The Mekong

R(

Publication of this book was made possible by the support of Obayashi Corporation.

The Mekong: Environment and Development

Hiroshi Hori

United Nations
University Press

TOKYO · NEW YORK · PARIS

The views expressed in this publication are those of the authors and do not necessarily reflect the views of the United Nations University.

United Nations University Press
The United Nations University, 53-70, Jingumae 5-chome,
Shibuya-ku, Tokyo, 150-8925, Japan
Tel: +81-3-3499-2811   Fax: +81-3-3406-7345
E-mail: sales@hq.unu.edu
http://www.unu.edu

United Nations University Office in North America
2 United Nations Plaza, Room DC2-1462-70, New York, NY 10017, USA
Tel: +1-212-963-6387   Fax: +1-212-371-9454
E-mail: unuona@igc.apc.org

United Nations University Press is the publishing division of the United Nations University.

Cover design by Joyce C. Weston
Cover photograph by Hiroshi Hori

Printed in Hong Kong

UNUP-986
ISBN 92-808-0986-5

I dedicate this book to my wife, Toyoko, who passed away just before this book was published.

The mighty Mekong in slumber

Late one afternoon in June
  A faint sound of thunder was heard
Somewhere in the blue sky
  Over the beautiful city of Vientiane.
Very soon dark clouds gathered over the Mekong
  And raindrops began to tap her placid face.
Hurriedly I took cover
  Near the earthen wall of an old temple
And watched enchanted
  The silvery raindrops splashed themselves on the emerald roof tiles;
Half an hour passed and the rain stopped.
  Suddenly from under a big tree
A noodle pedlar appeared
  Pushing his cart along the wet highway.
A fading sunset drew long silhouettes
  Of the pedlar, his cart, and the tall tree.
A scene which soothed so gently and softly
  My exhausted nerves
After such strenuous trips
  To so many dam sites of the mighty Mekong.

—Hiroshi Hori, *ESCAFE Staff Bulletin* No. 25 (1965)

Hiroshi Hori

Contents

Foreword

This informative and critical account of the experience in planning for development of the water resources of the Mekong Basin is unique in at least two major respects: it describes the Mekong experience; it also provides a concrete demonstration of how international thinking and action on water management evolved over the years following 1958. The author describes what was done, and why, in the Mekong, while encouraging thoughtful evaluation of possible lessons for use elsewhere.

The efforts to plan for the Lower Mekong in Cambodia, Laos, Thailand, and Viet Nam were among the earliest to engage the new United Nations Organization and its regional office. The Asian Development Bank, and teams of experts from France, Japan, the United States, and other countries were among those drawn in, and continued (notwithstanding difficulties) during the Viet Nam War. The political problems became even more complex as China and Myanmar participated.

Over four decades, the prevailing notions of what are important factors in water planning for the basin evolved significantly: the attention given to the social and economic condition of indigenous peoples grew notably; investigation of the inter-relations of human activity and the quality of ecosystem processes changed radically; methods of assessing the needs for hydroelectric power and the full consequences of proposed large reservoirs evolved drastically. For example, the policy positions of the US Bureau of Reclamation on the role of large reservoirs are stated as being quite different at two points in time.

While the author was directly involved in a few of the reported studies, he has been a critical observer of the whole history to date. He strives to state the views of the major participants accurately. Alongside that reporting he shares his own assessment of the actions of major groups and methods of estimating past and future needs.

The resulting book not only is a comprehensive description of the basin and of unfolding policies for its development but also provokes critical appraisal of evolving water-management theory and practices elsewhere in the world as they are reflected in the Mekong.

Few basins in the world have a more complicated history of investigation. Much of the projected construction work has not been undertaken, but the variety of policies and methods that have been proposed is representative of that of many other areas. Profoundly important decisions lie ahead in the Mekong; the lessons here illuminate problems of environment and development faced elsewhere in the world.

Gilbert F. White
University of Colorado at Boulder, USA

Preface

This book is an English translation of the original Japanese edition published by Kokon Shoin in April 1997, which took me 2 years and 9 months to write, from the spring of 1993 to the beginning of winter 1995. As indicated by the title, it is mainly concerned with dam development and its anticipated environmental impact on the lower Mekong Basin. Before embarking on this main theme, however, I would like to begin by briefly outlining the background behind my decision to write this book.

Background

In my 50-year working life, I have spent 3 years on a study of the environmental impact of developing the Kumano River in Japan and more than 20 years working overseas. Through this experience, I have become very familiar with the environmental problems of dam-development projects, not only in Japan but also overseas, including the lower Mekong Basin.

My experience of life abroad began when I went to the United States in 1955. After studying as a Fulbright Scholar at the University of Illinois in Urbana, I joined Harza Engineering Co., a water-resources consultancy in Chicago, where I took part in planning for a dam in Honduras in Central America and became acquainted with the system in the United States for providing aid to developing countries.

After returning to Japan in the fall of 1956, I went back to work on the development of the Tenryu River. Hoping to make use of my experience in the United States, I moved in 1958 to the head office of the Electric Power Development Company (EPDC), a semi-governmental organization, where I was put in charge of preparing the ground for overseas technical aid. At EPDC, I initiated the study of comprehensive development plans for the lower Mekong Basin in preparation for the secondment of a Japanese government reconnaissance team to the area. At the same time, I studied reports of consultants in the West concerning plans such as the Volta River hydroelectric power development project in the Gold Coast (Ghana), the Mantaro hydroelectric power project in the uppermost reaches of the Amazon in Peru, and the Malikina Dam project in the environs of Manila in the Philippines. As described in chapter 5, the Volta River dam project involved many environmental problems. The Malikina Dam was to be a high-arch dam built in a limestone region, and fears concerning the safety of the people of Manila and the possibility of water leakage gave rise to a long debate about the advisability of the project, which was finally abandoned.

From 1959 to 1960, while I was still at the head office of EPDC, I was appointed by the Ministry of Foreign Affairs as a member of the Japanese Government Reconnaissance Team (led by Yutaka Kubota) sent to investigate the major tributaries and the main stream of the lower Mekong Basin. We investigated the potential of dam and hydroelectric power development in the three countries of Indo-China (Cambodia, Laos, and Viet Nam) and observed the social situation and natural features of the river basin in this tropical continent prior to development.

At the end of 1960, I was seconded by EPDC to the United States as a member of the Japanese Government Science and Technology Agency's Disaster Prevention Study Team. After returning to Japan, I published an in-company report titled "Research on Flood Control in the USA," which even included themes relating to hygiene, such as malaria control by the Tennessee Valley Authority (TVA).

Thereafter, for 4 years until the spring of 1964, I took part in studies on development projects for rivers in Hokkaido and later worked mainly on survey and research on the environmental impact of the construction of the Nanairo and Komori dams in the middle and upper reaches of the Kumano River. There was local opposition to the development of the Kumano River, mostly due to fears that it would damage the natural beauty of the Doro Gorge, a tourist attraction downstream from the dam – a foretaste of the environmental issues that frequently arise today. I was responsible for the technical aspects of compensation and had to provide explanations concerning all the various environmental problems expected to result from the construction of these dams. These problems included reservoir sedimentation; inundation of houses; discharge of water for sightseeing to conserve the scenery at the dam site; compensation for the fishing industry downstream from the dam; conservation of the scenery of the Doro Gorge, which is located immediately downstream; the issue of propeller-driven boats for tourists between the coastal towns of Shingu and Doro; the need to move the water intake of the Kishu papermaking plant at the furthest point downstream to avoid damage by sea-water intrusion in the dry season; the danger of water leakage into the mine tunnels of Ishihara Sangyo Co., Ltd; processing of river pollution resulting from construction of the dam; the earthquake resistance of the Kumano River Basin, which is traversed by the Great Fault Line; and damage prevention in the event of flooding.

In 1964, I was sent by EPDC to serve as Chief Planning Engineer of the Secretariat of the United Nations Committee for Coordination of Investigations of the Lower Mekong Basin (referred to below as the Mekong Committee) and spent four and a half years in Bangkok at this post. I was responsible for the technical aspects of the formulation of the 1970 Indicative Basin Plan report and for "on-the-job" training of young engineers seconded to the Mekong Committee by the governments of the riparian countries. In 1966, while I was serving on the Mekong Committee, I visited the site of the Aswan High Dam and inspected the Nubian residents' resettlement area, which was still under construction. That year, the Mekong Committee received an award from the Magsaisai Foundation in the Philippines, which is viewed in Asia as equivalent to the Nobel Prize.

In 1969, while still at EPDC, I was assigned to the Overseas Technical Cooperation Agency (OTCA, the predecessor of the Japan International Cooperation Agency, or JICA) and appointed Chief Executive and Planning Director of the newly organized National Water Resources Council of the government of Tanzania in East Africa. For 2 years, I formulated a plan for water-resource development throughout Tanzania in collaboration with the ministers of government ministries such as economic development, agriculture, land, trade and industry, health and welfare, regional development, and finance. At the same time, I worked on a study for a project to construct the Stiegler's Gorge Dam in the lower reaches of the Rufiji River, the largest river in Tanzania (with the eightieth-largest basin in the world). Protected areas such as the Serengeti National Park are now well known in Japan, but at that time lions were frequently seen in the suburbs of Tanzania's capital Dar es Salaam and many wild animals inhabited the site where development of the Rufiji River was planned. This made me aware how essential it was to consider the impact on the animals and vegetation when planning dam development in tropical regions. I also began a study for the comprehensive development of the uppermost part of the Blue Nile – i.e. the Kagera

International River (running through Tanzania, Uganda, Rwanda, and Burundi) – in order to secure funds from the United Nations Development Programme (UNDP). Through these various activities, I came to realize that the African government officials concerned took little interest in preserving the environment while promoting development. Although the local governments would pay lip-service to the importance of environmental conservation, their greatest desire was to achieve the fastest possible economic growth. The situation may not differ much, even today. I was also made painfully aware that, although the construction of large dams on continental rivers always had to be carefully planned and implemented (bearing in mind the international situation, because of differences in the degree of disadvantages caused by development as well as different interests and viewpoints of neighbouring countries regarding socio-economic effects as well as the environmental impact), it is, in practice, extremely difficult to formulate a comprehensive development plan for an entire river basin.

From 1971 to 1973, EPDC seconded me to work as a resident technical adviser to the Asian Development Bank (ADB), where I was responsible for the technical aspects of the Temengor Dam Project in West Malaysia, the Tenom Pangi Dam project in East Malaysia, a hydroelectric power project in Western Samoa, and a project for the development of the Laguna Lagoon and surrounding area on the outskirts of Manila. Among these, the Laguna development project (completed in 1982) in particular involved various environmental problems. The aims of the project were to mitigate flood damage in the city of Manila, to enable the transportation of mining and agricultural products over the lagoon on the far side, and to pump water from the lagoon to irrigate the surrounding area. However, with the increase in the population living around the lagoon and industrialization in recent years, the disposal of agricultural, household, and industrial waste had resulted in the increasing eutrophication of the lagoon. In order to deal with this problem, water-quality experts were invited to take part in the preliminary studies.

In 1973, EPDC sent me to serve as a resident deputy representative of the UNDP in Turkey, where I dealt with wide-ranging developmental issues throughout the country. While I was in Turkey, I inspected the Keban Dam at the most upstream reaches of the Euphrates River and observed the problem of dam leakage. From 1973 to 1975, as Senior Technical Adviser at the UNDP Head Office in New York, I took charge of water-resources development and energy-development programmes which were to be promoted by the United Nations throughout the world.

After returning to Japan in 1975, I worked on overseas projects in such countries as the Philippines, Iraq, Egypt, South Africa, and Mexico.

While working at a consulting company (PCI) and a construction company (Nissan Construction Co.) as Overseas Director and Adviser, I was Vice-Chairman of the Association of Japanese Consulting Engineers (AJCE). In 1979, I attended the United Nations Water Conference in Argentina as adviser to the Japanese governmental representative. From 1980 to 1990, I served as Director and World Committee Chairman of the International Water Resources Association (IWRA). In 1985, I participated in the International Large Dams Conference at Lausanne as Chairman of the Japanese delegation on environmental problems.

As far as the Mekong River is concerned, I visited Bangkok in 1984 at the invitation of the Interim Mekong Committee and prepared guidelines and a budget for the revision of the Indicative Basin Plan of 1970. However, the Japanese government (Ministry of Foreign Affairs), wary of the political situation in the countries of the lower Mekong Basin, vacillated and in the end did not respond to requests for funding; without financial support, the proposed revisions could not be made. In 1985 and 1990, I attended general meetings of the Interim Mekong Committee as an observer. In 1988, I reviewed and evaluated the draft of the 1988 development plan ("Perspectives for Mekong Development") which had been drawn up by Swiss consultants with funding from the UNDP. In 1987–1988, at the request of the Governor of the province of Vientiane in Laos, I conducted a study for small-scale hydroelectric power projects to assist the people living in the mountains. In 1989, I was requested by the Institute of Energy Economics, Japan to head a survey team to study the energy situation in Viet Nam. From 1990, I spent a year and a half studying flood-control measures for rivers running through the capitals of three countries of the lower Mekong Basin and five other countries for the UN Economic and Social Commission for Asia and the Pacific (ESCAP), and made recommendations to the governments of these eight countries. From 1991 to 1993, I headed a JICA team conducting a prefeasibility study for small-scale hydroelectric power projects in the Se Kong Basin in southern Laos. In December 1994, immediately after the agreement was made to establish the new Mekong River Commission, I made a speech at the Mekong River Development Symposium at the United Nations University, in which I emphasized the need to formulate a comprehensive development plan for the whole Mekong Basin, taking into account the changed situation of the river basin and the need to consider carefully the environmental impact of dam construction. To achieve these objectives, I stated that it would be necessary to set up an international advisory subcommittee of experts and specialists from a wide range of fields as part of the Mekong River Commission. Unfortunately, however, the Mekong River Commission did not set it up; moreover, it asked a northern European country to formulate the comprehensive plan, but, even at the time of writing, such a formulation has not been accomplished.

For 2 years (from 1996 to 1998), JICA conducted a study to formulate strategies to balance development and the environment in the Mekong Basin. For that purpose, it established a committee to carry out the study. I was appointed as its Chairperson and completed the study in July 1998. In the study report, I emphasized the following points:

- Aid to the least-developed countries should be based on love and respect for those nations and their people and be self-driven by the local residents;
- Development assistance with priority regarding the environment should first take a comprehensive, wide, all-embracing approach and, next, incorporate micro-innovations that address more specific concerns.
- It is important to consider not only the physical effects but also the spiritual and psychological effects of development on the local residents.
- Emphasis must always be placed on conservation of the natural environment, including preservation of biodiversity. This entails the production of a comprehensive database of information on environmental resources.
- Large-scale dam-development projects must address the true needs of communities in the region. Assessment should be fair and comprehensive. Development should proceed in a gradual manner that does not impose undue stress on the target region and which, from start to finish, is open to public scrutiny. The development should not interfere with, and should be in conformity with, the surrounding scenery and regional culture. It should be modified or adjusted according to changing situations and conditions.

Since I had lived overseas for a long time, I did not have much contact with the academic world in Japan for many years; however, from 1982 to 1990 I served as a lecturer to postgraduate students from overseas in the Civil Engineering Department at the University of Tokyo, in courses for domestic students in the Educational Department of Waseda University and in the Engineering Department of Hosei University, and in a course on social engineering at Tsukuba University. This university tutorial work presented me with plenty of opportunities to explain the environmental problems of large-dam construction and the problems involved in development itself. I have also been a member of various academic societies, both in Japan and overseas, as Chairperson of the National Committee and (as already mentioned) as Board Director of IWRA since 1982. In 1986 I submitted a thesis entitled "The Influence of Dam Construction on the Natural Environment and Society in River Basins

Focusing on Dam Development on Rivers in Tropical Continents" to the University of Tokyo, and was awarded the degree of Doctor of Engineering in February 1987.

About this book

This book centres around a revised version of the section of my doctoral thesis dealing with dam construction in the lower Mekong Basin. It consists of six chapters which outline the main features of the lower Mekong Basin, the history of development plans, and their future outlook, as well as examining the development of the upper Mekong Basin, i.e. the Lancang River, which I inspected at the end of 1994.

Chapter 1, "The Mekong River and its Basin," and chapter 2, "Current Conditions and Future Possibilities for Resource Development in the Lower Mekong Basin," aim to give the reader a general idea of the developmental situation in the Mekong Basin. Although constantly feeling frustrated by the lack of basic material, I have done my best to give a complete overview, but I feel that I have succeeded in describing only one aspect of the vast panorama comprising the Mekong Basin. Furthermore, the basin is today undergoing very rapid and dynamic changes, and what is written in one month may not be applicable the following month. Although I have striven to provide an accurate description of the current situation and future outlook of the basin, making use of the most up-to-date information and knowledge, I fear that I have fallen short of this initial aim.

Chapter 3 deals with "Planning of Dam Development in the Lower Mekong Basin – Chronology and International Cooperation." Almost one and a half centuries have passed since France established French Indo-China and started to develop the region. However, this book focuses mainly on the modern history of development of the lower basin since the end of the Second World War. Chapter 3 looks back at the history of past development plans, which many readers may feel is hardly necessary, but I believe that it was my duty to describe the changing attitudes, and approaches to the water-resources development in the lower basin, of the dam-development planners from Japan and the rest of the world, and to tell the tale of their efforts, aspirations, and visions. I hoped that this would be not simply a record for the future but also a source of reference, showing the various plans conceived in the past, to which we could turn again and again in our search for new and better methods of development.

Chapter 4, "Dam-Development Projects in the Upper Basin and the Lancang River," outlines plans for projects for the development of a series of dams on the Lancang River in Yunnan Province, of which I made an inspection tour in 1994. For a long time, there was no way of knowing the development situation in the upper Mekong Basin; however, recently, the Ministry of Electric Power of the Chinese government has made this information available. With feelings of gratitude to the Ministry, I have tried to describe the situation in the upper basin in as much detail as possible.

Chapter 5, "Environmental Problems of Dam Construction in the Tropics," as explained above, constitutes the core of this book. With some revisions, it is essentially the doctoral thesis that I wrote from 1985 to 1987.

Finally, chapter 6, "Creation of the Mekong River Commission and Future Outlook," describes how, after development had stagnated after the end of the Viet Nam War in 1975 and came to a complete standstill with the outbreak of various problems in the upper and lower basins in the late 1980s, spring eventually arrived in the lower Mekong Basin and a new Mekong River Commission was set up, with all four riparian countries finally marching forward in step again. The Mekong Commission can be expected to publish a revised development plan for the lower basin, reflecting this new situation. At the end of this chapter, I set out my own strong expectations of the new Mekong Commission.

River-basin development and environmental issues: underlying arguments

The Mekong River Basin has long experienced flooding, salt-water influx, depletion of forests, deterioration of groundwater, water pollution and other problems. The agreement of the Mekong River Commission in 1995 concluded with the following aim: "To protect the environment, natural resources, aquatic life and conditions, and ecological balance of the Mekong River Basin from pollution or other harmful effects resulting from any development plans and uses of water and related resources in the Basin."

However, in reality, several environmental problems continue to occur in the river basin. These include the depletion of natural resources, overpopulation of the cities, resettlement of refugees, and social problems such as the relocation of inhabitants and the securing of their employment accompanying the establishment of infrastructure. It is not an easy task to balance the need for increased food production and other developmental needs against the maintenance and management of resources in the Mekong River Basin, where the population is growing rapidly.

The challenge of satisfying both the need for development and the requirement of preservation of the environment can be seen not only in the Mekong but also in every river basin. The search for some form of balance is an ongoing task for all those concerned. In March 1997, supported by JICA, the Japanese Ministry of Foreign Affairs held an open symposium entitled "Free-talking Discussion Meeting on Development and the Environment in the Greater Mekong Subregion". This included a director-general of the Asian Development Bank, who was in charge of the Bank's Mekong activities; a professor of the Civil Engineering Department of the Tokyo University who was known in Japan as the "Mekong expert"; a representative of the Japan International Volunteer Centre, who was a social worker and environmentalist, and the author. The author had the role of key person, and a prominent television personality took the chair.

Stimulating comments and opinions were expressed and different viewpoints were exchanged during the discussion. During the meeting, questions and comments were received from the audience, who came from a wide range of fields related to Japanese international cooperation.

At the end of the meeting, it was concluded that we should approach the development/environment projects or programmes in the Mekong River Basin not merely from technical or physical standpoints but with much wider points of view, such as the socio-economic, political, archaeological, and other quite fundamental aspects, at all levels. Furthermore (and more importantly), we should always remind ourselves of the fact that the genuine heroes of the development/environment projects or programmes are undoubtedly the ordinary people who live in the target area.

Acknowledgements

This book is the product of my whole life's work and I regard it as a sort of testament. While writing it, I have constantly been aware of the great debt of gratitude to all the people without whose assistance, both direct and indirect, it could not have been written. I would like to express my heartfelt thanks to all of these friends and colleagues both in Japan and overseas, including some who have, sadly, now passed away. I wish to extend my warm thanks to Dr Juha I. Uitto, formerly Senior Programme Officer on the Environment and Sustainable Development, and Dr Manfred Boemeke, Head of the United Nations University Press, for so readily agreeing to take on the publication of this English edition; my thanks are also due to Ms Makiko Yashiro for her editorial assistance. I am also very grateful to Mr Ted Manuel, an American from Chicago and my very close friend since our days in the University of Illinois in the 1950s, for his selfless and dedicated assistance with the translation. I am also deeply indebted to the River Planning Division of the River Bureau of the Japanese Government Ministry of Construction for providing both reference materials and funding for this project. Last, but by no means least, I would like to thank my wife Toyoko Hori who stood by and warmly supported me for over 15 years from the writing of my doctoral thesis to the completion of this book, but who passed away in January 1999.

I cherish the hope that this book may make some contribution to the long-term sustainable development of the Mekong Basin and other river basins, and to the happiness and prosperity of the people who live in them.

Figures and Tables

Figures

1

The Mekong River and its basin

The Mekong River originates in Tibet and flows through the Chinese province of Yunnan before continuing on a long southward journey ending in the South China Sea. In the course of this southern journey, the extensive Mekong Basin touches the territories of six countries – China, Myanmar, Laos, Thailand, Cambodia, and Viet Nam. The area of the basin itself amounts to 795,500 km^2 with a total length of 4,620 km.[1] Both figures place the Mekong among the world's largest rivers. Although different figures have recently been announced by the Chinese,[2] the author has retained the values quoted here.

A journey down the Mekong River

At their origins, the waters of the Mekong are fast moving as they rush down the steep Tibetan slopes, but soon the flow takes on an unhurried pace which becomes gradually slower as the river travels south. Having passed through China and Myanmar, the Mekong comes to Laos and Thailand, after which its speed is further reduced. The entire trip to the estuary takes three weeks to complete when the river is flooded, while the same course requires more than three months during the dry season.

Let us take a journey down the winding path of the Mekong (fig. 1.1).

Figure 1.1. Watershed of the Mekong Basin
Source: UN 1957

The Chinese section (2,000 km)

The Mekong originates in the Tanggula Shan Mountains of the Tibetan Plateau, more specifically in the Rup-sa Pass Swamps (elevation 4,968 m) located on the northern side of the eastern section of the Tanggula Shan Range. At the outset, the river takes a south-eastern route and is initially referred to as the Za Qu River in the section between its source and the large town of Quamdo. Further downstream, the river takes the name of the Lancang River and continues southward sandwiched between the Jinsha River, an upper branch of the Yangtze, to the east and the Nu River, an upper branch of the Salween River, to the west. Further southward, the Mekong eventually moves away from the Jinsha River while continuing to flow almost parallel to the Nu River. Maintaining a southerly direction, the Mekong eventually finds its way into the province of Yunnan and continues on to the southern extremity of the province (Mekong Secretariat 1988). In other words, during its journey through Chinese territories, the Mekong is initially called the Za Qu River and later the Lancang River. Throughout this section, it flows through deep ravines while being fed by the melting snow of the Tibetan Highlands and the rainfall which flows down the mountains of Yunnan. The Chinese sector of the basin is covered by forests and the river discharge is abundant. The average annual discharge in the lowest portion of the Lancang River amounts to roughly 2,000 m³/s, which remains relatively constant from year to year. (According to a report in the *Yunnan Geographic Environmental Research Journal* published in March 1999, this figure is 2,410 m³/s.) The banks of the river here are sparsely cultivated and the population density is low; this area is therefore highly suited to hydroelectric power generation. The combined length of the Za Qu and Lancang rivers is about 2,000 km (Kunmin Hydroelectric Investigation and Design Institute 1985), or 2,161 km according to the above-mentioned report. From the headwaters to the southern border of China (altitude approx. 400 m), the altitude of the river falls by approximately 4,570 m. The approximate gradient of the river bed between the source and the northern end of Yunnan Province is 1:290 and declines to 1:660 as it flows through the province. This is roughly the same gradient as that frequently seen in the rivers of Japan.

From the Chinese Border to the Golden Triangle (220 km)

Immediately before leaving China, the Mekong serves as the border between China and Myanmar over a distance of roughly 25 km. After this short interval, the Mekong begins to mark the border between Laos and Myanmar as it heads toward the trilateral border area where Myanmar,

Laos, and Thailand meet. This is the "Golden Triangle," notorious for its production of opium; the mountain slopes of this area are densely covered with the poppy plant.

From the Golden Triangle to Luang Prabang (Laos) (360 km)

Leaving the Golden Triangle, the Mekong follows an eastward course through Laos over a distance of 350 km before making a sharp turn. The beautiful city of Luang Prabang is located on the left bank of the river as it comes out of this turn.

Luang Prabang was the capital of the Laotian kings between the late sixteenth century and the early eighteenth century and was built on the elevation overlooking the levee immediately below the confluence of the Mekong and one of its major tributaries, the Nam Khan. Surrounded by rich greenery, the city is home to many historic Buddhist temples and has maintained the flavour of its past elegance. Upstream from Luang Prabang there are numerous navigation aids (markers), which were erected by the French many years ago. Here, immediately upstream, many local people can be seen collecting sand gold along the left bank of the Mekong in the dry season.

From Luang Prabang to Vientiane (430 km)

After passing through Luang Prabang, the Mekong heads due south and picks up speed as it passes over the shallow river beds of this area. The river can be navigated by boat between Jinghong in Yunnan Province and Luang Prabang, and this route is now being served by hydrofoil vessels. However, south of Luang Prabang the river becomes much less inviting over a 220 km stretch ending at the town of Pak Lay. During the dry season, boats have to be pulled over the shoals by rope, while during the peak period of the rainy season all river traffic is brought to a stop by the treacherous current and rocks. Throughout this entire area, the Mekong flows through deep valleys and ravines.

Travelling downstream from Pak Lay, we reach Chiang Khan after a distance of 70 km. The Mekong again makes a sharp turn over this distance and takes an easterly direction as it begins to etch the boundary between Laos and north-east Thailand.

Chiang Khan itself is a north-eastern Thai town situated on the right bank of the Mekong. A further 110 km from Chiang Khan, the mountains on both sides of the river start to become smaller and lower. Not far from here is the proposed site for the Pa Mong Dam, which has been considered the most promising candidate for a dam-construction project on the main stream in the lower Mekong Basin. Here the topography around the

Mekong is squeezed between steep hills on both sides and the flow is interrupted by numerous outcrops of hard rocks which rise up from the river bed. Navigation aids erected on these outcrops during the French colonial period can still be seen in the middle of the river and remain useful in directing the river traffic in this area.

The left bank of the Pa Mong Dam site opens to the expansive Laotian plain known as the Vientiane Plain. Some 20 km downstream from the dam site is the capital city of Vientiane which stands on the northern (left) bank of the river with its back to this plain. This is a sleepy little city which makes one wonder whether it can really be the national capital. Yet, with a population of 400,000 people, Vientiane accounts for roughly 10 per cent of the total population of Laos. Across the river from Vientiane, the view opens on to north-east Thailand.

At this point, the width of the Mekong is roughly 800 m. Both sides of the river bank are vulnerable to erosion and are protected by carefully reinforced levees. The river bed on the Vientiane side is covered by piles of boulders and pebbles. In April 1994, a bridge linking Laos and Thailand was completed a short distance from Vientiane. Named the "Friendship Bridge," this elegant structure is the first bridge built across the Mekong in the lower basin area (the area of Thailand, Laos, Cambodia, and Viet Nam). The length of this reinforced concrete bridge is 1,170 m.

From Vientiane to Savannakhet (460 km)

The Vientiane Plain is spread out on the left bank of the river as it flows by the city of Vientiane. The mountains of the Annam Cordillera, which rise to heights of from 1,500 m to more than 2,000 m, are visible in the distant north. In stark contrast, the southern vista which opens on to north-east Thailand is flat as far as the eye can see, with no mountains to break the monotony of the plain. This area is known as the Korat Plateau. It is said that north-east Thailand was covered by the sea in ancient times but later emerged as a tray-shaped plateau.

The Korat Plateau in north-east Thailand is one of the poorest regions of this country. Annual rainfall comes to only 1,000 mm and the land is not fertile; irrigation and manuring are therefore prerequisites for the agricultural development of this region. The farmers do not have the financial means to undertake these improvements: because agricultural productivity is low, they cannot afford to make the necessary investments.

Shortly after passing Vientiane, the Mekong comes to its confluence with the Nam Ngum River which flows down the Vientiane Plain to meet the Mekong. A further 200 km downstream from this confluence the Mekong adjusts its eastward direction and begins to follow a south-easterly course. This turn of the river marks the confluence of the Nam Theun River, another major tributary of the Mekong. The Nam Theun is spanned by a bridge at this point. Both the Nam Theun and the Nam Ngum rivers will certainly be very important for Laos in the near future: this is because the Nam Theun holds the promise of large-scale hydro-electric power output, while the Nam Ngum, which is already generating 150,000 kW of power, is scheduled for various new hydropower genera-tion and irrigation projects in the near future.

After combining with the Nam Theun, the Mekong takes a south-eastern course and flows between the Laotian towns of Thakhek and Nakhon Phanom on the Thai side. After this, the Mekong heads due south before eventually arriving at Savannakhet, the principal city in central Laos. Throughout this entire section, the river channel is marked with numerous navigation aids.

Savannakhet is the second-largest city in Laos. It serves as the hub of the economy in central Laos and is also important as a political centre. As it stands now, Savannakhet is a small country town but it has the poten-tial to develop into an important junction in the future transportation network of this region. Mountain roads running through the Annam Cordillera connect Savannakhet to the old Vietnamese capital of Hue. Foreign goods unloaded on the Vietnamese coast must pass through Savannakhet on their way to Bangkok. The goods are carried over the river from Savannakhet to Mukdahan on the Thai side before making their way through the Korat Plateau to Bangkok. In other words, Sav-annakhet constitutes an important link in the land-based trade route linking Viet Nam and Thailand. At the present time, the only means of crossing the river is by ferry; however, the Asian Development Bank is currently working to realize the construction of a bridge spanning the Mekong between Savannakhet and Mukdahan. When this bridge is com-pleted, Savannakhet will emerge as the leading city in this area and one of the most important growth centres in the entire lower Mekong Basin.

From Savannakhet to Pakse (260 km)

Passing beyond Savannakhet, the Mekong continues on its southerly course to arrive at the southern Laotian town of Pakse. (The section of the Mekong between Savannakhet and Keng Kabao was difficult to navigate because of the many outcrops in the river; however, these were removed some thirty years ago and navigating conditions were improved significantly.) The Nam Mun River, the largest Mekong tributary in north-east Thailand, meets the Mekong at a point some 50 km upstream from Pakse at its right bank. Pakse is the principal transportation junction in southern Laos where a diverse cross-section of people gather, including Vietnamese crossing from the Annam Cordillera, Cambodians travelling north on the Mekong, Thais crossing from the east, and numerous highland tribes of this region. As such, Pakse functions as a very lively border town in the southernmost part of Laos.

As it flows from Savannakhet to Pakse, the Mekong cuts deeply into the sandstone layers which cover many parts of the plateau areas of the river. In these sandstone regions, the river is contained by sheer cliffs on both sides. The current is fast and reaches speeds in excess of 4 m/s when the water level is high; consequently, boats can navigate this section only when the water is at normal levels. However, as the river approaches Pakse, the current becomes relatively calm as the river gets wider, allowing the operation of regular ferry services.

From Pakse to the Khone Falls (155 km)

In the section between Pakse and the Khone Falls, the Mekong turns into a calm river flowing from north to south; nevertheless, its passage is studded by a number of outcrops which rise above the water level. The Khone Falls are located at the southern end of this section where the river slopes steeply downwards over a 20 km stretch to create a 30 m head. The waterfall is quite broad and the passage is separated into narrow strips by numerous rocky islands in the river. In the early twentieth century, the French devised a system for lifting cargo over these falls to open the north to trade; the rails and reloading facilities built at that time still remain.

The Khone Falls constitute the greatest impediment to the use of the Mekong for navigation. It is reported that, when these falls were discovered in the 1860s, the French abandoned their plan to trade with the interior of China using boats sailing up through the Mekong Estuary.

From the Khone Falls to Stung Treng (55 km)

A Thai developer is promoting the development of a small run-of-the-river type hydroelectric power station using a section of the falls, while also turning this area into a tourist attraction. Leaving the magnificent view of the falls behind, the Mekong crosses over the border from Laos to Cambodia and travels a further 50 km to arrive at the north-eastern Cambodian city of Stung Treng. Near this point, the Se San River – the largest Mekong tributary, which flows from the vicinity of the central Vietnamese city of Kon Tum – meets the Srepok River which travels down from the Darlac Plateau of central Viet Nam. Shortly after this confluence, this combined river takes in waters of the Se Kong River (which flows down the eastern side of the large mountain ranges of the

Boloveng Plateau located east of the Laotian town of Pakse) and finally meets the Mekong at Stung Treng.

At the end of the 1950s, the attention of the Japanese reconnaissance team which investigated the major tributaries of the lower Mekong was drawn to this very important juncture at Stung Treng. At the time, they formulated a development concept which featured the building of a dam immediately downstream from Stung Treng. This dam, equipped with giant navigation locks, would adjust the flow of the Mekong. An additional advantage of the dam was that the Khone Falls would be submerged by the reservoir extending northwards. The Japanese proposal was thus aimed at solving the problem of navigation in this region. In addition, the dam could serve to generate a huge amount of hydroelectricity, support irrigation in the region, and make an important contribution to flood control in the Mekong Delta. Furthermore, it might be possible to construct a canal system on the western edge of the giant Stung Treng Reservoir to irrigate up to a million hectares of land on the expansive plains north of the Great Lake. Building a reservoir at Stung Treng would mean, of course, that the town of Stung Treng itself and a large surrounding area would be submerged; fortunately, population densities in the upstream areas and the area of the proposed reservoir lake are relatively low. However, owing to the Viet Nam War and subsequent strife in Cambodia, the development plan was left untouched for a long time. Nevertheless, the idea has remained alive as a difficult but highly attractive plan which is worth reconsidering.

From Stung Treng to Kratie (135 km)

Travelling down from Stung Treng, the Mekong approaches another Cambodian town, Kratie, also situated on the left bank of the river. Some 20 km north of Kratie, the current becomes treacherous as it enters the Sambor Rapids, a challenge to river navigators past and present. During the dry season, grass-covered boulders and shoals are found scattered along this stretch of the river. A proposal, known as the Sambor Dam Project, was made in the early 1950s to build a dam immediately below these rapids, the aim of this construction being to submerge this difficult stretch. The dam was to be equipped with a series of locks to allow boats to travel upstream, while also contributing to power generation and irrigation. In the late 1960s the Japanese government conducted a serious prefeasibility study on this project. However, as in the case of the Stung Treng Project, the Sambor Dam has not yet materialized. Nevertheless, the new Cambodian government has shown keen interest in resurrecting this project and holds high hopes of its realization.

From Kratie to Phnom Penh (215 km)

South of Kratie, the current of the Mekong becomes slow and calm enough for the passage of vessels of 200 tons. The town of Kompong Cham located on the right bank of the river has, as its name indicates, long been inhabited by the Cham Tribe. After it passes Kompong Cham, the Mekong finally enters its delta.

In order to travel by land from Phnom Penh to north-eastern or eastern Cambodia, the traveller must cross the river at Kompong Cham by ferry to begin the land route from the left bank of the river at this point. As of 1996, the construction of a bridge 1,300 m long across the Mekong at this spot is being considered by the Japanese government, which is also planning another bridge, the Thakhek Bridge, further upstream on the Mekong between Nakhon Phanom (Thailand) and Thakhek (Laos). When the Kompong Cham Bridge is realized, the entire situation in the area will be dramatically changed and the improvement of transportation from Phnom Penh will considerably expedite the development of the remote, somewhat backward, regions of north-eastern and eastern Cambodia.

From Phnom Penh to the estuary (330 km)

Immediately downstream from Phnom Penh, the Mekong divides into two rivers: the river on the east side is the main stream of the Mekong and that on the west is the Bassac, a branch of the Mekong; however, the two rivers join together again at Ban Nao, some 50 km south of the Cambodia–Viet Nam border, but separate again within a short distance. After this second division, the Bassac River on the west flows directly to the sea, while the Mekong further divides into five branches after flowing through Sadec before finally finding its way to the South China Sea.

Directly downstream from Phnom Penh, where the Mekong separates from the Bassac, it takes in the waters of the Tonle Sap River. Standing on the river-bank at Phnom Penh where the two rivers meet, one can clearly see the contours of the courses of four rivers – those of the Mekong flowing in from the north, the Bassac River, the Tonle Sap River and the flow of the Mekong towards its delta. This is why the French called this area Quatre Bras (four arms).

During the monsoon season, ocean-going vessels (not exceeding 3,000 tons) can sail from the South China Sea directly to Phnom Penh. Smaller ships of up to 500 tons can travel further upstream as far as Kratie. During the dry season, the river route to Phnom Penh is passable for ships of nearly 2,000 tons; in the same period of the year, the influx of salt water from the South China Sea travels up the Mekong and Bassac rivers.

In the delta, the South China Sea repeats a cycle of two major daily high and low tides.[3] At certain places in the delta, the difference in water level between high and low tides reaches 4 m. The influx of salt water during the dry season poses a difficult challenge to farming in coastal areas of the delta, as well as creating the (no less difficult) task of securing reliable sources of drinking-water. A fundamental solution to this problem requires the construction of a dam on the Mekong to adjust water levels during the dry season in order to push back and flush out the encroaching salt water. Another important need is to dredge both the Mekong and Bassac rivers, which are gradually becoming shallower and narrower owing to the deposition of silt.

The recent global-warming trend is causing some concern that the level of the South China Sea is gradually rising in this coastal area. (According to one theory, the water level will rise by 15–65 cm over the next few decades.)

Outline of the Mekong River Basin

As explained at the beginning of this chapter, the source of the Mekong is located in the eastern Tibetan Highlands. The headwaters of the Jinsha and Salween rivers are located in these same Tibetan Highlands. For the first 1,000 km of its journey to the sea, the Mekong flows roughly parallel to the Jinsha and Salween rivers. The distance between the three rivers varies from as much as 150 km to as little as 20–30 km.

The source of the Mekong is located at roughly latitude 33°N; however, on its southward course, the river eventually cuts across the demarcation between the temperate and torrid zones at 23° 30′N (at this point, the Mekong is still in Chinese territory). In other words, the Mekong is a river which traverses the temperate and torrid zones and its source is located in the swamps which constitute the watershed between the upper reaches of the Jinsha River and the flow which develops into the Mekong. With the exception of a brief period during August and September, this entire area is covered by deep snow for most of the year.

At its upper reaches, the Mekong Basin covers a relatively narrow strip of land which gradually begins to expand after the river passes the joint boundaries of China, Myanmar, and Laos (located at roughly latitude 22° 15′N). The basin reaches its maximum width at latitude 13°N where it stretches from east to west to cover a width of 700 km; this is located at a relatively short distance from the estuary (9° 30′N). Thus, the shape of the entire basin resembles a Japanese lute or a Western violin.

The length of the river located north of the borders of Thailand, Myanmar, and Laos (i.e. the upper reaches of the river) is 2,200 km. The

Table 1.1 Land areas and populations (1995) in the six riparian countries

Country	Land area ($\times 10^3$ km^2)	Total population ($\times 10^6$)	Land area within Mekong Basin ($\times 10^3$ km^2)	Population within Mekong Basin ($\times 10^6$)
China (Yunnan)	395.0	39.0	165.0	9.6
Myanmar	676.6	46.5	24.0	0.9
Thailand	513.1	58.8	184.2	28.1
Laos	236.8	4.9	202.4	4.5
Cambodia	181.0	10.2	154.7	8.5
Viet Nam	331.7	74.5	65.2	22.7
Total	2,334.2	233.9	795.5	74.3

Source: Mekong Secretariat (1988).

drainage area is 189,000 km^2 and the mean breadth of the upper basin is 118 km. On the other hand, the length of the lower reaches is about 2,400 km and its mean breadth is 250 km. The breadth of the lower basin is therefore more than twice that of the upper basin.

The total area of the Mekong basin as distributed among the six countries of this region is as follows. Chinese territory accounts for roughly 165,000 km^2 (approx. 21 per cent) of the entire basin. Myanmar accounts for 24,000 km^2 (3 per cent), Laos for 202,400 km^2 (25 per cent), Thailand for 184,200 km^2 (23 per cent), Cambodia for 154,700 km^2 (20 per cent), and Viet Nam for 65,200 km^2 (8 per cent). The upper basin (China and Myanmar) covers an area of 189,000 km^2, while the lower basin (Laos, Thailand, Cambodia, Viet Nam) covers 606,500 km^2. Of the total basin area of roughly 795,500 km^2, approximately 24 per cent is located in the upper basin (China, Myanmar) with slightly more than 75 per cent belonging to the lower basin (Laos, Thailand, Cambodia, Viet Nam).

Table 1.1 shows the land areas and populations of the six countries of the Mekong Basin. It is interesting to note the proportion of the Mekong Basin in the total area of each of these countries. In China,[4] the Mekong Basin accounts for 1.7 per cent of the country's total land mass, accounting for 41.8 per cent of the province of Yunnan, which has an area of 395,000 km^2; in Myanmar, the basin covers roughly 3.5 per cent of the total land mass; the Mekong Basin thus covers only a relatively small share of the land mass of China and Myanmar, the two upper-basin countries. The situation is reversed in the four riparian countries: the percentage share of the basin in the land mass of these countries is 85.3 for Laos, 35.9 for Thailand, 85.6 for Cambodia, and nearly 20 for Viet Nam.

According to Table 1.1, the population of the upper basin (China, Myanmar) amounts to 10.5×10^6, while the lower basin (Laos, Thailand, Cambodia, Viet Nam) has a population of 63.8×10^6; this puts the population of the entire basin at 74.3×10^6.

The topography, soil, and vegetation of the Mekong Basin (figs. 1.2, 1.3)

The region of the northern mountains

The 189,000 km^2 of the upper basin covering China and Myanmar, as well as the northern Laotian portion of the lower basin, consist of an uninterrupted region of mountains with an elevation of 400 m or more which join the soaring heights of the Annam Cordillera to the east.

Travelling north by road from the Laotian capital of Vientiane (elevation 170 m), one traverses the Vientiane Plain with a clear view of the high mountains which run from north to north-east. These distant mountain ranges rise as high as 2,800 m, while never dropping below the 1,500 m mark. Covered by thick forests and deep valleys, this is the region of the northern mountains.

The area to the south of the upper Mekong and the middle reaches of the Lancang River comprises a fertile low-lying plateau which covers a relatively large belt of land extending from the central-southern part of the upper basin to Hsishuangpanna. Villages of houses on stilts (with elevated floors) dot the landscape and the paddy fields yield good harvests of rice. Sugar cane, peanuts, and other products are also grown. Aside from this fertile belt, the region of northern mountains is sparsely populated by villagers who cultivate the narrow strips of land on the banks of the rivers and streams which flow down these hills. The villagers also slash and burn the slopes of the hills to plant upland rice. The entire region holds much promise for hydroelectric power generation but the opportunities for improving agricultural output by irrigation are limited because of the lack of flat land.

The northern mountains are mostly covered by evergreen tropical rain forests and the topsoil consists of a thin deposit of weathered sandstone and igneous rocks. These tropical rain forests are receding as a result of logging and the slash-and-burn agricultural practices of the highland tribes in this region.

The Korat Plateau

Flying north-east from Bangkok, we pass over a range of low mountains. These are the Phetchabun Mountains (also known as the Dong Praya Yen Mountains) which run from north to south, dividing the Korat Plateau of central Thailand into its eastern and western sections.

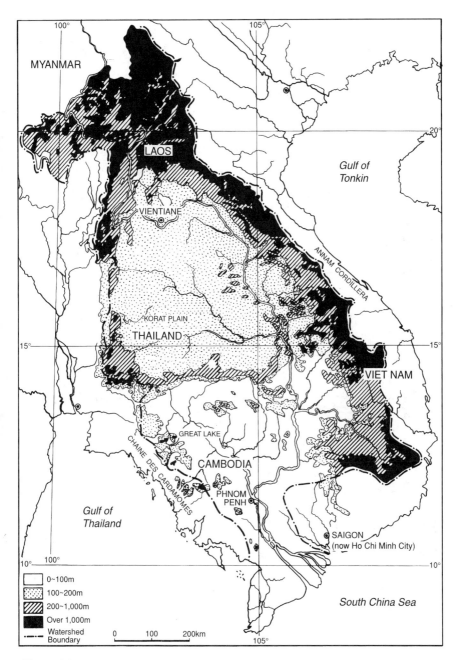

Figure 1.2. Topographical map of the lower Mekong Basin
Source: UN 1957

13

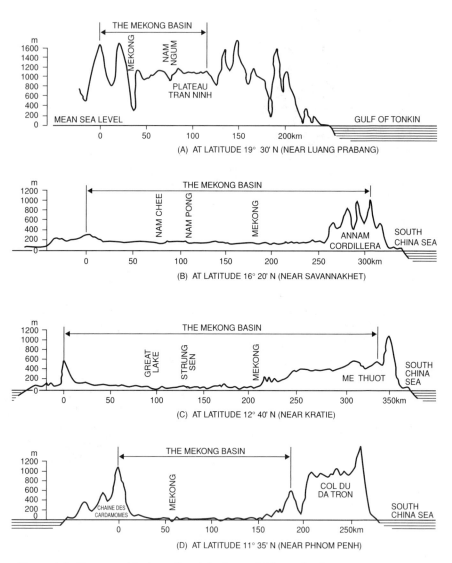

Figure 1.3. Topographical profile of the lower Mekong Basin
Source: UN (1957)

Referred to as Ihsahn (north-east Thailand), the area of the Korat Plateau is the poorest region in Thailand. The elevation of the plateau ranges between 100 and 200 m above sea level. This is a very mildly undulating area which slopes gently in a south-westerly direction. The Mun River flows from west to east through the middle of the plateau to join the mighty currents of the Chi River before completing its journey across the plateau and into the main stream of the Mekong. The Korat Plateau is not particularly fertile, but the alluvial regions of the Mun River and its tributaries have been developed for rice cultivation and pumped irrigation is spreading rapidly in this area.

South of the Korat Plateau lies the range of the Dangrek Mountains which run from east to west, separating this region from the lowlands of Cambodia to the south. The Korat Plateau is thus surrounded at its west and south by mountain ranges. During the monsoon, damp seasonal winds from the south-east blow towards the plateau but are stopped short by the mountains to the south. As mentioned previously, the plateau receives relatively little precipitation for this reason.

The Mekong flows through the rim of the Korat Plateau, tracing a lazy curve from north to east. Several of the Mekong tributaries flow through the Vientiane Plain and the Laotian flat lands which adjoin that plain; the alluvial regions of these tributaries are used for rice cultivation (as in the case of the Korat Plateau). Pumped irrigation can be expected to spread throughout this area in the future.

Because of its limited precipitation, the Korat Plateau is mostly arid. In the past, the plateau was studded with forests but most of these have disappeared as a result of deforestation and all that remains are grasslands and clumps of small woods. As a result, the rice fields of the Korat Plateau are now threatened by both floods and dry spells.

The entire plateau rests on a horizontal stratum of Mesozoic sandstone and shale. The surface of the ground is undulating and has a pinkish-grey hue. One look at the ground reveals that this is not an area suited for agriculture. However, there are some pockets of higher productivity, such as in the vicinity of the town of Loei located in the northern section of the plateau where rivers flow through the narrow strips of flat land sandwiched between the hills. The alluvial regions of these rivers and streams contain sand of fairly good quality which can support a relatively higher level of agricultural output.

The area in the vicinity of the town of Udon in the Korat Plateau rests on a bedrock with a high salt content. Salt water seeps through the topsoil from place to place and salt is produced during the dry season. Needless to say, the subterranean water sources in such areas are too brackish to be used.

The soil throughout the Korat Plateau is infertile and cannot adequately support agricultural activities. In addition, the average precipitation of 30 mm/month during the dry season makes it very difficult to maintain crops. Precipitation during the monsoon season reaches no more than 220 mm/month, which is insufficient for achieving stable harvests; the agricultural development of this area must, therefore, await the completion of the irrigation facilities noted earlier. However, irrigation may result in a rapid increase of salinization because of the rock-salt stratum on which the plateau rests. To avoid salinization, it will be necessary to develop adequate drainage facilities.

The region of the eastern mountains (Annam Cordillera)

The Annam Cordillera stands like a giant screen on the eastern edge of the lower Mekong Basin. Starting at its juncture with the northern mountains, the Annam Cordillera takes a south-easterly course as it slides down the spine of Viet Nam before coming to rest in southern Viet Nam. The mountains rise to elevations of more than 2,000 m while never dropping below the 1,000 m mark. Southern Laos and north-east Cambodia are separated from Viet Nam by these mountains. In their highest area, the Annam Cordillera mountains rise to more than 2,800 m. These summits are located in the upper reaches of the Nam Nhiep and Nam Ngum, two major tributaries which join the Mekong on the eastern side of Vientiane.

The Annam Cordillera gradually loses its elevation as it goes towards the south-east, but this downward slope is interrupted at the upper reaches of the Se San River (a major tributary of the Mekong) where the range rises again to an elevation of more than 2,000 m. This height is maintained up to the famous resort town of Dalat in southern Viet Nam. At their highest point, the mountains in this region rise to elevations of more than 2,500 m.

The region of the Annam Cordillera which adjoins the northern mountains is mostly covered by tropical rain forests. However, here, too, the forests are suffering from the excessive slash-and-burn agricultural practices of the mountain tribes and the encroachment of commercial logging. As a result, the forests of the Annam Cordillera are receding very quickly. The areas left behind by loggers and slash-and-burn farmers turn into grasslands with occasional shrubs and tall trees; only patches of forests remain in the folds of the valleys. Whereas the regions between an elevation of 600–1,000 m are covered by deciduous monsoon forests, those above the 1,000 m mark are covered by tropical evergreens. Both regions have only a thin topsoil layer composed of weathered sandstone and igneous rock. Agricultural productivity is therefore low, as in the case of the northern mountains.

The Boloveng Plateau of southern Laos which adjoins the Annam Cordillera covers an area of 100,000 ha and has an elevation of 700–1,000 m. The soil is a fertile laterite. Annual precipitation exceeds 2,500 mm and the climate is temperate, making this a favoured region for grazing and for the cultivation of coffee, tea, and other products.

The Vietnamese part of the Mekong Basin is located on the western side of the Annam Cordillera. Between the towns of Ban Me Thuot and

Kontum, the basin consists of a plateau with an area of roughly 3.3×10^6 ha. This plateau has an elevation of 500–600 m and is characterized by a gently rolling topography. The headwaters of two major Mekong tributaries, the Se San and Srepok rivers, are located in this plateau. The soil is fertile, consisting of a mixture of basalt and red soil (*terre rouge* or *terra rocca*) which is capable of supporting high agricultural productivity; for this reason, this region has many rubber plantations. Over the years, various proposals have been made for the irrigation of this plateau.

The southern section (lower Mekong Basin) and the lowlands
(area of the Great Lake and Mekong Delta)

The Khone Falls mentioned previously are located at a point where the Mekong meets the eastern edge of the Dangrek Range. The falls mark the border between Laos and Cambodia, the northern face of the falls being in Laotian territory while the southern side is in Cambodia. Going downstream from the Khone Falls, one approaches the town of Stung Treng and the confluence of the three mighty tributaries which have travelled down the slopes of the Annam Cordillera – the Se Kong, Se San, and Srepok rivers. The town of Kompong Cham is located downstream from the rapids at Sambor, which follow the confluence of these three rivers with the Mekong. The city of Phnom Penh is located further downstream where the Mekong meets the Tonle Sap River on the west to join with the waters of the Great Lake (figs. 1.1, 1.2).

Geological studies of the Mekong Delta reveal that this area was part of the ocean until the early alluvial era (roughly 6,000 years ago). During this period, the Great Lake and the lowlands in its vicinity formed a bay which was gradually filled by sand and silt carried by the Mekong. Eventually, sand bars formed in the area of Kompong Chhnang on the Tonle Sap River and this gave shape to the Great Lake.

The Tonle Sap River, which joins the Mekong from the west in the Quatre Bras region, is a relatively short river through flat terrain which carries the waters of the Great Lake a distance of about 130 km before meeting the Mekong. Some 90 km west of Phnom Penh is the busy fishing port of Kompong Chhnang.

During the dry season, water from the lake flows out through the Tonle Sap River and on to the Mekong. The direction of the flow is reversed during the rainy season, when approximately one-fourth of the flow of the Mekong is taken up by the Tonle Sap River and carried back into the lake. The lake thus acts as a temporary reservoir for the Mekong during the rainy season. This function of the Great Lake is one of the salient features of the lower Mekong Basin.

Because sand and silt from the Mekong are deposited in the Great Lake in large quantities, combined with the sand and silt directly carried

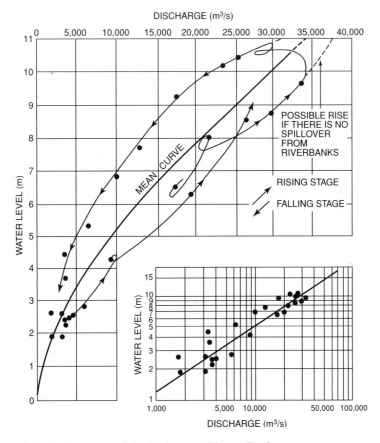

Figure 1.4. Rating curve of the Mekong at Phnom Penh
Source: UN 1957

into the lake from tributaries, it is feared that both the lake and the Tonle
Sap River have become much shallower in recent years. It now appears
that these fears are well founded.

   The water level of the Mekong begins to rise with the onset of the rainy
season in May and June. In the lower basin areas, the water level reaches
its maximum heights during September and October. With the end of the
rainy season between October and December the water level quickly
subsides and continues to decline gradually between January and April.
Although differences in precipitation affect the water level of the Mekong
from year to year, this annual cycle is repeated without fail. Because the
Great Lake is indirectly connected to the Mekong, its water level follows
this cycle: thus, between June and October of every year, when the level
of the Mekong rises from 2–3 to 10–11 m (fig. 1.4), the Great Lake also

rises gradually. During the dry season, the Great Lake covers an area of approximately 3,000 km^2. At this time, the lake has a maximum length of 150 km and a width of not more than 320 km. In the rainy season, the lake expands to cover approximately 10,000 km^2. With the inundation of the surrounding flood forests, the lake becomes passable to ships of considerable size. The waters recede at the beginning of each year to reveal a rich bed of newly fertilized soil.

It is estimated that about 60×10^9 m^3 of water flow into the Great Lake annually, making the lake an important factor in reducing the flooding in the delta. On the other hand, during the dry season, the lake replenishes the Mekong and increases the discharge in the delta. (It is estimated that the Great Lake receives a total of 85×10^6 m^3 of water annually from its tributaries.) This contributes to maintaining irrigation and navigation in the delta as well as to stopping the influx of sea water.

The Great Lake is surrounded by a low-lying valley (with an area of roughly 6,500 km^2) which is covered with forests. During the rainy season, the swollen waters of the lake reach the height of the branches of these trees and the entire area turns into an excellent breeding ground for freshwater fish. These flood forests are in turn surrounded by paddy fields which extend into the shallow portions of the lake. Beyond the rice fields is a sandy plain with sparse groves of trees that continue to the base of the surrounding mountains. Battambang, Cambodia's second-largest city, is located on the north-western fringes of the lake in the middle of a rice-growing area. National highways encircle the lake, and the cradle of the Angkor civilization is located beyond these roads at the north-western end of the lake.

A vast plain extends toward the south from the town of Kompong Cham and continues unhindered to the South China Sea. This is the Mekong Delta, an alluvial region formed by the sand and silt carried by the Mekong. The delta can be more exactly defined as a triangular plain starting at Kompong Cham or Phnom Penh and gradually expanding as it approaches the South China Sea. The delta is bound on the south-west by the Gulf of Thailand; on the south-eastern side it is hemmed in by the eastern branches of the Vaico River, itself a branch of the Mekong in the lowest regions of its basin. In total, this delta covers an area of 49,520 km^2. The northern end of the delta is Cambodian territory, while the southern side is located in Viet Nam. The Cambodian portion accounts for 26 per cent of the delta, with the remaining 74 per cent located in Viet Nam.

The water level of the rivers in the delta is high between September and November. During this period, the flood waters overflow the levees and inundate an area of more than a million hectares. In years of extreme flooding, the inundated region extends to between 3 and 4×10^6 ha. At such times, the water covering the area between Kompong Cham at the top of the delta and the city of Can Tho in Viet Nam can reach a depth of 4 m. The silt carried by the flood waters consists of relatively large particles and, when deposited along the banks of the Mekong and Bassac rivers, tends to form ridge-shaped natural levees. This combination of natural levees surrounded by flood plains can be seen throughout the upper delta in a highly developed form: here the natural levees of the upper delta are relatively high and the flood plains are deep. Towards the lower reaches of the delta the levees become significantly lower and the flood plains tend to be shallower. In the lowest part of the delta there are no natural levees and the floods form a shallow expanse over wide areas.

Along the coast of the South China Sea, the difference in sea level between high and low tides reaches as much as 4 m, giving the coastal lands excellent drainage even though the rivers in this area have very gentle slopes. The coastal area is marked with bands of outcrops formed when the sand carried by the Mekong and Bassac rivers is pushed back by the sea.

Because of the westerly direction of the coastal currents in this area, the sand, whose movement is stopped by the sea, gradually drifts toward the west where it accumulates. For this reason, the southern coast of the Ca Mau Peninsula in southernmost Viet Nam is slowly expanding into the sea and mangrove trees dot the landscape. Many lagoons have been created on the inside of the low ridges facing the coast and most of the area is covered by swamps.

Rice cultivation in the Mekong Delta can be characterized as follows. Long-stalked "floating" rice is generally grown in the area south of Phnom Penh and extending over the Cambodian border to as far as Long Xuyen in Viet Nam. As mentioned above, the Mekong in this area features relatively high natural levees. Flood waters reaching over the levees can reach depths of 3–4 m. To adapt to these conditions, floating rice with stems of up to 6 m is cultivated. Further downstream from Long Xuyen, rice cultivation is based on transplantation to wet paddy fields. In the area between Long Xuyen and Vinh Long, double transplantation is practised; this changes to single transplantation in the area between Vinh Long and the coast.

In certain parts of the delta, the soil is acidic and ill-suited to agriculture. This is the area of the Joncs Plain (a low-lying swamp area located in the Vietnamese region of the delta, sandwiched between the Mekong as it flows south from the Cambodian border and Vaico Occidental on the east). This area is completely barren. Another pocket of barren land is located in the interior of the Ca Mau Peninsula located in the southwestern section of the delta. Apart from these barren fields, rice is extensively cultivated throughout the entire delta. Irrigation facilities have been improved in recent years, with the result that yields have risen significantly. Because of these advances, Viet Nam has regained its pre-war position as a major exporter of rice; nevertheless, export capacity varies from year to year.

As mentioned earlier, it is believed that the trend towards global warming and the rising sea level may cause severe social and economic losses in the delta. There is much concern that the rising waters may do extensive damage to the densely populated coastal regions and their production of food.

The alluvial plains surrounding the Great Lake are not particularly well suited for agriculture: the sandstone base is lacking in low-acidic nitrogen and humus and is low in available lime, phosphates, and potassium. Nevertheless, these alluvial plains have a far higher agricultural potential than the area of the Korat Plateau. Furthermore, there are various topographical advantages: for instance, there is good potential for the development of irrigation dams and the future irrigation of these plains. The Great Lake is surrounded by a belt of swampy forests.

The Mekong Delta is completely flat, with the exception of a single isolated mountain in Tay Ninh, located in the north-eastern part of the delta, and the low hills situated near Long Xuyen. (The delta has an average slope of roughly 0.0085, or 1:118) This means that gravity irrigation is impossible and the entire area must depend on pumped irrigation. The soil of the delta consists of the fine silt sedimentation of the Mekong – clayey soil of medium fertility. The soil in certain sections of the delta is strongly acidic or contains toxic levels of sulphuric acid. The loss of water through evaporation increases the level of acidity in the soil and leads to the oxidation of surrounding land and water. This means that the water resources in these areas must be carefully monitored, even in areas where there is no irrigation. These areas are mostly low-lying swamps where the flood waters of the Mekong cannot be properly drained. As a result, the water level of the swamps declines mostly through evaporation, which leaves behind a residue of alum. This deposit obstructs the growth of vegetation and renders these lands unsuitable for agriculture. As mentioned above, these barren swamps are referred to as the Joncs Plain, which covers approximately 1×10^6 ha of land in the lower Mekong Basin.

During the period of French colonial rule of Viet Nam from the latter half of the nineteenth century through the first half of the twentieth century, the French authorities worked on the development of a canal system in the Mekong Delta. Their initial objective was to use the canals for transporting military equipment to counter the guerrilla forces, but subsequently the goal became the creation of irrigation and drainage systems for the rice fields and the development of agricultural land. Currently, these waterways not only are used for navigation and agriculture but also have an important role in the fishery industries; in addition, various sections are being used for shrimp farming. The canals of the delta thus perform the three functions of providing a means of transportation, raising agricultural production, and supporting fisheries and shrimp farms.

Apart from the unusable areas of the delta, the soil of the delta is generally of medium fertility. The productivity of most of these tracts can be greatly improved by providing for adequate drainage, effectively controlling floods, and undertaking proper fertilization. However, as mentioned above, the low and flat delta has an elevation of no more than 2–3 m. During the dry season, the water channels of the coastal area are thus threatened by sea-water influx. For this reason, wetland rice cultivation must be suspended during the dry season in areas which are relatively close to the coast, which is mostly covered by mangrove trees and swampy forests.

The mountains of the south-eastern section

The south-eastern section of Cambodia is covered by mountains and hills which are extensions of the Cardamom and Elephant ranges, these two mountain ranges being almost linked. The Battambang and the Stung Pursat rivers flow down the Cardamom Range and flow northwards to the Great Lake, while the Prek Thnot River flows from the Elephant Range, joining the Bassac River below Phnom Penh, and the Kampot River flows down to the Gulf of Thailand. All these rivers have the potential to support hydroelectricity generation and irrigation projects. During the French rule, an irrigation dam was constructed on the Battambang River.

The southern slopes of these southernmost mountains have very high levels of annual precipitation, which, in some areas, exceeds 4,000 mm. Agriculture has not made significant advances in this area because of the dense jungles and sparse population. Nevertheless, these mountainous regions are very promising in terms of water-resource development.

Geological features of the basin

From a macroscopic perspective, the Mekong Basin was created by the movement of the earth's crust dating back to before the Palaeocene period in the Cenozoic era. In modern times, this area has been relatively stable.

The Annam Cordillera on the eastern rim of the basin was formed during the Palaeozoic era, while the northern mountains were formed during the Archaeozoic, Proterozoic, Palaeozoic, and Mesozoic eras. The western side of the basin was formed during the Mesozoic era and Palaeocene period. The southern delta is said to have been formed as a result of the floods which rapidly eroded the valleys of the northern mountains during the Palaeocene period. The flood waters carried with them fine sand particles, silt, and clay, which created a wide-ranging area of diluvial and alluvial strata.

It is interesting to take a second look at the upper Mekong Basin north of Vientiane from the perspective of these theories. If they are correct, the main current of the Mekong and its tributaries can be viewed as a related group of rivers created when the steep slopes of the mountains of this region were eroded to form deep ravines and valleys. The fact is that these ravines reveal a complex mixture of phyllite, limestone, shale, clay slate, quartzite, schist, and igneous rock; furthermore, the area is covered with a highly fissured metallic accumulation. On the other hand, the Korat Plateau, situated south of Vientiane, rests on a stratum of sedimentary rocks from the Mesozoic era. The plateau was thus formed from the silt and sandstone deposits which accumulated in this region.

As briefly mentioned previously (p. 17), one theory about the formation of the Great Lake holds that the lake and much of the Mekong Delta were part of the sea until the early alluvial era (roughly 6,000 years ago) when the oceans of the world were in an expansive phase. In later times the sea receded and the Mekong covered the newly exposed land with sand, silt, and clay, causing the sea gradually to become shallower. In its present configuration, the boundary between the plains surrounding the Great Lake and the Mekong Delta is marked by the flow of the Tonle Sap River. In the area of Kompong Chhnang, which is situated along the Tonle Sap River, there is a low hill composed of granite and volcanic rocks. As the sea grew shallower over time, this rocky hill anchored the sand bars which were created in this area; this process culminated in the formation of the Great Lake. In other words, the Great Lake is the remnant of a huge marine lake. This lake is gradually shrinking in area and is becoming shallower on account of the large amounts of sand and silt (mostly silt) which flow into it during the rainy season from the Mekong via the Tonle Sap River. At the same time, the sand and silt deposits which flow back out of the lake during the dry season are accumulating near the junction of the Tonle Sap River, giving rise to a reverse delta phenomenon which is attracting wide attention.

Temperature, humidity, wind, and rain in the basin

Various comments have already been made, throughout this chapter, on climatic conditions. Observations on temperature, humidity, wind, and rain in the basin are summarized in this section.

The source of the Mekong is located in the snowcapped Tibetan Mountains; however, further downstream, where the river runs close to the border between China and Myanmar, it enters a temperate zone where the seasons are affected by rainfall and temperature cycles. These conditions persist until the river crosses the border between China and Myanmar. Beyond this point, the river flows through the torrid zone.

In the lower Mekong Basin, variations in temperature are slight throughout the year. Nevertheless, certain areas of the lower basin are subject to seasonal changes which can be attributed to differences in elevation. For instance, during the dry season the Boloveng Plateau in southern Laos (roughly 600 m above sea level) is noted for its mist and cool evening temperatures: before dawn, temperatures fall low enough to cause frost damage to crops. Further downstream, the delta and coastal regions are affected by the marine climate but show very little seasonal variation. Throughout the entire area, daytime temperatures are high. A brief respite from the heat is enjoyed between the months of November and February, when monsoon winds blowing in from central Asia bring somewhat cooler temperatures.

Humidity is at its highest during September, when it exceeds 80 per cent; by March, humidity falls below the 60 per cent mark. The rate of evaporation is lowest during January at approximately 80 mm and is highest in July at roughly 160 mm.

The lower Mekong Basin is swept by monsoon winds blowing in from the south-west between the end of April and the end of October; these winds reverse their direction between November and March, when they blow in from the north-east. Needless to say, the period of these wind cycles varies from year to year.

The summer monsoon winds from the south-west are wet and warm, and generally begin blowing through the lower Mekong Basin in April. In part, these winds are held back by the tall mountains of south-eastern Cambodia. Because of this natural barrier, the Great Lake and its vicinity – located to the north-east of these mountains – have no more than 1,500 mm of annual rainfall (fig. 1.5). On the other hand, the areas south-west of these mountains, which lie beyond the limits of the lower Mekong Basin, enjoy as much as 4,000 mm of annual precipitation. Most of the south-western monsoon winds remain unhindered until they reach the Dong Praya Yen Mountains, which divide north-east Thailand (part of the Mekong Basin) from central Thailand, and the Dangrek Mountains, which separate north-east Thailand from the area to the north of the Great Lake in Cambodia. Although both of these ranges are relatively low, the south-western monsoon winds lose much of their force at these points. Thus, the Korat Plateau and the area around the Great Lake receive a meagre 1,000 mm of rain annually. Any warm and humid winds

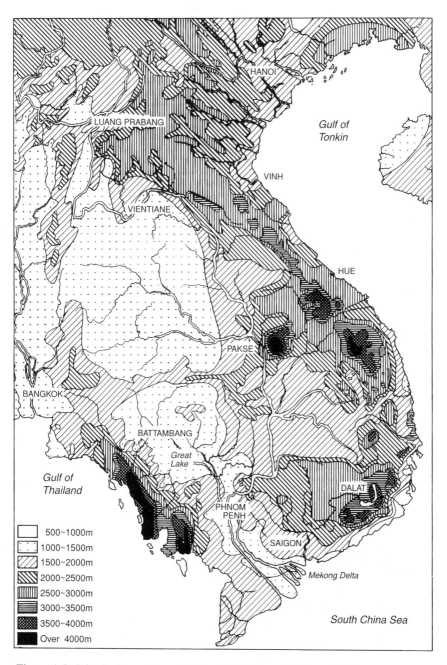

Figure 1.5. Distribution of precipitation in the lower Mekong Basin
Source: UN (1957)

Table 1.2 Average monthly rainfall in the lower Mekong Basin

Month[a]	Rainfall (mm)	Percentage of total
1	8	0.5
2	15	0.9
3	40	2.4
4	77	4.6
5	198	11.8
6	241	14.4
7	269	16.1
8	292	17.5
9	299	17.9
10	165	9.9
11	54	3.2
12	14	0.8
Total	1,672	100.0

Source: Ohya and Kuwahara (1974).
a. 1 = January; 12 = December.

which pass over these low mountains eventually collide with the western slopes of the Annam Cordillera and provide more than 1,500 mm of annual rainfall to the western slopes. The annual rainfall in the higher regions of the Annam Cordillera exceeds 2,500 mm. The Boloveng Plateau situated east of the southern Laotian town of Pakse receives more than 3,000 mm of rain in certain years.

The monsoon winds from the north-east, which blow through the lower Mekong Basin between November and March, bring dry air into the lower basin with the result that there is hardly any precipitation during this period. The lower Mekong Basin therefore experiences a rainy season which extends from April to October (with conditions of high temperatures and high humidity) and a dry season which lasts from November through March (relatively low temperatures and low humidity). This cycle has an important influence on the seasonal patterns of activity of the population of the basin.

As shown in table 1.2, the average annual precipitation for the entire lower Mekong basin is approximately 1,600 mm. This means that there is just enough water[5] to sustain rice cultivation in a year of average rainfall. However, in years of below-average rainfall, rice production faces a critical shortage of water. On a regional basis, the Korat Plateau and the vicinity of the Great Lake are perennially short of water. Irrigation is an essential requirement for rice cultivation in many other areas of the basin, owing to fluctuations in rainfall during the rainy season.

Throughout the rainy season, the lower Mekong Basin experiences frequent rain which is often accompanied by thunder and lightning.

However, the rains do not continue for long and tend to be local. A serious problem for this region is that it experiences violent typhoons from time to time. These tropical storms typically follow one of two courses: in the first instance, typhoons originating over the waters of the Pacific Ocean east of the Philippines take a westerly course and hit the Mekong Basin; the second type of typhoon originates in the South China Sea. On average, the Mekong Basin is hit by one typhoon every other year. The period between July and November is the most prone to typhoons and an approaching storm often causes torrential rains in the northern regions of the Mekong Basin. The worst of such rains was recorded in early September 1966, when a record-breaking flood swept through the Laotian capital of Vientiane. This flood was triggered by typhoon "Phyllis," which bore down on northern Viet Nam in early August and proceeded to make its way through northern Laos and on to the region of the Mekong Basin in northern Thailand and Myanmar. The heavy rains following the typhoon continued for about four weeks, while Vientiane itself was lashed by particularly severe torrential rainstorms. (Areas downstream from Vientiane did not experience serious flooding during this incident.)

Water level of the Mekong

The French authorities began recording the water level of the Mekong at Vientiane in 1897 and added other observation points (fig. 1.6), for example at Luang Prabang, Savannakhet, and Pakse. These observations have provided an excellent long-term record of the water level of the Mekong which is rarely matched in other developing countries.

The water level in the lower basin is determined by the amount of precipitation. To outline once more the water-level cycle of the main stream and tributaries of the Mekong, the rivers begin to rise in May and June as the south-eastern monsoon winds start blowing. In the upper reaches of the rivers, maximum water levels are registered during August and September, whereas in the lower reaches the maximum water levels are registered during September and October. Thereafter, the water level declines rapidly until December. After January, the level continues to decline gradually until April when the minimum levels are registered. The water-level cycle at various points in the lower Mekong Basin is as follows. At points upstream from Kratie, the water level is primarily influenced by the inflow from the various tributaries and is subject to sudden and wide fluctuations. These fluctuations follow the same pattern throughout most of the points in this region. However, the situation differs significantly after the Mekong passes the city of Kompong Cham, where the pattern of fluctuation becomes much more smooth and gentle. This is attributed to the holding reservoir function of the Great Lake

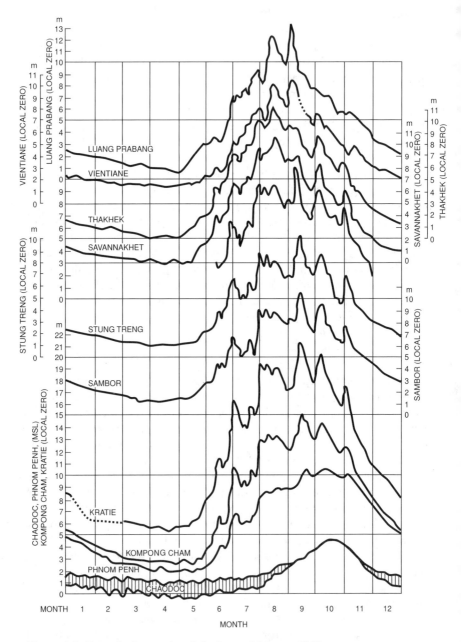

Figure 1.6. Stage hydrographs of the lower Mekong, 1950
Source: UN (1957)

and the overflow of water from both banks of the Mekong below this point.

River discharge

The upper Mekong Basin (the territories of China and Myanmar) covers an area of 189,000 km^2 and accounts for 24 per cent of the entire Mekong Basin. The total volume of water which annually flows out of the upper basin amounts to 80×10^9 m^3; this is roughly equivalent to 18 per cent of the annual discharge of the entire Mekong Basin (China accounts for 16 per cent and Myanmar for 2 per cent), which totals approximately 450×10^9 m^3 (although the discharge varies from year to year). The remaining 82 per cent, or roughly 370×10^9 m^3, originates in the lower basin (35 per cent from Laos, 18 per cent from Thailand, 18 per cent from Cambodia, and 11 per cent from Viet Nam).

Figure 1.7 shows the total area and the average annual discharges of the main tributaries of the lower Mekong Basin and their average shares (per cent) in the discharge of the entire basin.

Throughout most of the lower basin, the left bank of the Mekong faces mountainous regions (covering 28 per cent of the entire basin) which account for roughly 55 per cent of the inflow of water. On the other hand, the north-eastern regions of Thailand (covering 19 per cent) which are located on the right bank of the river contribute only 10 per cent of the water. The remaining areas (covering 29 per cent) supply approximately 17 per cent. In other words, more than half of the water which flows through the entire basin originates on the left bank side (Laos, Viet Nam, and north-east Cambodia) of the Mekong.

Let us define any year in which the discharge exceeds the average annual discharge by more than 10 per cent as a "wet year"; similarly, let us define a "dry year" as any year in which the discharge falls more than 10 per cent below the average annual level. A review of water-level observations made over a 40-year period at Vientiane and Kratie reveals that, on average, Vientiane experiences one wet year and one dry year in every 4-year period, whereas Kratie has one wet year and one dry year in every 5-year period. The records also reveal that a dry year is frequently followed by one or two more consecutive years of relatively low water levels. This pattern holds true for both Vientiane and Kratie.

This is a very important factor for the people of the lower basin, particularly from the viewpoint of agricultural production. As is discussed later in greater detail, the situation is especially serious in the delta where, during the dry season of an average year, the influx of sea water (salt content of more than 4 g/litre) extends nearly 50 km inland from the coast. Therefore, the people of the delta are seriously threatened by any

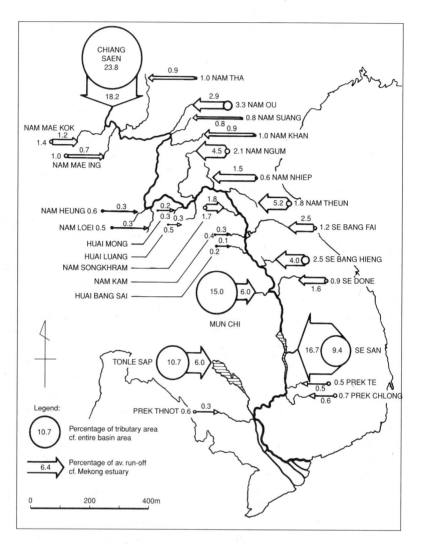

Figure 1.7. Run-off from tributaries in the lower Mekong Basin
Source: Kawai 1993

diversion of the waters of the Mekong during the dry season for purposes of irrigation or otherwise. This problem will certainly become even more serious in the coastal regions of the delta if global warming results in higher sea levels in the future.

Next, let us examine the problem of flooding in the lower basin. Of the total area of the Mekong Basin (795,000 km²), approximately 81 per cent, or 646,000 km², is located above the town of Kratie which is situated 545 km from the estuary. Of the 450×10^9 m^3 of water which annually flow through the Mekong, roughly 93 per cent pass through Kratie.

When the Mekong is flooded, it is generally possible to obtain measurement data relating to the flood at Kratie, but accurate measurement is not possible after Kratie because flood water starts to overflow both banks of the river beyond this point. In addition, the period for which water-level records are available varies from place to place. Table 1.3 presents the discharge (maximum, minimum, and average discharges) recorded at the principal observation points on the Mekong. Among the observation points cited in table 1.3, the river used to overflow both banks of the river at Vientiane. (The levees at Vientiane have now been reinforced to withstand the floods expected to be the greatest for a quarter of a century.) In the case of Vientiane, the maximum recorded flood volume is not very accurate because it includes the estimated overflow. On the other hand, the flood-volume data for Kratie are reliable because the river at this point is held in firmly by the banks. The highest recorded flood volume at Kratie was 66,700 m^3/s (recorded in 1939) and this can be used as an important benchmark for the development of the Mekong. (Based on the records of observations at Kratie, the author has estimated the flood volume of the largest projected floods in 100-year and 200-year periods to be 73,900 and 77,500 m^3/s, respectively.)

A review of the minimum discharges shown in table 1.3 indicates that the discharge falls to as low as 1,250 m^3/s near the Mekong Estuary. The ratio between this figure and the maximum discharge of 66,700 m^3/s is merely 1:53. The ratio between minimum and maximum discharges is normally referred to as the run-off coefficient of the river regime. This coefficient is 53 at Kratie, which is relatively high compared with the coefficient for the Nile at Cairo (30) or the coefficient of the Mississippi (3–21). However, this figure is very much smaller than the coefficient for many of the rivers of Japan[6] and indicates that the Mekong can be relatively easily controlled.

In the case of the Chao Phraya River in Thailand, the coefficient stood at approximately 30 when the river was in its natural state. However, it was reduced to 3 by the construction of various dams and sluices in the river and its tributaries. The problems of flooding and drought were effectively solved in this manner. In addition, these projects yield 2×10^6 kW of power and have made possible the irrigation of 2×10^6 ha of land. These figures should serve as an important reference in future development projects along the Mekong. Specifically, the construction of several large-scale and medium-scale dams at various points on the Mekong should reduce damage from flooding and drought in the Mekong Delta.

Standing on the banks of the Mekong at Phnom Penh, which is located

Table 1.3 Flow volume of the main stream of the Mekong River

Location	Drainage area (10^3 km^2)	Period of record	Flow volume (m^3/s)			Av. annual flow (10^9 m^3)
			Max	Min	Av	
Chiang Saen	189	1961–92	23,500	540	2,790	88
Vientiane	299	1913–92	26,000	700	4,020	146
Pakse	545	1934–92	57,800	1,060	10,400	327
Kratie	546	1924–68	66,700	1,250	14,000	441
Phnom Penh	663	1960–73	49,700	1,250	13,130	414

Source: Kawai (1993).

Table 1.4 Run-off coefficients of major world rivers

River	Country	Measuring site	Run-off coefficient
Kurobe	Japan	Unazuki	1,164
Yoshino	Japan	Chuohashi	658
Shimanto	Japan	Gudo	662
Nile	Egypt	Cairo	30
Ohio	USA	Metropolis	86
Colorado	USA	Grand Canyon	181
Thames	UK	London	8
Donau	Austria	Vienna	4
Elbe	Germany	Dresden	82
Rhine	Switzerland	Basle	18
Seine	France	Paris	34

Source: Sakaguchi et al. (1986)

at the head of the delta, the observer can view the following panorama during a flood. As the flood approaches its peak – say, at a discharge of more than 15,000 m^3/s immediately upstream from Phnom Penh – part of the flood water flows into the Great Lake via the Tonle Sap River, while the remainder continues down the Mekong to flood the banks of the Mekong and its branch, the Bassac. The overflowing waters find their way into the low-lying areas of the delta. During a flood of this magnitude the water levels of the two rivers increase gradually, rising at a rate of several centimetres a day; on the other hand, the flood water which flows into the delta remains trapped there for several months.

During the dry season, the influx of sea water extends to roughly 50 km from the coast (salt content of more than 4 g/litre). However, by the end of the flood season, the invading salt water is pushed back to a distance of about 15 km from the sea.

Water quality and temperature

Water quality in the lower Mekong Basin has been systematically monitored since 1972. Table 1.5 is based on figure 1.8, which depicts the average levels of ionized elements noted during the period 1985–1993.

The water quality of the Mekong is changing from year to year as a result of the development projects in the upper-basin regions of Yunnan Province, the expansive logging operations in north-east Thailand and parts of Laos, and the development of the delta. Nevertheless, until now, water quality has remained essentially uniform between Chiang Saen at the lower edge of the upper basin and Phnom Penh in the lower basin. In this stretch of the river, the mineral content of the water remains low. Moreover, during the rainy season, the average concentration of dis-

Table 1.5 Ionic components of river water in the lower Mekong Basin

Site no.	Av. percentage, 1985–1992/3					
	Ca + Mg	Alk	SO$_4$	Cl	Na	K
10501	40.2	36.0	8.6	5.3	9.1	0.9
11201	41.9	36.8	8.4	3.7	7.2	1.0
11901	41.9	36.8	8.3	4.0	7.2	1.0
13101	39.3	38.7	8.7	6.4	9.8	1.0
13801	40.4	38.0	8.4	5.4	8.9	0.9
13901	40.3	36.1	8.5	5.6	8.3	1.1
19802	39.8	38.7	2.5	3.6	8.5	1.8
19801	40.8	39.6	5.6	4.0	8.7	1.4
19803	39.1	38.3	5.7	5.9	9.7	1.3
39801	39.3	38.1	6.1	5.6	9.5	1.4
19804	39.6	38.1	6.2	5.4	9.4	1.2
39803	39.1	38.1	6.0	5.9	9.7	1.3
19805	20.2	10.8	8.0	31.4	28.6	1.0

Source: Mekong Secretariat (1995).

solved minerals falls to one-third of the level registered during the dry season.

According to data compiled in 1975, there was no evidence of pollution in the vicinity of Chiang Saen at the head of the lower basin; however, pollution caused by development in north-east Thailand was already in evidence below the confluence of the Mun and Mekong rivers. Nevertheless, the waters of the main stream of the Mekong in both the upper and lower basins (with the exception of the delta) can be utilized without any problem for irrigation, drinking, and industrial use. In any case, the pollution and level of impurities in the Mekong are said to be comparable to those in the other rivers of South-East Asia and these impurities can be easily removed through high-speed filtration.

The present-day data (table 1.5 and fig. 1.8) indicate declining levels of hardness (Ca + Mg) and alkalinity as the water flows from the upper to the lower regions of the Mekong. Regarding the concentrations of SO$_4$, Cl, Na, and K, there are no significant differences in observed levels throughout the upper and lower regions of the Mekong (with the exception of the delta).

According to a report written in 1975, the water temperature of the main current of the Mekong falls as it flows down the lower basin: the average annual water temperature at Chiang Saen is 27°C, compared with 22°C in the delta. This is an interesting pattern which runs contrary to the trend in atmospheric temperature. It may be possible to explain this drop in water temperature by arguing that the lower reaches of the river are more directly affected by the water temperatures of the inflow-

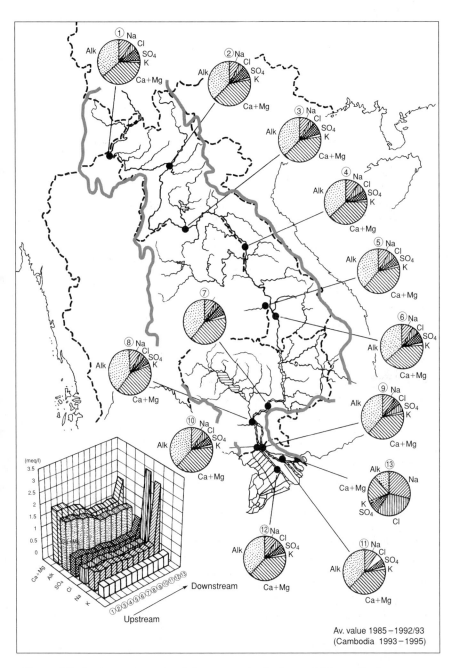

Figure 1.8. Ionic components of river water in the lower Mekong Basin
Source: Kawai 1993

ing tributaries. In any case, these figures may prove to be misleading and must be checked.

Notes

1. According to the ADB Project Profile Compendium (February 1995 edition), the total area of the basin is 795,000 km² and the total length is 4,880 km (Asian Development Bank [ADB] 1995).
2. According to a report by He Daming, of the Yunnan Institutes of Geography, Kunming, China, in the *Journal of Chinese Geography* Vol. 7, No. 4, 1997, entitled "Facilitating Regional Sustainable Development through Integrated Multi-objective Utilization Management of Water Resources in the Lancang–Mekong River Basin," the total length is 4,889 km.
3. Water-level differences at Paolo Condre located outside the Mekong Delta reach 3.9 m. However, water-level differences between high and low tides in the delta exceed this level by roughly 0.15 m.
4. The total land area of China is 9,597,000 km² and the total population is 1.2×10^9 (1995 figures).
5. The minimum precipitation required for rice cultivation is estimated to be 1,500 mm during the five months of the rainy season. Monthly precipitation during this period exceeds 300 mm (United Nations [UN] 1957).
6. The run-off coefficients of some of the major rivers in Japan up to 1980 are as follows: Tone River at Kurihashi: 1782; Shinano River at Ojiya: 117; Tenryu River at Kashima: 1430; Kiso River at Inuyama: 384; Kurobe River at Unazuki: 5075.

BIBLIOGRAPHY

Asian Development Bank (ADB) (1995). *A Compendium of Project Profiles.* Prepared for Forum of Comprehensive Development of Indochina, ADB, Tokyo, February 1995.

Davies B., and K. Walker (1986). *The Ecology of River Systems*. Dr. W. Junk, Amsterdam.

Hopkins J., and A. Hopkins (1991). *The Mekong*. Post Publishing Co., Bangkok.

Kunmin Hydroelectric Investigation and Design Institute (1985). *Cascaded Hydropower Projects on the Lancang River in Yunnan Province*. Kunmin Hydroelectric Investigation and Design Institute, Kunmin, Yunnan Province.

McKay, D. (1956). *ICA Reconnaissance Report on Lower Mekong River Basin.* US Bureau of Reclamation, March.

Mekong Secretariat (1970). *Report on Indicative Basin Plan.* Mekong Committee, Bangkok.

Mekong Secretariat (1976). *Environmental Effects of the Mekong*, Mekong Committee, Bangkok.

——— (1988). *Perspectives for Mekong Development.* Mekong Committee, Bangkok, April.

——— (1995). *Ionic Composition of the Mekong River.* Mekong Committee, Bangkok, July.

O'Neill T. (1993). *Mekong River*. National Geographic, Washington D.C., February.

The Times (1989). *The Times Atlas of the World*. The Times Inc., London.

United Nations (UN) (1957). *Development of Water Resources in the Lower Mekong Basin*. Flood Control Series No. 12, United Nations, Bangkok.

Viet Nam Government (1995). *Major Issues in Water Resources Development in the Mekong Delta*. SIWRPM, Ministry of Water Resources, Viet Nam Government, April.

2

Current conditions and future possibilities for resource development in the lower Mekong Basin

The regions of the upper Mekong Basin are covered in chapter 4 which includes a discussion of the dam-development projects on the Lancang River in Chinese territories. Chapter 2 outlines the current conditions and future possibilities for the development of various resources in the lower basin. The first section of this chapter focuses on the development of agricultural resources. Next, forestry, fishery, mineral resources, and navigation are considered in turn. In the final section of chapter 2, hydroelectricity generation is discussed.

Agriculture and irrigation

From a macroscopic perspective, it goes without saying that the principal crop of the entire lower basin is rice. With the exception of the delta, rice cultivation in all areas of the lower basin is mainly based on rain-fed paddy fields which do not depend on irrigation. This means that the agricultural methods tend to be traditional and the yields per unit relatively low. Productivity is also held back by the cycle of flooding and drought which affects not only the delta but also all areas of the lower basin. A further drawback is that the soil is relatively infertile. Although the progress of irrigation and drainage projects in the delta is supporting triple-cropping of high-yield rice in increasingly large areas, the coastal agricultural lands continue to suffer the ill-effects of salinization caused by the intrusion of sea water.

Table 2.1 Paddy cultivation in riparian countries, 1985

Annual production of unhulled rice	Kampuchea (Cambodia)	Laos PDR	Thailand	Viet Nam	All
National					
Area ('000 ha)	1,600[a]	700	9,600	5,900	17,800
Yield (t/ha)	1.1	1.9	2.1	2.8	2.2
Basin					
Area ('000 ha)	1,600	700	4,200	2,000	8,500
Yield (t/ha)	1.1	1.9	1.7	3.7	2.1

a. About half the area that could be cultivated.

Table 2.1 presents the annual production of unhulled rice in the four countries of the lower Mekong basin according to the *Perspectives for Mekong Development* (Mekong Secretariat 1988). The figures at the top differ somewhat from those in Britain's Economist Intelligence Unit third-quarter 1993 profiles of the three countries of Indo-China (Economist Intelligence Unit 1993) (table 2.2). However, the 24 June, 1991 issue of the *Nihon Keizai Shimbun* contains the following statement: "The rice exports of Viet Nam have increased very rapidly, rising from 100,000 tons in 1988 to 1.4 million tons in 1989 and 1.7 million tons in 1990. This is due to the success of Viet Nam's Doi Moi policies. Total rice production in Viet Nam passed the 20 million ton mark in 1989." These figures also differ from those in table 2.2; nevertheless, a general idea of the agricultural output of the four countries can be gained from tables 2.1 and 2.2.

Laos

Agriculture currently accounts for two-thirds of the Laotian gross domestic product (GDP), while three-quarters of the population (4.17×10^6 as of 1990) are farmers. For the most part, Laotian agriculture consists of traditional family farming, with an average of two hectares of cultivated land per household. Agricultural cooperatives were established in some areas but their numbers were limited, and the few that did exist had been dissolved by the mid-1990s.

Although Laos is known for its highland plateaus, a significant portion of the country is suited to paddy-field cultivation. These areas are located on the banks of the Mekong (this area is relatively limited) and in the middle and lower sections of the Mekong tributaries. Large tracts of arable land are available in the northern districts in the vicinities of Luang Prabang and Sayaboury. The central regions of Laos contain the fertile

Table 2.2 Production and imported volumes of unhulled rice in Viet Nam, Laos, and Cambodia

Country	Volumes ('000 t)	Year								
		1984	1985	1986	1987	1988	1989	1990	1991	1992
Viet Nam	Production	15,506	15,875	16,003	15,550	17,000	18,990	19,225	20,500	18,500
	Imported	284	422	500	450	200	20			
Laos	Production	1,321	1,395	1,449	1,207	1,003	1,404	1,491	1,223	1,653
	Imported	38	23	24	22	35	42			
Cambodia	Production	1,970	2,100	2,000	1,855	2,400	2,500	2,155	2,550	2,254
	Imported	100	55	80	88	105	50	20		

Vientiane Plain which draws its water from the lower reaches of the Nam Ngum River. The low-lying plains of southern Laos, which are watered by the Se Bang Fai, Se Bang Hieng, Se Kong, and Se Dong rivers, also constitute an area of extensive rice cultivation.

Laos has roughly 5×10^6 ha of arable land which accounts for 21 per cent of the country's total land mass (approximately 236,800 km^2). However, the area currently under cultivation amounts to about 700,000 ha, or only 14 per cent of the total arable land. The land is used primarily for rice cultivation and a variety of other secondary crops. At the present time, a little more than 120,000 ha are irrigated. Indications are that irrigation will reach 600,000 ha in the future. In addition, Laos has 750,000 ha of pasture land and 50,000 ha of reservoirs, led by the Nam Ngum Dam reservoir, all of which have been stocked with fish. On the other hand, slash-and-burn farming practices persist in northern Laos. As of 1995, slash-and-burn practices were in use in an area of 190,000–200,000 ha (Boupha 1995).

The Vientiane Plain constitutes the most important region for Laotian agriculture. Although certain regions of these plains are covered by marshes with poor drainage, approximately 100,000 ha are used as paddy fields. Japanese and Australian aid has been used to install irrigation pumps serving an area of 8,000–13,000 ha. In addition, gravity-based irrigation systems are in place in two regions of the plains. These two irrigation systems draw their water from two reservoirs, each with a capacity of approximately 40×10^6 m^3. Beyond the Vientiane Plain, various small-scale irrigation systems have also been installed, which serve a combined area of roughly 50,000 ha. These systems provide supplementary irrigation using small-scale weirs. It is reported that approximately 20,000 ha are served by relatively well-developed irrigation systems with good reliability.

The Boloveng Plateau in southern Laos is another promising agricultural region. As mentioned in the previous chapter, this region has fertile soil and ample precipitation and has been the site of coffee and tea cultivation for many years. It is also said that the Boloveng Plateau is the most suitable region for dairy production throughout the entire lower basin of the Mekong.

As we have seen, 75 per cent of the Laotian population is involved in agriculture, and agriculture is by far the largest source of production in the Laotian economy. A closer look reveals that, with the exception of those in the Vientiane Plain, individual farms tend to be very small. Modern farming methods have made little inroads into Laotian agriculture and the country faces difficulties in feeding itself. Furthermore, agricultural output tends to be seriously impaired in years of drought. Thus, the government of Laos is doing its best to improve this situation.

According to the *Basic Statistics (1975–1990)* of the Laotian government (Finance Ministry of Laos 1990), the productivity of rice cultivation has risen quite rapidly in the 17 Laotian provinces where efforts for agricultural improvement have been the most successful. In these regions, the average per-unit yield of rice (t/ha) stood at 1.27 in 1976 and 1.45 in 1980, but has continued to increase to 1.84 in 1988, 2.36 in 1989, and 2.30 in 1990. According to government statistics for 1990, the total production of rice in unirrigated low-lying paddy fields, irrigated fields, and slash-and-burn fields amounted to 1,068,000, 45,500, and 204,000 t, respectively.[1] Notable crops other than rice include coffee, tea, vegetables, corn, cotton, tobacco, peanuts, soybeans, and sugar cane. The total tonnage of crops harvested during the fiscal year 1990 came to 5.1×10^6.

In March 1989, the government made a significant reversal in its policy in order to promote more effective utilization of agricultural lands. A new law was enacted at this time which affirmed that "all land belongs to the state, but the state will do no more than administer the land." Under this system, the Laotian people were given the right to long-term use of the land. This law also contains a strict provision to the effect that farmers with low productivity must transfer their land to others.

Government-supported agricultural cooperatives were all dissolved by the mid-1990s. In place of these cooperatives, joint-procurement systems were instituted to facilitate the smooth distribution of fertilizers, seeds, pesticides, and other materials. Such developments as "the right to land use" and "the expansion of joint-procurement systems" indicate that the government is pursuing an active policy for fundamental reform in the agricultural sector. The government has also been promoting livestock farming in a bid to improve the nation's nutritional intake. Statistics for total livestock production in 1987 and 1990 are shown in table 2.3. The government's objective has been to maintain this rate of increase beyond 1990. However, it appears that the government is committed to a policy of reducing the number of water buffaloes.

Table 2.3 Livestock production in Laos

Livestock	No. of animals ('000)		Annual percentage increase, 1990–1995
	1987	1990	
Water buffalo	1,000	997	−1.6
Cattle	700	830	8.8
Swine	1,400	1,300	9.7
Sheep and goats	82	116	8.9
Domestic animals	8,000	8,039	5.9

Thailand

In 1981, agriculture accounted for 21.4 per cent of the GDP of Thailand. This ratio has followed a steady downward trend, reaching 12.4 per cent in 1990. By the same token, the share of agricultural products in total exports has dropped from slightly over 50 per cent in 1982 to 21.2 per cent in 1990. The primary reason for the declining share of agriculture in the Thai economy is that there is little room for developing new farm-lands. Another significant factor is that market prices for Thailand's primary agricultural exports have failed to show a positive trend.

The total area of agricultural fields in north-east Thailand amounted to 4.2×10^6 ha in 1985. This was equivalent to 44 per cent of the country's total area of agricultural land, which stood at 9.6×10^6 ha. However, agriculture in north-east Thailand is based on traditional methods and this district has a far lower ratio of irrigation (12 per cent or 500,000 ha) than the national average of 36 per cent. As shown in table 2.3, the farming villages of north-east Thailand report very low levels of productivity and remain impoverished to a great extent.

Agricultural activity in Thailand winds down during the dry season and farmers migrate to other areas in search of work. The only viable way to ensure a stable livelihood for these farmers, while obviating this seasonal migration, is to improve irrigation facilities in order to allow double-cropping. With this objective in mind, various large-scale irrigation projects have been planned and implemented in north-east Thailand, with irrigation reaching 140,000 ha in the basins of the Mun River and its tributary, the Chi River. In addition, various European countries have provided economic assistance for the completion of several irrigation projects in the Mun and Chi rivers as well as in some of the smaller river systems in this area, such as the Hoei Mong pumped irrigation project. As a result of these advances, during the 1980s the rate of irrigation grew at an annual rate of 9 per cent in north-east Thailand. This was nearly double the national rate of irrigation growth, which stood at 5 per cent.

Plans have been formulated for raising the height of some of the existing low dams on the Mun and Chi river basins and building new dams in order to take water from the Mekong during the rainy season to be stored for irrigation use during the dry season. Although some progress was made on these projects, further work has been suspended owing to opposition from Viet Nam, which is located downstream from these Thai projects. The Vietnamese position is as follows: "The use of the waters of the Mekong must be carefully considered from long-term and international perspectives. Work on any project must be preceded by an international agreement. National requirements do not provide adequate justification for embarking on a project without international consensus." The two countries have failed to reach a compromise on this issue and Vietnamese opposition to the Thai projects has continued. On the other hand, very strong local opposition to the construction of dams has made it virtually impossible for Thailand to undertake large- and medium-scale dam projects.

There is no doubt that the future of agriculture in north-east Thailand depends on the continued expansion of irrigation. However, in addition to the international issue noted above, there are various fundamental problems which loom as obstacles. The first problem is one that applies equally to the Vientiane Plain in Laos – that irrigation and the promotion of modern methods of rice cultivation do not in themselves constitute an adequate agricultural policy. Concrete steps must be taken toward promoting crops other than rice. Although the importance of this approach is appreciated, in practice it has proved very difficult to diversify the agricultural production of this region.

Turning to Thailand as a whole, although it is true that the dependence on rice cultivation is unchallenged, in terms of gross tonnage, the production of cassava compares favourably with that of rice. During fiscal 1991, a year of poor rice crops, the cassava harvest of nearly 20×10^6 t actually outstripped the production of rice (17×10^6 t). Other significant crops include corn, sorghum, mung beans, and oil seeds such as soybeans, peanuts, castor-oil plant, sesame, coconuts, and balm oil. Thailand also boasts a substantial production of cotton, kenaf, kapok, sugar cane, tobacco, rubber, pineapples, and coffee. The wide variety of vegetables cultivated include leeks, garlic, and onions. However, most of these crops are concentrated in areas outside north-east Thailand: for instance, northern Thailand accounts for 80 per cent of the soybeans grown in Thailand, in addition to 50 per cent of the peanuts and a sizeable portion of all vegetables. Thus, the situation in north-east Thailand remains poor and unaffected by this diversity.

Fortunately, it is reported that the soil and climate of north-east Thailand is suitable for the production of peanuts, soybeans, mung beans, and other crops. Indications are that these non-rice crops, which are being cultivated in other parts of the country, can be grown in north-east Thailand without great difficulty, shedding a ray of hope on the future of this region. However, low productivity is not the only problem in north-east Thailand: another important drawback is that the marketing system in this region remains generally underdeveloped.

Although the diversification of agricultural production, the improvement of irrigation technologies, and the proper development of market systems are issues common to the entire lower basin, the question is what specific measures can be taken to overcome these problems in north-east

Thailand and the Vientiane Plain in Laos. For the resolution of these problems, the Mekong Committee has been promoting the implementation of an action programme under the slogan of "the promotion of sustainable irrigated agriculture." This action programme can be outlined as follows:

1. Organizing farmers in areas suited to irrigation;
2. Strengthening the irrigation-related departments of responsible government agencies;
3. Promoting regional cooperation for the improved economic efficiency of irrigated agriculture;
4. Raising agricultural productivity in irrigated areas.

If the Thai government is able to resolve the impasse over the pumping of water from the Mekong which has currently been suspended, and if a comprehensive development programme for the agricultural sector in north-east Thailand – featuring the diversification of crops and the improvement of marketing systems – can be successfully implemented, there is reason to believe that agriculture in north-east Thailand will show dramatic progress.

Viet Nam

According to *Perspectives for Mekong Development*, Viet Nam has a total arable area of 11×10^6 ha; however, only 7×10^6 ha are currently under cultivation, 4.9×10^6 ha of which are paddy fields.[2] Total arable land in the Vietnamese portion of the Mekong Delta amounts to 3.9×10^6 ha with 2.4×10^6 ha currently under cultivation. Of this, irrigated paddy fields account for 2×10^6 ha.[3] The delta is thus of extreme importance to Viet Nam.

As mentioned previously, nearly one-half of the total area of the Mekong Delta remains inundated during the rainy season, which continues for 4–6 months every year. The peak of the inundation occurs in October (a major flood occurs approximately every 10 years). In the northern reaches of the delta near the Cambodian border, rice cultivation depends on long-stalked "floating" rice, which has a low yield.[4] Because of the predictability of the annual floods and dry spells, the farmers in this region lead a relatively stable existence. Nevertheless, it must be remembered that between 1.2 and 1.4×10^6 ha of farmland remain under water for an extended period every year. The most reliable way of raising productivity in this region would be to reduce the flooding by introducing effective flood-control systems on the Mekong and its tributaries. In this regard, Viet Nam has no choice but to await the development of dam projects in the upstream territories of its neighbouring countries.

Some areas of the delta do not suffer from flooding and inundation.

However, as we have seen, the coastal parts of these areas are subject to extensive salinization due to the intrusion of sea water. The water level of the Mekong in the delta begins to subside as the end of the rainy season approaches in December. By April and May, the flow rate falls to roughly 2,000 m^3/s and the sea begins to encroach on these areas as the level of the river falls. During the dry season, the effects of salt water are felt over an area of $1.7–2.1 \times 10^6$ ha for a period of 1–8 months. Normally, a flow rate of 2,000 m^3/s would be sufficient to sustain irrigation in the dry season. The problem of the delta is that up to 1,500 m^3/s of water must be used to push back the influx of salt water, leaving only 500 m^3/s for irrigation needs. Thus there is a constant shortage of water for irrigation during the dry season. Moreover, the efficiency of cultivation and the harvest yield in the coastal fields steadily decline as the salt level rises.

Certain regions of the Mekong Delta face yet other problems. Approximately 1×10^6 ha of reed fields and low-lying marshes which extend from the environs of the city of Long Xuyen to the sulphate-bearing land of the Ca Mau Peninsula cannot be cultivated. As mentioned in the preceding chapter, when the acidic soil of the reed fields and areas near Longshan becomes soaked during the rainy season, the effluence of acidic water from these areas raises the acid level of nearby lands and renders the entire region unsuitable for cultivation.

Although the Ca Mau Peninsula is not affected by flooding, poor drainage in the area results in partial inundation from heavy rains. Roughly $1.0–1.2 \times 10^6$ ha remain submerged for periods of 2–4 months, rendering them unsuitable for cultivation. Although it is very important for the future development of agriculture in the Mekong Delta that these problems of the reed fields, the environs of Long Xuyen, and the Ca Mau Peninsula be effectively removed, the fact remains that no easy solution is available.

With the exception of these difficult regions, agricultural development in the Mekong Delta can be effectively promoted through dry-season irrigation. As mentioned earlier, of the total 2.4×10^6 ha currently under cultivation in the delta, approximately 2×10^6 ha have already been irrigated. Nevertheless, the construction of irrigation facilities lags behind in some of the most promising agricultural areas of the delta.

Fortunately, large portions of the delta are unaffected by salinization and sulphate-bearing soil and are blessed with sufficient water resources. One example of prime agricultural land is the area sandwiched between the banks of the Mekong and Bassac rivers, extending from the Cambodian border southward to within 40 or 50 km of the coast. Superior conditions and extensive irrigation result in very high yields in such areas. However, not even these lands are free from the threat of flooding and various protective measures, such as water pumps and low dykes, are in place to reduce the damage from flooding.

In the Mekong Delta, rice is harvested during the periods of May–June and October–November. Of the 2×10^6 ha of irrigated rice paddies of the delta, roughly 900,000 ha are amenable to double-cropping and 100,000 ha are capable of supporting triple-cropping. The per-hectare yield averages 3.7 t of unhulled rice per harvest (1991). However, it should be noted that this figure results from the use of high-yielding varieties of rice in irrigated fields with adequate use of fertilizers and pesticides. In addition, as is often observed, high-yielding varieties are inferior to conventional varieties of rice in terms of taste, and require further improvement. The rice harvests of fiscal 1993 were particularly good, while prices remained stable. (Export prices for unhulled rice in 1991 stood at 280 US$/t for Vietnamese rice and 302 US$/t for Thai rice.) Because of these favourable conditions in 1993, Viet Nam's combined rice exports from both the north and south stood at 1.1×10^6 t as of the end of August (roughly a 9 per cent improvement over the previous year).

Although Viet Nam continues its efforts to increase the production of rice in the delta, rice is not the only export crop of this region: such products as coffee, sugar cane, pineapples, and bananas are also being grown and exported from the delta. Compared with the northern and central regions of the country, the delta has the advantage that flooding and other natural disasters are very predictable and tend to occur in the same place from year to year. This means that farmers in areas other than these trouble spots can predict and respond very smoothly to changing climatic conditions, such as the water level of rivers, amount of rainfall, temperature, and humidity. Nevertheless, farmers are not completely secure from the threat of flooding during the rainy season and the shortage of water during the dry season.

The key to easing the delta's shortage of water during the dry season finally lies in pursuing the development of upstream dams and improving the utilization of the waters of the Great Lake and Tonle Sap River in Cambodia. The building of dams will contribute to agricultural development by facilitating effective flood control. These dams will provide an added bonus in the form of hydroelectricity. However, as has been repeatedly stated, the future construction of large-scale dams faces many difficult obstacles. Moreover, none of these prospective dam sites are located within Vietnamese territory.

Cambodia

Cambodia's agricultural soil can be roughly divided into the following five categories: sandy, neo-alluvial, proto-alluvial, *terre rouge*, and *terre noire*.

Sandy soil can be found in the provinces of Kampot, Takeo, Svey Rieng, Kompong Chhnang, Kompong Thom, and Siem Reap. Poor in nutrients and high in acidity, these soils can support only very low yields of rice and are used for other crops such as sugar palms, kapok, pepper, coconuts, and peanuts.

The neo-alluvial soil is found in the natural levees of the Mekong and its tributaries and in the clay layers situated beyond these levees. The levees are widely used for the cultivation of vegetables and fruit. The areas behind the levees are covered by mud containing lime and phosphates carried by the river. When not inundated, they are used for growing such crops as tobacco, beans, persimmons, and sesame.

The proto-alluvial soil contains clay deposits and covers large tracts of land located at a distance from the rivers of the basin. Deprived of fresh deposits of mud, these lands are low in nutrients and high in acidity. The fertility of the clay content can support a relatively high level of rice production, but the land is poorly suited to other agricultural uses.

Terre rouge (red soil) containing basalt is found in part of the provinces of Kompong Cham, Kratie, and Ratana Kiri. Together with the *terre noire* (black soil), which is found in parts of the province of Battambang, this soil is suitable for dry-land cultivation.

Rice is, of course, the mainstay of Cambodian agriculture. In Cambodia, rice cultivation is concentrated on the banks of the Mekong in the provinces of Kandal, Takeo, Prey Veng, and Svey Rieng and along the shores of the Great Lake in the provinces of Battambang, Siem Reap, and Pursat. Cultivation of long-stalked "floating" rice is commonly seen along the Great Lake and in the frequently inundated areas along the Mekong and the Bassac near the Vietnamese border. In addition, pockets of dry-land rice cultivation can be found in the hilly areas. Because of the lack of water control, the great majority of Cambodian rice fields are limited to a single crop per year, which is cultivated during the rainy season from May to October. Farming methods are primitive and fertilizers are not used, resulting in relatively low yields.

During the 1970s, the South-East Asian Research Center of Kyoto University conducted an extensive investigation of the Mekong Delta conducted under the aegis of the Japan International Cooperation Agency (JICA). According to the report of this study, which was published in March 1975, paddy rice cultivation accounted for 2.8×10^6 ha, 85 per cent of Cambodia's total area under cultivation. Of a total national population of 6.5×10^6 people, 62 per cent were engaged in agriculture for a total of 800,000 farming households. Even in years of poor rice harvests, such as in 1956, the country managed to export 320,000 t of polished rice, earning 36 per cent of its total export revenues from rice shipments to foreign markets. Although it is not known what the nation's total rice production was at that time, other sources state that Cambodia registered a record-breaking rice harvest of 2.5×10^6 t of unhulled rice in the 1968–69 season. However, as indicated in table 2.2, Cambodia subsequently became a chronic importer of rice, owing to the long period of hostilities and incessant internal turmoil. Nevertheless, as the table shows, Cambodian rice production in 1989 reached 2.5×10^6 t, to match the peak levels of the past. This achievement was the result of improved yields per unit; it appears that the total area under rice cultivation has fallen below 2×10^6 ha. The 21 June 1992 edition of the Japanese newspaper *Mainichi Shimbun* reported that Cambodia would have a 150,000 t shortage of rice in 1992. Table 2.2 shows that the output of unhulled rice dropped to 2,254,000 t in 1992, a 300,000 t decline from the previous year. Because Cambodian rice cultivation is dependent upon rainfall, production levels are unstable. In addition, Cambodia is second only to Viet Nam in terms of annual flood damage.[5] Furthermore, there has been a marked deterioration in the management of water resources as a result of the ill-conceived waterways and reservoirs which were built by the Pol Pot regime.

Table 2.4 presents the recent production figures for Cambodian agriculture, livestock farming, fisheries, and forestry. This table is derived from the June 1992 report of the World Bank; unfortunately, the figures for rice production differ from those shown in table 2.2, which is based on the estimates of the Economist Intelligence Unit.

As outlined earlier in this section, Cambodia harvests several other grains which are cultivated to suit the terrain and soil of various regions of the country. Corn constitutes a major crop and is cultivated behind the natural levees. Normally, corn is rotated with green beans. Other agricultural products include cassava, kapok, sweet potatoes, and pepper. Kapok is a fine, fibrous substance surrounding the seeds of *Ceiba pentandra*, which is a tree generally found in sandy soils and which is cultivated extensively throughout Cambodia; kapok is used to provide stuffing material for cushions, pillows, and mattresses. Pepper is mostly grown in the coastal regions of the province of Kampot and is of a very high quality. Other common crops include sugar palms, peanuts, and sesame.

Cambodia was formerly well known for its rubber production, which used to be one of the country's leading industries. Until the Viet Nam War, large-scale rubber plantations owned by the French and Chinese were in operation in the red-soil regions of the provinces of Kompong Cham and Kratie. Cambodian rubber was noted for its high quality and uniform texture and was almost totally exported. These plantations fell into ruin during the war.

Forestry resources

The forests of the lower Mekong Basin can be divided into two broad categories – deciduous and evergreen. The former can be further divided

Table 2.4 Production volume of agriculture, livestock, fisheries, and forestry in Cambodia

Product	Unit ('000)	1980	1981	1982	1983	1984	1985	1986	1987	1988	1989	1990	1991
Cereals													
Rice	t	1,717	1,490	1,949	2,039	1,260	1,812	2,093	1,815	2,074	2,278	2,150	2,400
Corn	t	101	85	51	43	48	42	51	38	45	54	36	56
Cassava	t	152	182	76	42	31	17	62	46	72	64	58	75
Sweet potatoes	t	45	59	31	16	14	16	27	17	40	24	34	54
Vegetables	t	106	290	131	172	141	143	156	146	193	193	170	249
Sugar	t	23	56	240	300	190	169	154	162	139	245	150	304
Rubber	t	1	4	7	9	13	18	25	25	31	33	23	24
Livestock													
Dairy cattle	Head	772	917	1,143	1,271	1,436	1,560	1,705	1,852	1,947	2,098	2,234	...
Water buffalo	Head	375	404	482	50	603	613	635	659	675	741	737	...
Swine	Head	131	223	723	824	1,009	1,203	1,161	1,251	1,531	1,741	1,516	...
Domestic animals	Head	2,442	2,883	4,779	4,595	5,430	6,398	7,347	7,164	9,171	8,720	8,164	...
Fish													
Fresh	t	20	52	69	68	64	71	74	82	87	82	111	118
Dried	t	...	4	5	3	3	3	3	3	3	2	1	3
Timber													
Logs	m³	0	11	68	90	73	97	214	306	283	225	257	322
Board	m³	13	1	25	27	23	13	26	27	19	16	16	...
Charcoal	t	4	2	9	11	5	4	4	6	19	7	7	...

Source: World Bank (1992).

into the deciduous monsoon forests, which are found in areas of relatively high precipitation at elevations of less than 1,000 m, and savannah-type growths which appear in dry river valleys with low precipitation. The category of evergreen forests can be divided into the following five sub-groups: evergreen forests on hilly terrain at elevations of more than 1,000 m; tropical rain forests growing in filtered moist soil in high-precipitation areas (more than 2,000 mm/year) at elevations below 1,000 m; coniferous forests growing in sandy soils in medium-precipitation areas (600–1,400 mm/year); freshwater swamp forests found in the vicinity of the Great Lake and parts of the Mekong Delta; and the mangrove forests of the Mekong Delta.

The hilly terrain evergreen forests are located in northern Laos and north-east Thailand, where they are constantly threatened by the slash-and-burn farming practices of the Meo and other tribes. Furthermore, the coniferous forests are of limited scale and distribution. On the other hand, the coastal regions are covered by mangrove forests which provide ideal breeding grounds for crabs, shrimp, and various fish. These mangrove forests also play an important part in the prevention of coastal erosion. The freshwater swamp forests around the Great Lake make a vital contribution to the maintenance of fishery resources. In the preceding section, brief mention was made of the rubber plantations, located near Snoul in eastern Cambodia. The World Bank production figures (table 2.4) indicate that the output of these plantations has been slowly recovering in recent years.[6]

Deforestation

Brief reference has already been made to deforestation in the section in chapter 1 on the topography, soil, and vegetation of the Mekong Basin (p. 12). According to the 1988 report of the Mekong Committee entitled *Perspectives for Mekong Development*, approximately 50 per cent of the entire lower basin was covered by forests until around 1970; however, the forests have been pushed back by the significant increase in population. In 1985, roughly 8.5×10^6 people of a total population of 46.3×10^6 in the lower basin depended on slash-and-burn agriculture, the effects of which were apparent over an area of 175,000 km^2, equivalent to 30 per cent of the lower basin. As a result, forests now cover no more than 27 per cent of the lower Mekong Basin.

According to a study published by the Food and Agriculture Organization (FAO) in 1980, the forests of Thailand are receding at an annual rate of 330,000 ha, while reforestation is reclaiming 13,000 ha. In net terms, Thailand is therefore losing nearly 320,000 ha every year. In Laos, 125,000 ha are destroyed annually, with 1,000 ha reclaimed through reforestation, leaving a net loss of 124,000 ha/year. The ratio of deforestation is higher in Laos than in the other countries of this region.

According to *Perspectives for Mekong Development* (Mekong Secretariat 1988), the four riparian countries of the lower Mekong Basin produced a total of 79×10^6 m^3 of logs in 1985; in contrast, these countries jointly consumed 115×10^6 m^3 in 1982; as much as 100×10^6 m^3 appear to have been used for fuel. In overall terms, Cambodia and Laos continue to have access to surplus wood and should be able to continue the export of processed lumber until at least the year 2000. On the other hand, north-east Thailand and the Mekong Delta regions of Viet Nam were experiencing a net shortage of wood as early as 1985. The report indicates that, in the year 2000, Thailand and Viet Nam will register net shortages of 33×10^6 and 19×10^6 m^3, respectively. In particular, a serious shortage of fuel wood is expected.

The marked increase in logging, which dates back to the 1970s, has resulted in serious problems involving soil erosion and the depletion of fishery resources. These issues currently have become the subject of very close scrutiny. There is a real danger that the lower basin will be denuded of its forests in the future. To stop this process, it will be necessary to rectify the view that forests are readily exploitable resources. By the same token, the casual acceptance of slash-and-burn agriculture must be reconsidered. The countries of the lower basin will have to decide whether priority should be given to the environment or to economics: in other words, they will have to choose between long-term advantages and transitory gains. In any case, there is a urgent need to halt the ongoing process and to reconsider the conventional values which have governed the exploitation of forests. However, it is very difficult to overlook the immediate and pressing needs of poverty in these countries.

Thailand

The rapid pace of deforestation in north-east Thailand has given rise to a sharp increase in flood damage. For instance, it is believed that the record-breaking floods of 1950 were primarily caused by unregulated logging. The primary reasons for deforestation have been analysed as follows in the 1982–83 edition of the *Summary of the Thai Economy* published by the Bangkok Office of the Japanese Chamber of Commerce and Industry:

1. The development of new roads has resulted in a rapid increase in cassava and maize cultivation;
2. The rapid population expansion has resulted in increased demand for firewood in rural areas;
3. Commercial loggers in Thailand have not shown any great interest in reforestation;
4. Measures for forestry preservation are not properly implemented.

In order to stop the deforestation, since 1977 the Thai government has banned the export of logs and permits only the export of processed lumber.

Laos

In the case of Laos, the forests which cover nearly 50 per cent of its total area represent the most valuable natural resource available to the country. Traditionally, Laotian farmers have combined agriculture and forestry to support themselves. However, a sizeable proportion of the farming population, particularly in northern Laos, is engaged in slash-and-burn agriculture.[7] Furthermore, rapid deforestation was triggered by the entry of large-scale logging operations from Japan, Thailand, and other neighbouring countries in the 1980s.[8] Faced with the looming crisis, the Laotian government followed the example of Thailand in banning the export of unprocessed logs. In October 1987, the government took the first step to institute a partial ban on log exports. This measure had a twofold purpose – to protect the country's forestry resources and to promote the development of the domestic wood-processing industry. In 1988, the government introduced a total ban on the export of logs. However, this measure evoked a strong reaction from Thailand and, in January 1989, Laos agreed to lift the ban on log exports to Thailand.

As a result of these developments, among all major export products, the export of wood products now constitutes the largest single source of revenue for the Laotian government. Thus, the export of electric power to Thailand, which previously held the top spot, has now fallen to second place after wood products. Although the government welcomes the growth of export revenues, since 1989 it has levied a high tax on the logging of trees and other materials in a bid to preserve the valuable resources of the forests. The government is hoping that these measures will restrain the expansion of logging operations and that the growth in the volume of logging during the 5-year period 1990–1995 will remain below 20 per cent. In any case, it is clear that the government finds itself in the difficult position of having to balance the two conflicting demands of "the forests as a source of income" and "the protection and preservation of forests."

Viet Nam

Viet Nam has suffered severe damage to its forests as a result of the various military operations of the American forces during the war. In March 1992, Prime Minister Kiet ordered the Ministry of Forestry and the Ministry of Tourism and Trade to implement a total ban on the export of logs. This measure was aimed at stopping the deforestation which was proceeding at a particularly alarming rate in central and southern Viet Nam. Before March 1992, the export of logs constituted the third most important source of foreign exchange for Viet Nam, following closely on oil and rice exports; however, the government was forced to opt for the protection of the environment over the pressing needs of economic reconstruction. Prime Minister Kiet ordered the introduction of severe restrictions on logging operations, together with the ban on log exports. It is reported that the central government is now urging the regional and local governments to undertake a review of the licensing of lumber mills.

Cambodia

More than 70 per cent of the Cambodian land mass was covered by forests before the outbreak of hostilities in that country. Up to one-third consisted of dense forests of choice and first-class timber and 20 per cent consisted of semi-dense forests containing building materials and timber for general use. Another 40 per cent consisted of sparse forests containing trees for general use, charcoal, and firewood. The remaining 10 per cent consisted primarily of mangrove forests.

Cambodian logs are of high quality and were being exported to many countries of the world, including Japan. The Mekong and the other rivers of the region were used in transporting the logs to Phnom Penh and the seaport of Sihanoukville. However, the river transport of logs has come to a halt. Even after the end of hostilities, the forests of Cambodia remain devastated and unattended because of the lack of security in jungle areas. This is exacerbated by the fact that 90 per cent of the entire population depends on firewood and charcoal as its main fuel (table 2.5). One of the most serious concerns of deforestation is that the resulting soil erosion seriously threatens to choke the Great Lake. The lake has been becoming shallower and this has already had an adverse effect on its fishery resources.

Fishery resources

Fishery resources play an important part in the livelihoods of the people of the lower Mekong Basin, particularly the people of Cambodia and the delta region of Viet Nam.

It is reported that more than 400 species of freshwater fish inhabit the lower Mekong Basin, of which approximately 150 species are brought to market. Of these marketable species, it is said that up to 50 are economically viable, while a total of about 20 species account for the bulk of the fish traded in the markets. Some 150,000–200,000 t of fish are brought to market annually, while another 100,000 t are consumed locally, bringing the annual catch to 250,000–300,000 t. In addition, the coastal regions of

Table 2.5 Wood production and exports in Cambodia

Production/export volumes ('000 m³)	Year												
	1966–68	1980	1981	1982	1983	1984	1985	1986	1987	1988	1989	1990	1991
Products													
Log	347.9	0.24	11.03	67.7	90	73.28	96.53	213.55	306.16	282.95	224.83	257.35	308.89
Board	–	0	1.44	2.99	3.41	4.63	8.58	9.99	11.78	12.75	16.37	15.71	–
Firewood	331	26	30	82	200	164	84	99	57	96	123	105	–
Charcoal	–	3,500	8,000	8,500	10,620	21,170	53,100	4,272	7,428	9,376	6,980	6,951	–
Exports													
Log	78.8	0	0	1.75	15.16	15.17	25.12	25.47	84.68	106.59	78.72	93.75	131.79
Board	7.57	0	0	0	0	0	0	0	0	0	0.98	5.52	5.33

Source: World Bank (1992).

Table 2.6 Estimated fish production in the lower Mekong Basin, 1990

Country	Total fish production (t)	Natural fish catch (%)	Fish farming (%)
Cambodia	60,000	80	11
Laos	20,000	96	4
Thailand	180,000	50	50
Viet Nam	250,000	51	49

Source: Mekong Secretariat (1990).

the South China Sea yield an annual catch of 250,000 t. The total annual catch of the lower basin area thus comes to approximately 550,000 t (table 2.6). A detailed survey of the types of fish marketed is presented in a section on the impact of dam development in chapter 5 (pp. 267–279). Regarding the freshwater catch, carp and catfish make up the bulk of the fish brought to market. Also included in the freshwater catch are various species which migrate over long distances in the middle reaches of the Mekong.

To cite some old statistics, in the mid-1960s the fishing industry in the lower basin area accounted for 2.0, 8.8, and 3.0 per cent of the GDP of Thailand, Cambodia, and Viet Nam, respectively. At the time, it was reported that residents of the lower Mekong Basin depended on fish for 50–75 per cent of their total intake of animal protein. It was generally held that the annual per capita consumption of fish in Cambodia and Thailand amounted to 20 kg, placing these countries among the top consumers of fish in the world. During the same period, annual per capita consumption of fish in the delta region of Viet Nam was estimated to be 12 kg, which itself is a relatively high figure. It can be safely assumed that there has been no dramatic shift in these figures from the mid-1960s and that the same basic trends persist today. Nevertheless, it is noted that consumption has declined in certain areas because of resource exhaustion caused by environmental destruction and the rapid growth of population resulting in over-fishing.

Needless to say, the largest single source of freshwater fish in the entire lower basin area is the Great Lake (Lake Tonle Sap) in Cambodia (table 2.7); this is followed by the Nong Hang marshlands in north-east Thailand.

The Great Lake boasts an extremely high level of fish production, with a projected annual catch of 66,000 and 72,000 t for the 1981–82 and 1982–83 fiscal terms, respectively. It is notable that, with the exception of some of the Chiam tribes, the Cambodian fishing industry is primarily controlled by immigrant Vietnamese and Chinese populations to whom the lion's share of the profits of this industry accrue. Various reasons are

Table 2.7 Commercial fish yield (metric tons) in Cambodia

Yield (t)	Year											
	1980	1981	1982	1983	1984	1985	1986	1987	1988	1989	1990	1991
Catch of freshwater fish	18,400	50,780	65,700	58,681	56,703	59,400	66,381	64,654	65,800	56,038	71,500	81,400
Catch of river and Great Lake fish	18,400	50,780	65,700	58,681	55,093	56,400	64,181	62,154	61,200	50,500	65,100	74,700
Stock of lake fish	–	–	–	–	1,610	3,000	2,200	2,500	4,600	5,538	6,400	6,700
Catch of sea fish	1,200	814	3,015	9,444	7,721	11,178	7,247	17,417	21,000	26,050	39,900	36,400
Total	19,600	51,594	68,125	68,125	64,424	70,578	73,628	82,071	86,800	82,088	111,400	117,800

Source: World Bank (1992).

given for this, such as Cambodian people's tendency to avoid occupations which involve the taking of animal life because of their belief in Hinayana Buddhism. Also, it cannot be denied that the Cambodians cannot easily compete with the Vietnamese, who have access to adequate capital for investing in fishing boats and nets.

The greater part of the catch from the Great Lake and other Cambodian sources is either consumed locally or sent to Phnom Penh and other large cities. Distant farming communities are unable to benefit from the fishery resources of the Great Lake because of the lack of salting, drying, and other processing facilities; instead, the people of these communities eat various small fish, crabs, and frogs which inhabit the paddy fields.

Fed by the flood waters and irrigation projects of the lower Mekong Basin, the numerous canals, waterways, and swamps of this area provide an excellent environment for the growth of fry. These young or newly hatched fishes not only provide an important dietary source but also contribute to the fertilization of agricultural lands. Thus, for the people of the lower basin, fishery is a vital resource second only to rice cultivation. For this reason, the proper management of fishery resources is of critical importance to the welfare of the entire population.

Proper production- and resource-management systems have already been instituted in the Nong Hang marshlands of north-east Thailand and other marshlands. The Nong Hang marshlands, which are linked to the Mekong, cover an area of more than 100 km^2, while the Kawang Payao marshlands cover approximately 45 km^2. Such marshes have truly outstanding levels of productivity. The same holds true for the various artificial lakes and reservoirs of this area, such as the Nam Pong Dam lake in north-east Thailand and Lake Nam Ngum in Laos. Following the completion of the dams, these artificial lakes were stocked with fish and, at one time, their fish populations reached very high levels; however, subsequently, a gradual reduction was observed, which points to the importance of proper resource management to avoid such declines.

Fishing in the tributaries and main stream of the Mekong is also very widespread, with particularly good results in the area between Vientiane and Thakhek and between Kratie and the Mekong Estuary. The Tonle Sap River, which connects the Great Lake to the Mekong, and the Bassac River, a tributary of the Mekong, are also very rich in fishery resources. The coastal regions of Viet Nam provide excellent fishing grounds for squid, lobsters, and prawns, which are in large part directed to export markets. At the same time, shrimp farms are growing rapidly in this area. Statistics show that 35,000 t each of sea prawns and shrimp from shrimp farms were harvested during 1990 for a total haul of 70,000 t. In this connection, it is being argued that the rush to clear mangrove forests to make way for artificial shrimp-farming ponds is resulting in environmental destruction; this, too, underlines the need for effective resource management.

Fishing methods throughout the lower basin area remain essentially primitive. However, there are certain pockets of modernization: for instance, the 1970 report of the Mekong Committee (Mekong Secretariat 1970) states that modern fishing methods were already being employed in parts of north-east Thailand and in the vicinity of the Great Lake, and that this was contributing to significant increases in the total catch. Shortly after this report was compiled, the local populations of the Great Lake and the Tonle Sap river were caught up in the heavy fighting which covered this area. As mentioned above, many of the fishermen in this region are of Chinese and Vietnamese descent. Given their tradition for the aggressive pursuit of business interests, it is quite possible that these people will try to revive the large-scale and modern fishing methods which they had previously established.

A major source of concern for the fishing industries in this area is that the Tonle Sap River, as well as the mouth of the Great Lake, are becoming shallower with the influx of large volumes of silt in recent years. It is feared that this trend may undermine the development of large-scale fishing operations in these waters. Whereas there are long-standing bans on gill-net and dragnet fishing in the Kawang Payao and Nong Hang marshlands in north-east Thailand, it appears that, except for these areas, no restrictions on fishing methods are in force throughout the region. Some observers fear that, unless Cambodia implements such bans and restrictions in the future, the resources of this area will be rapidly depleted and the fishing industry undermined.

Although there is no doubt that first priority must be given to protecting and developing the fishery resources of the region, there is a no less critical need to establish refrigeration and transportation facilities in order to bring greater stability to the markets.

The release of fish in natural marshlands and irrigation ponds is a common practice throughout north-east Thailand and Viet Nam. As previously stated, the marshlands and artificial lakes of north-east Thailand have been extensively stocked and a more stringent management of resources is required in this area. Nevertheless, fishery productivity is extremely high in this area because of the tropical climate: for instance, Viet Nam reports a yield of 1,500 kg/ha and experimental fishing farms in north-east Thailand boast annual yields of 2,000–3,000 kg/ha.

The people of the Mekong Basin have traditionally used their paddy fields to stock fish. However, with the spread of modern agricultural techniques and the use of large amounts of chemical fertilizers and pesticides, this practice now appears to be doomed. Other environmental changes are also resulting in declining catches in some areas.

There is little doubt that both the upper and lower Mekong basins are today undergoing very significant social and environmental transformations. For instance, the rapid increase in the population of northeast Thailand is triggering very substantial changes in socio-economic conditions. Basically, the same can be said for the Laotian plains and the delta region of Viet Nam, where the transition toward a market economy is now beginning to affect the fishing industry. Cambodia is no exception to this trend as the prospects for fully fledged progress toward economic reconstruction become brighter. These developments indicate that the time is now ripe to embark on a thorough and far-reaching reorganization and improvement of the fishing industry in the Mekong Basin. The key to success in any such effort lies in the ability to formulate and promote comprehensive and integrated development plans which take into consideration the realities of the entire region.

Mineral resources (fig. 2.1)

During 1990–1991, the Asian Development Bank (ADB) financed a systematic survey of the mineral resources of Laos conducted by the British Geological Survey (BGS). A survey of this nature had yet to be implemented in any other part of the lower Mekong Basin, with the exception of north-east Thailand. However, it is known that Laos and other areas of the lower basin hold rich deposits of mineral resources, the best-known of which include tin, copper, and iron ores. If properly linked with the future development of energy sources in this region, these mineral resources will no doubt emerge as an important driving force in the socio-economic advancement of the countries of the lower basin.

Up-to-date materials and maps outlining the location of mineral deposits in the entire lower basin area are not readily available. This means that there is no choice but to use the information appearing in some relatively old sources (UN 1957). This section outlines the information gathered by the author from various sources. These findings should be considered in conjunction with the data in figure 2.1.

Laos

Various reports have been compiled concerning the mineral resources of Laos. The mineral deposits identified in reports published prior to 1991 can be summarized as follows.
- Northern Laos and area near Myanmar border: lead, lignite, antimony;
- Phong Saly Province area near Chinese border: copper, lead, tin, zinc, magnesium, gypsum, brown coal;

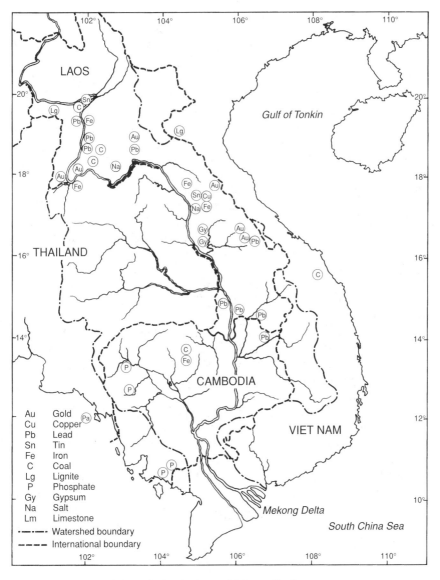

Figure 2.1. Mineral deposits in the lower Mekong Basin
Source: UN (1957)

- Luang Prabang area: lead, iron, copper, tin, coal;
- Tran Ninh Plateau area: lead, iron, tin, zinc, manganese, brown coal;
- Area north of Vientiane: limestone, coal;
- Mekong banks upstream from Vientiane: gold, iron;

- Kammouan and Savannakhet provinces downstream from Vientiane: gold, lead, copper, iron, gypsum, limestone, salt;
- Cham Pasak area: gold, silver, copper, lead, iron, graphite, salt;
- Attapu Province: gold, silver, copper, lead.

In addition to these known deposits, it is said that the province of Se Kong has deposits of silver, copper, gold, zinc, and tin. This is, indeed, a very diverse and rich list of mineral resources. The accuracy of this information was confirmed in detail by the BGS survey conducted during 1990–1991.

Notwithstanding its wealth of mineral resources, Laos has made little progress in the exploitation of its endowments. One exception is the tin mines located 30 km north of Thakhek, which are operated by the provincial government of Kammouan: at one time before the Second World War, the mines boasted an annual output of 1,800 t of tin concentrate; in later years, the two strip mines producing tin were both nationalized; thereafter, annual production (with Soviet assistance) has stood at 350 t (including 100 t of tin). The output from these mines was exported to the Comecon countries until recently. The estimated tin reserves of these mines amount to more than 130,000 t. The future electrification of Kammouan should lead to a dramatic increase in tin production. In other areas, the largest Laotian mining operation is located at the Dong Hen gypsum mines in the province of Savannakhet. The annual output amounts to 100,000 t, which is being exported to Viet Nam. Finally, a small-scale coal mine in the province of Vientiane produces about 1,000 t of coal per year for domestic consumption.

It is reported that high-grade copper deposits are located in the region of Cham Pasak, but this area has not been sufficiently investigated.

Iron ores constitute one of the largest mineral reserves of Laos. The numerous iron ore deposits which are scattered throughout the country are estimated to hold reserves of 1×10^9 t. One of the most extensive deposits is located in the Xieng Khouang region; however, access to this region will pose a serious challenge to prospective developers.

Another promising mineral resource is bauxite, which has, at the time of writing, remained untouched. There are indications of significant deposits in areas north-west of Luang Prabang and in the Boloveng Plateau.

Laos is also rich in precious minerals and gems: sapphire mining is currently being carried out in a joint venture with a Thai company which has achieved an annual yield of 15,000 carats. It is reported that the banks of the Mekong immediately upstream from Luang Prabang and from Vientiane hold rich deposits of gold. The author himself has seen local people panning for sand gold dust in this part of the river; however, the economic viability of these primitive methods is questionable. The processing and manufacture of gold handiwork can be seen throughout the

Table 2.8 Trends in mineral production (t/year) in Laos

Year	Product			
	Tin	Gypsum	Salt	Coal
1981	255	40,000	3.3	–
1982	356	–	5.1	111
1983	362	70,000	6.5	750
1984	430	84,000	7.8	830
1985	520	100,000	9.1	1,000
1986	559	98,000	8.6	1,556
1987	450	70,000	13.8	1,550

Source: Economist Intelligence Unit (1990–1991).

country. It is known that the region of the Annanese Range produces various gems and precious stones, but the central government apparently has not been able to determine the value of the gems being excavated.

Table 2.8 shows the trends in the exploitation of mineral resources in Laos during recent years. As can be seen from this table, the production of all mineral resources, with the exception of salt and coal, has traced a very flat curve.

It should be noted that there are known reserves of oil and gas in an area of more than 20,000 km^2 in the province of Savannakhet. Likewise, oil has been discovered in an area of 26,000 km^2 in the provinces of Saravan and Cham Pasak. Exploration agreements have already been concluded: the Savannakhet fields have gone to a British and French consortium, while the Saravan and Cham Pasak fields have been assigned to Hunt Oil.

In April 1994, the Prime Minister of Australia attended the opening ceremony of the first bridge to span the Mekong in its lower basin. Immediately before the ceremony at this bridge, which links Laos and Thailand, the Premier visited Vientiane and signed an agreement promising Australian aid for a new survey of Laos's mineral resources. The findings of this investigation are eagerly awaited.

Thailand

The Korat Plateau of north-east Thailand sits on top of hydrogenous rock formations dating from the Mesozoic period. As such, this area has not been affected by mineralization processes. However, high-grade iron ore deposits with an iron content of more than 60 per cent have been found at Chiang Khan in the Roi Et Region. The scale of these deposits has been estimated at 30×10^6 t. With regard to copper, some years ago the US Geological Survey announced the discovery of copper deposits in the

Roi Et Region. At the time, it was estimated that these deposits held 78×10^6 t of ore with an average copper content of 0.8 per cent.

The area stretching from north-east Thailand to Laos is studded with vast deposits of rock salt. However, the Association of South-East Asian Nations (ASEAN) soda ash project, which was designed to exploit these salt deposits, was suspended in 1985 because of its poor economic viability.

Cambodia

The iron ore deposits in the vicinity of Phnom Penh have been well known for many years. These deposits hold ores of magnetic iron ore and haematite which are relatively free of impurities. A large-scale development of these deposits was attempted but ended in failure because of transportation difficulties. The Phnom Penh area also has some coal deposits which could be exploited for use in steel-making.

The areas of Battambang and Kampot have phosphorus deposits with a phosphate content of 33 per cent. In its report of 1982, the Food and Agriculture Organization (FAO) stated that the phosphate production of the region could be increased significantly if the phosphate mines at Kampot, which had been destroyed during the Cambodian Civil War following the war in Viet Nam, could be restored to operation. It is reported that bilateral aid projects are being planned for this purpose. On the other hand, in the same report, the FAO states that the reconstruction of the phosphate fertilizer factory in Battambang has been completed and that the plant is producing 6 t of fertilizer per day.

The province of Kampot is also known for its rock salt, which is indispensable in the production of pickled vegetables and salted fish products. However, the transportation of rock salt out of Kampot has posed a challenge. Kampot was also the site of Cambodia's single cement plant; however, the 1982 FAO report states that this plant has yet to be restored.

Other reports of mineral deposits in Cambodia include those of gold, copper, and iron ore in Kompong Tom; of lead, copper, and zinc in Kompong Speu; and of gold, lead, copper, and various precious stones in Stung Treng. There are unconfirmed reports of oil seepage in the vicinity of the Great Lake, but serious investigations have yet to be undertaken.

Navigation

As described in chapter 1, the European powers were initially attracted to the Mekong for its potential as a regional artery for navigation. By the middle of the nineteenth century, the British were making plans for opening up a trade route passing through Burma to Yunnan in China.

Out of their keen trade rivalry with the British, the French were anxious to initiate trade with Yunnan before British interests became established there. The route of choice for the French was to sail directly up the Mekong; for this purpose, a small expeditionary force of naval officers was sent to Phnom Penh in 1866. Shortly after sailing out of Phnom Penh, the expeditionary force was obstructed by the shoals at Sambor. Having passed through the shoals with great difficulty, the advance of the group was soon stopped by the Khone Falls located above the shoals; shortly thereafter, the group was again impeded by the rapids at Khemarat and Keng Kabao; finally, all hopes of finding a route to Yunnan through the Mekong were abandoned. The French eventually had better luck with the Red River in northern Viet Nam, which was used successfully to establish a trade route with China.

The French experience proved what is only too well known today: the main Mekong with its powerful currents, waterfalls, and rapids does not provide an uninterrupted waterway running its entire length; at best, it can sustain several local and regional waterways. As this is dealt with fully in the first sections of chapter 1, the discussion in this section focuses primarily on shipping.

The lower part of the Mekong (all sections below the border of Myanmar) has two navigable segments: the first consists of a 547 km stretch between the South China Sea estuary and the town of Kratie in Cambodia; the second measures 733 km and runs between Kinak in Laos and the town of Chiang Saen on the border of Myanmar. Kinak is the name of the village located immediately above the Khone Falls; an old length of rails laid here by the French to carry cargo around the waterfalls still remains.

The Khone Falls thus divide the waterways of the Mekong into two separate segments – an upper segment which reaches the border of Myanmar and a lower segment which starts below the falls and the shoals at Sambor and winds through the delta.

Navigation above the Khone Falls and the east–west corridor (fig. 2.2)

The Mekong above the Khone Falls is tightly sandwiched between the territories of Thailand and Laos. Clearly, Laos is far more dependent on the Mekong than is Thailand. For the Thai people, the Mekong remains a distant reality located far from the capital of Bangkok. Unlike the Chao Phraya River, which skirts the capital, the Mekong fails to attract the serious attention of the people residing in the heart of the nation.

It is estimated that, during the 1980s, approximately 100,000 t of goods crossed the Mekong to be imported into Laos from Thailand. To this day, the trade in the opposite direction consists almost solely of Laotian ex-

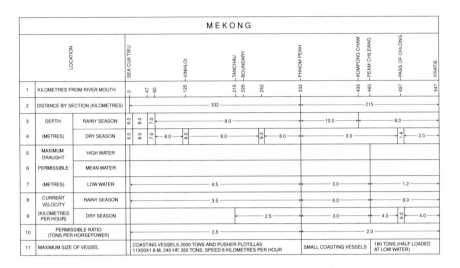

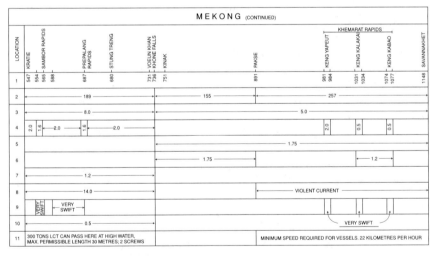

Figure 2.2. Navigability of the lower Mekong main stream and Tonle Sap River
Source: UN (1957)

ports of logs and lumber. In addition to importing various goods from Bangkok via the Mekong, Laos has access to a land route which stretches eastward through the Annanese Mountains to reach Da Nang and other seaports on the eastern coast of Viet Nam. This land route serves as a vital conduit for the importation of foreign goods into Laos. Cargo arriving in Laos from Da Nang amounts to about 50,000 t/year; of this total volume, approximately 10,000 t are unloaded at Savannakhet at the border between Laos and Thailand; the remaining 40,000 t is transported via

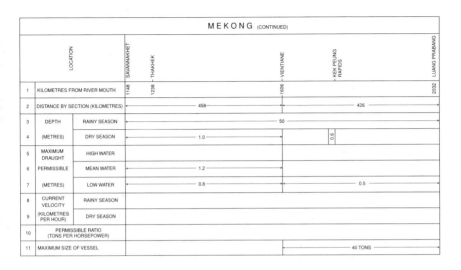

	LOCATION		SAVANNAKHET	THAKHEK		VIENTIANE	KEK PEUNG RAPIDS	LUANG PRABANG
	MEKONG (CONTINUED)							
1	KILOMETRES FROM RIVER MOUTH		1148	1238		1606		2032
2	DISTANCE BY SECTION (KILOMETRES)		← 458 →		←	426 →		
3	DEPTH	RAINY SEASON	← 50 →					
4	(METRES)	DRY SEASON	← 1.0 →				0.6	
5	MAXIMUM DRAUGHT	HIGH WATER						
6	PERMISSIBLE	MEAN WATER	← 1.2 →					
7	(METRES)	LOW WATER	← 0.8 →		←	0.5 →		
8	CURRENT VELOCITY	RAINY SEASON						
9	(KILOMETRES PER HOUR)	DRY SEASON						
10	PERMISSIBLE RATIO (TONS PER HORSEPOWER)							
11	MAXIMUM SIZE OF VESSEL				←	40 TONS →		

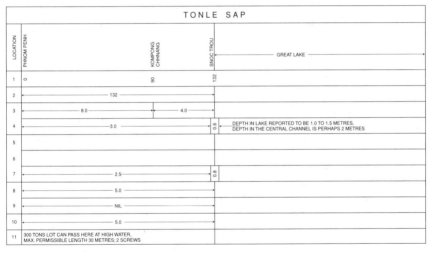

	LOCATION	PHNOM PENH	KOMPONG CHHNANG	SNOC TROU	GREAT LAKE
	TONLE SAP				
1		0	90	132	
2		← 132 →			
3		← 8.0 →	← 4.0 →		
4		← 3.0 →		0.6	DEPTH IN LAKE REPORTED TO BE 1.0 TO 1.5 METRES, DEPTH IN THE CENTRAL CHANNEL IS PERHAPS 2 METRES
5					
6					
7		← 2.5 →		0.8	
8		← 5.0 →			
9		← NIL →			
10		← 5.0 →			
11		300 TONS LOT CAN PASS HERE AT HIGH WATER, MAX. PERMISSIBLE LENGTH 30 METRES; 2 SCREWS			

Figure 2.2 (cont.)

Savannakhet to Vientiane. Approximately 75 per cent of the goods carried on to Vientiane are transported over land via Laotian Highway 13; the remaining 25 per cent, amounting to 8,000–10,000 t, is transported along the Mekong by ships. To facilitate navigation over this stretch of the river, wider passages have been blasted through the treacherous shoals at Keng Kabao with American and British assistance in 1973.

The land route which leads from Bangkok to Vientiane consists of a friendship highway (built by the Americans after the Second World War) and a railway link. Previously, the eastbound cargo had to be unloaded at the Thai river-port of Nong Khai and ferried over the Mekong to Vientiane on the opposite side. However, with the completion of the "Friendship Bridge" in April 1994, the land route has become far more efficient. The completion of the bridge, however, means that foreigners are no longer able to cross the river into Laos by ferry: at present, visitors must take a bus over the bridge, an arrangement which may have a bearing on the fact that the number of tourists entering Vientiane through Thailand has not increased in recent years.[9] Nevertheless, the important point is that cargo from Bangkok now arrives directly without the need for transshipment. In addition, the new bridge is encouraging more active trade between Thailand and Viet Nam via Laos.

There are several available trade routes linking Bangkok to Viet Nam, all of which pass through north-east Thailand, over the Mekong and through Laos to arrive finally at their destination in Viet Nam. The most active of these routes runs along Highway 9, which links Bangkok to the town of Mukdahan in north-east Thailand. At Mukdahan, the cargo is loaded on to ferries which cross the Mekong to Savannakhet in Laos. From there, the cargo is again carried by trucks which traverse the mountain roads (Laotian Highway 9) into Viet Nam and the port city of Da Nang. Westbound cargo retraces this route from Da Nang in the opposite direction.

A new bridge spanning the Mekong between Mukdahan and Savannakhet is scheduled to be constructed in the near future with financing from the Asian Development Bank. The completion of this link is sure to result in a dramatic increase in land-based transportation between Thailand and Viet Nam. In addition to this bridge, possibilities are now arising for the construction of bridges between Thakhek (Laos) and Nakhon Phanom (Thailand), as well as between Pakse (Laos) and the opposite banks of the Mekong (leading into Thailand).[10]

Turning next to the river-link between Vientiane and the ancient capital of Luang Prabang located to the north, approximately 4,000 t of cargo per year are transported by ships in both directions between these cities. Some ten years ago, a Japanese grant was used to upgrade the river port at Vientiane; however, the series of smaller ports of call – at Houei Sai, Pak Lay, Pak Beng, and other points south of Luang Prabang – require extensive improvement. Vientiane is also connected by ferry service to Savannakhet in the south with three scheduled departures per week.

Navigation in the Mekong Delta

After separating from the Bassac River at a short distance south of Phnom Penh, the Mekong splits into three branches as it completes its long journey to the South China Sea: these three branch rivers are the

Cua Co Chieng, the Song Ham Long, and the Song Cua Dai. The so-called Saigon River (which takes the name of the Cua Soirap River as it flows further downstream) flows beyond the eastern edge of the Mekong Basin and thus does not belong to the Mekong Basin.

The waterways of the delta which connect the coast to Phnom Penh and Kratie, located further upstream from Phnom Penh, are discussed in chapter 1 and are not discussed again here. Suffice it to say that the waterways of this area criss-cross between the Mekong and Bassac to create a highly complex network with the two rivers at its core. These waterways connect Phnom Penh to various Vietnamese cities, such as My Tho, Vinh Long, Cao Lanh, Can Tho, and Long Xuyen. These also serve as important links between various Vietnamese port cities, such as Ha Tien and Rach Gia, and the coastal cities of Soc Trang, Bac Lieu, Ca Mau, Tra Vinh, and Ho Chi Minh City (formerly Saigon).

A total of 7×10^6 t of cargo per year was being transported through the waterways of the Mekong Delta at the end of the 1980s.[11] This figure is equivalent to roughly one-half of the gross cargo tonnage in the region. Throughout the 1980s, the gross tonnage transported through the waterways of the delta was growing at an annual rate of 10–12 per cent. Of this total, the tonnage carried by ocean-going vessels stood at 150,000–200,000 t/year.

The waterways of the Mekong Delta clearly have an outstanding potential for future development. However, this potential can be fully realized only with the construction of large dams on the Mekong with massive reservoirs which would take in and submerge the Khone Falls and the many treacherous shoals along the river. Another advantage of these large-scale dams would be that, by establishing effective flood control throughout the lower basin, the water level of the waterways in the delta could be maintained at adequately high levels throughout the entire year.

However, as already mentioned, there are no immediate prospects for the construction of such large dams. With the exception of the project to build a low dam at Pa Mong immediately upstream from Vientiane, the dam-development projects for the Mekong have failed to attract serious discussion. The completion of the low dam at Pa Mong will have little effect on flood control in the region, nor will this dam be able to play a part in maintaining the water level in the waterways of the delta.

There would be a major difference if, instead of the Low Pa Mong Dam, the public would give its support to the construction of the high dam which was originally planned for Pa Mong in the 1960s. In addition, if large dams were to be erected at Stung Treng and other critical sites on the Mekong and its tributaries, and if these dams were to be furnished with the appropriate locks (the building of locks might be economically feasible if trade between the lower basin area and Yunnan Province and eastern Myanmar increased), this would result in the creation of a vast waterway linking the Mekong Delta to distant sections in the upper reaches of the river. However, the construction of such dams cannot advance until international agreements are reached on such crucial matters as the overall direction of development, environmental protection, and the resettlement of large populations which would be displaced by the massive reservoir lakes. Because of these obstacles, the construction of such a monumental waterway will remain a dream with little hope of realization in the foreseeable future.

Nevertheless, water-level conditions in the Mekong Delta can be expected to improve significantly with the completion of the serial dams on the Lancang River (which China is surveying and implementing), the construction of the Nam Theun 2 Dam in Laos, and other dams on the major tributaries of the Mekong. As the Lancang Jiang and Nam Theun projects are already making headway, this would be a good time for the governments of this region to review the future role of the Mekong Delta waterways in the transportation of rice, petroleum products, cement, and other materials.

It was reported in April 1994 that the Australian government had reached an agreement with Viet Nam for undertaking a feasibility study for the construction of a bridge at My Thuan, which is located at the mouth of the Mekong. (As in the case of the "Friendship Bridge" in Vientiane, Japan had been involved in planning for this bridge since the 1960s before the above agreement was reached.) It remains to be seen how the construction of this bridge will affect the balance between land and navigation in the Mekong Delta. The answer to this and other questions will be found in the results of the Australian study.

Hydroelectric power generation

Hydroelectric power potential

During the author's years of service on the Mekong Committee in the 1960s, he attempted to estimate the gross theoretical potential of the lower Mekong Basin for power generation, among other possibilities, using a map of the lower basin drawn to a scale of 1 : 2,000,000. This was used to create a rainfall map on which the elevations from sea level for key points were noted (the sea level at Ha Tien in Viet Nam was taken as the point of reference). This information was combined with precipitation data and presumed run-off coefficients for various points in the individual river basins of the lower Mekong Basin. This exercise yielded some very rough estimates of the lower basin's theoretical hydroelectric power potential. The related table was included in the Indicative Basin Plan report published by the Mekong Committee (Mekong Secretariat 1970). To the best of the author's knowledge, these estimates have not been modified in the intervening years. The relevant data are shown in tables 2.9 and 2.10.

Table 2.9 Estimated hydroelectric potential of the Mekong Basin, 1970

Region of Mekong Basin	Potential hydroelectric power (10^9 kWh)	Potential installed capacity (10^6 kW)
Upper	665	76
Lower	505	58
	1,170	134

Source: Mekong Secretariat (1970).

Table 2.10 Estimated hydroelectric potential of the countries of the lower Mekong Basin, 1970

Country[a] and region[b]	Catchment area (km^2)	Run-off per year (10^6 m^3)	Hydro-electricity (10^6 kWh)	Potential capacity (10^3 kW)
Cambodia				
A	93,825	49,534	27,063	3,094
B	47,608	22,630	4,676	534
C	(141,433)	(72,164)	(31,739)	(3,628)
Laos				
A	160,745	180,835	325,466	37,195
B	41,655	31,915	40,829	4,662
C	(202,400)	(212,750)	(366,295)	(41,857)
Thailand				
A	147,310	35,555	16,653	1,903
B	36,935	19,363	8,981	1,027
C	(184,245)	(54,918)	(25,634)	(2,930)
Viet Nam				
A	29,850	41,797	80,950	9,256
B	Nil	Nil	Nil	Nil
C	(29,850)	(41,797)	(80,950)	(9,256)
Grand total for entire lower Mekong Basin apart from delta	557,928	381,629	505,000[c]	58,000[c]

Source: Mekong Secretariat (1970).
a. Current names given to avoid confusion.
b. Regions are as follows: A, major tributaries; B, along the main stream and small tributaries; C, total (A + B).
c. Totals rounded up.

According to these computations, the gross theoretical hydroelectric potential of the lower basin area comes to 505×10^9 kWh (58×10^6 kW), while that of the entire Mekong basin amounts to $1,170 \times 10^9$ kWh (134×10^6 kW).

Table 2.11 indicates that it was technically feasible to develop roughly 30 per cent of the gross theoretical hydroelectric potential of the lower Mekong Basin. This amounts to 150×10^9 kWh (17.5×10^6 kW). After this estimation had been completed, the author, as Chief Planning Engineer of the Mekong Committee, undertook to identify the technically feasible dam-development projects in almost all the tributaries of the lower basin. This was done using precipitation and flow data and a map drawn to a scale of 1:50,000 (covering the entire lower basin), which had been acquired during the 1960s. On the basis of this study, the author was able to identify roughly 150 development projects (with a total installed capacity of 186×10^9 kWh per year). The author concluded that these projects merited further consideration on technical grounds (but with no reference to economic viability). This figure vindicates the earlier estimate of 150×10^9 kWh of technically exploitable hydroelectric potential in the lower basin. By applying the same methodology, the technically exploitable hydroelectric potential for the entire Mekong Basin was estimated to stand at around 350×10^9 kWh (40×10^6 kW). Needless to say, these tables and their estimates are based on the assumption that the construction of large-scale dams is, in fact, possible.

Electricity supply and demand

Perspectives for Mekong Development (Mekong Secretariat 1988) contains estimates of the gross production of energy and projected supply and demand in Laos, Thailand, and Viet Nam between 1986 and 2010, as shown in table 2.12. A comparison of these figures with the estimates generated by the Snowy Mountains Engineering Corporation (SMEC) of Australia (published in the feasibility study report of the Nam Theun hydroelectric power project; SMEC 1990) reveals a wide discrepancy, as shown in table 2.13. This fact is indicative of the great difficulty in projecting future demand for energy and electricity.

Domestic conditions in Cambodia and Viet Nam began to change significantly at the start of the 1990s. Thailand, in order to meet its rapidly growing demand for electricity, concluded an agreement with Laos in 1993 for the long-term importation of between 1.0 and 1.5×10^6 kW of power. These developments indicate that the estimates shown in tables 2.12 and 2.13 both need to be broadly revamped to reflect more accurately the growth in the production and demand for energy.

At the time of the Tokyo Forum held in February 1995, the ADB re-

Table 2.11 Comparative hydroelectric potential of the Mekong Basin

River basin or country	Catchment area (km²)	Estimated (10⁶ kWh)	Potential			
			Technical		Economic	
			(10⁶ kWh)	(Percentage of estimate)	(10⁶ kWh)	(Percentage of estimate)
Upper basin[a]	186,000	665,000[e]	—[f]	–	–	–
Lower basin[b]	557,928	505,000[e]	150,000	29.7	–	–
Entire basin	743,928	1,170,000[e]	–	–	–	–
Sweden[c]	450,000	199,000	–	–	85,000	42.7
Finland[c]	337,000	46,500	–	–	18,000	38.7
Switzerland[d]	41,000	144,000	70,000	48.6	27,000	18.7
Austria[d]	84,000	152,500	–	–	30,000	19.7
France	547,000	314,000	138,000	44.0	60,000	19.1
Czechoslovakia[d]	128,000	39,000	–	–	7,000	17.9
East and West[d] Germany	356,000	111,000	26,000	23.4	18,000	16.2
Poland	313,000	32,000	–	–	5,500	17.2
Hungary[d]	93,000	7,000	2,000	28.6	1,500	21.4

Source: Mekong Secretariat (1970).
a. Region upstream of Chiang Saen.
b. Not including delta region.
c. Report of group of experts on hydroelectric power resources, ECE, Geneva (1964).
d. Hydroelectric Potential in Europe and its Gross Technical and Economic Limits. (E/ECE/EP/131). United Nations (May 1953).
e. Reference level is mean sea-level (at Ha Tien).
f. Dashes indicate data not available.

Table 2.12 Anticipated energy production and demand in three riparian countries, 1986–2010

Annual energy production/demand (10^6 kWh)	1986	1990	1995	2000	2010
Production, Laos	867	890	943	943	943
Export, Laos to Thailand	698	585	389	158	349
Production, Thailand	24,780	33,719	43,862	55,732	63,613
Production, southern Viet Nam	2,980	4,480	4,725	5,835	5,835
Demand, Laos	169		554	785	1,292
Demand, Thailand	24,780		43,862	55,732	90,206
Demand, southern Viet Nam	2,200		4,045	5,672	1,285

Source: Mekong Secretariat (1988).

Table 2.13 Projected domestic energy requirements (GWh) in Laos

Energy recipients	1986	1990	1995	2000	2005	2010	2020
Cities							
Vientiane	128.0	159.9	273.8	384.2	490.3	625.4	1,018.0
Luang Prabang	3.3	7.5	12.1	19.5	29.3	39.2	65.1
Thakhek	3.2	20.5	36.1	50.7	65.3	83.3	135.7
Savannakhet	11.1	17.2	25.8	34.6	44.1	56.3	91.7
Regions							
Northern	26.9	32.8	44.6	63.0	86.3	115.5	188.1
Central	13.6	16.5	22.4	31.6	43.3	57.9	94.4
Southern	21.1	25.7	34.9	49.3	67.5	90.3	147.0

Source: SMEC (1990).

leased a series of basic data (table 2.14) presenting its estimates of the exploitable hydroelectric potential of the six riparian countries (including territories not located within the basin). It now appears that the ADB estimates must also be revised at an early date.

Laos

As of 1994, Laos had a total installed capacity of 212,000 kW. In contrast, domestic demand is said to be no more than 60,000 kW. Nevertheless, the small-scale factories located in Vientiane and its environs require power, and many towns and villages of the country are eagerly awaiting the supply of electricity by the government for lighting and for operating television sets. Likewise, army facilities require power, and the electrifi-

Table 2.14 Exploitable hydroelectric potential of the subregion

Country	Estimated exploitable hydroelectric resources	
	Total resources (TWh/yr)	Developed resources (TWh/yr)
Cambodia	41	–
Laos	102	1.1
Myanmar	366	1.1
Thailand	49	4.6
Viet Nam	82	5.8
Yunnan	450	7.9
Total	1,090	20.5

Source: ADB (1995).

Table 2.15 Presumed energy-generation potential in Laos

Location of drainage area	No. of sites	Installed capacity of power station (MW)	Energy generated (GWh/year)
Upstream from Luang Prabang	7	1,175	8,826
Luang Prabang– Paksan	13	2,975	14,858
Paksan–Pakse	10	3,511	15,664
Se Kong	14	3,131	15,614
Total	44	10,792	54,962

Source: ADB (1995).

cation of mountain villages is an important element in improving the standard of living of the highland minorities. Thus, considerable potential demand for power does exist throughout the country. The problem, however, is that local pockets of demand are too small to render a hydroelectricity project economically viable. Current projections indicate that long-term power demand will grow at a particularly fast pace in Savannakhet, Pakse, and other regions of southern Laos. In response to this projection, the government is planning to proceed with the construction of several dams on the Se Kong River system and is hoping to receive financial assistance for these projects from Japan and Australia (see table 2.15).

The largest hydroelectric power station in Laos currently in operation is located at Nam Ngum (installed capacity of 150,000 kW), north of Vientiane. More than 80 per cent of the energy generated by this station is exported to Thailand. The Saravan Hydroelectric Power Station supplies the needs of Pakse and other parts of southern Laos. Southern Laos is also served by the Se Set Hydroelectric Power Station (installed capacity of 45,000 kW), the excess power from which is being exported to Thailand. Work on this station was carried out in parallel with the repair and expansion of the Saravan station and was completed in 1991. For Laos, which is currently ranked among the least-developed nations of the world, the export of power from the Nam Ngum and Se Set hydroelectric power stations, together with the export of lumber, constitutes one of the most important sources of government income.

With an eye to future economic development, the government of Laos is concentrating on the electrification of the three provinces located south of the Vientiane Plain (Savannakhet, Champasak, and Saravan). In April 1994, it was reported that the Australian Prime Minister had visited Vientiane to attend the signing of an agreement for the provision of A\$1 $\times$ 10^9 in private funds for financing the development of large-scale hydroelectricity projects at Nam Theun and Se Kamang (in the Se Kong Basin). It can be expected that any excess power generated at these hydroelectricity facilities will be sold to Thailand, Viet Nam, and Cambodia.

Thailand

A brief look at table 2.12 is enough to show that production and demand for energy in Thailand are far greater than those in its neighbouring countries. Furthermore, as indicated in the reports of the Electricity Generating Authority of Thailand (EGAT, Thailand's public power corporation), the recent growth in power demand has outstripped all previous projections. During the 30-year period between 1961 and 1990, annual demand for electricity increased more than 90-fold, rising from 460×10^6 to 43.2×10^9 kWh. Growth in demand was particularly strong in 1989 and 1990 as a result of booming economic conditions and the large number of foreign enterprises rushing to invest in Thailand. During these two years, demand grew by 16.2 and 16.8 per cent, respectively. The same dynamic growth was seen in peak generating demand, which grew by 14.7 per cent in 1989 and 15.4 per cent in 1990. The daily peak load is recorded between the hours of 7 p.m. and 8 p.m. when household demand for lighting and electric appliances is highest. Power demand in Thailand is 46.8 per cent for industrial use, 31.3 per cent for commercial use, 21.1 per cent for household use and others 0.8 per cent. Industrial demand has thus come to account for nearly half of the total power consumption in Thailand.

The rapid growth in demand in recent years can be attributed to the ongoing electrification of the countryside and the steady progress in industrialization in both the environs of the capital city and the outlying cities. Current projections indicate that power demand will continue to grow at an accelerated pace, rising from 69.6×10^9 kWh in 1994 to an estimated 117×10^9 kWh in the year 2000 (Kanou 1995).

In response to this accelerated growth in electricity consumption, the government of Thailand has been working intensively on the construction of new generating facilities and transmission lines. As a result of these endeavours, the nation's total installed capacity, which stood at 173,600 kW in 1960, had expanded dramatically to 8,720,000 kW by 1990. In 1960, Thailand was obtaining two-thirds of its electricity from diesel generators; however, in the ensuing years, much effort was put into the development of hydroelectric and oil-fired plants. The first such project to be completed was the Bhumiphol Dam, which was constructed on a tributary of the Me Nam River. With the 1963 completion of this dam, which took the name of the king, Thailand came to depend on a combination of hydroelectric and diesel generation for the bulk of its power. However, a major shift has occurred in generating modes during the past three decades: by 1990, the share of hydroelectricity in total power generation had fallen to 27 per cent, while oil- and coal-fired plants accounted for 53 per cent; at the same time, gas turbines and hybrid gas turbines contributed 17 per cent, and imports from Laos accounted for 3 per cent.

Thailand's hydroelectricity-generating potential is far smaller than that of Laos, and most of this potential has already been developed. Vocal opposition movements bar the progress of work on any of the remaining possible dam sites where the problems of resettlement and deforestation are thought to be serious.

In 1993, EGAT announced that Thailand needed an additional 17.70×10^6 kW of installed capacity by the year 2000 and 31.75×10^6 kW by 2010 (Kanou 1995). No clear indications have been given as to how the additional capacities are to be achieved. These estimates obviously place the joint Thai–Laotian project for the construction of the Pa Mong Dam, as well as the other hydroelectricity development projects on the Mekong, in the spotlight. However, EGAT has taken the position that construction of the Pa Mong Dam will be difficult because of the problems of inundation and other environmental concerns. Against this background, EGAT commented in 1993 that Thailand would not only increase the import of power from Laos but would also henceforth provide active support for the long-term development of the electric power industry in Laos. As part of this effort, EGAT has exchanged a memorandum with EDL, the Laotian power-generating corporation, concerning the joint development of seven hydroelectricity facilities in Laos. Given its vast hydroelectric power potential (which Laos itself cannot properly develop), Laos has welcomed the Thai initiative as a guaranteed long-term source of national income.

The construction of the Nam Theun 2 Dam is one of the projects taken up as part of this agreement. The first phase of the project, scheduled to be completed in the year 2000, will yield 210,000 kW. Almost all of the output will be exported to Thailand, making this Laos's third power-export project after the hydroelectric power facilities on the Nam Ngum and Se Set rivers.

Viet Nam

Electric power conditions in Viet Nam vary significantly in the northern, central, and southern sections of the country. At present, demand and supply are highest in the north, followed by the south; however, in terms of potential demand, the south is a far larger market. Owing to the shortage of generating facilities, the great potential of demand in the south has been suppressed. Such potential demand is also very high in central Viet Nam, where the towns and villages hunger for more electricity. (In fiscal 1989, the total electricity generated in Viet Nam amounted to approximately 7×10^9 kWh, of which 55 per cent was produced in the north, 37 per cent in the south, and 8 per cent in central Viet Nam.) To alleviate this imbalance, the government completed a 500 kV transmission line connecting the north to the central and southern regions in 1994. Through this power line, Ho Chi Minh City now has access to enough power to meet its current needs. The same year witnessed the completion of the Hoa Binh Dam (2.1×10^6 kW) constructed on the Black River (a major tributary of the Red River which traverses the northern regions of Viet Nam). The dam provides sufficient capacity to satisfy Viet Nam's current levels of demand. (Completed in March 1994, the Hoa Binh Dam increased Viet Nam's total installed capacity to 3.2×10^6 kW.) In the future, Viet Nam is planning to construct the Son La Dam (3.6×10^6 kW) to be sited upstream from the Hoa Binh Dam. Both dams lie beyond the Mekong Basin and are not discussed in detail here.

The upper basins of the Se San and Srepok rivers, both tributaries of the Mekong, are located in the central regions of Viet Nam. Plans for the construction of a hydroelectric power station at the Yali Falls in the upper reaches of the Se San River have been discussed over the past several decades, but no action has been taken until recently.

The construction of the Sambor Dam in Cambodian territory at the furthest point downstream on the Mekong would inspire new expectations for the development of Ho Chi Minh City and the Vietnamese portion of the Mekong Delta as a massive base for power consumption. The overall demand for power in southern Viet Nam stood at 626,000 kW and 2.2×10^9 kWh in 1986. Current projections indicate that this will increase to 1,175,000 kW and 5.67×10^9 kWh by the year 2000, and to 2,395,000 kW and 12.8×10^9 kWh by 2010.

Table 2.16 Presumed energy-generation potential in Cambodia

Projects	Installed capacity (MW)	Energy generated (GWh/year)
Mainstream projects on the lower Mekong	6,500	30,800
Projects in lower Mekong River Basin	1,100	5,500
Projects beyond the lower Mekong Basin	1,000	5,000
Total	8,600	41,300

Source: ADB (1995).

Cambodia

Topographical conditions and rainfall levels provide a limited level of hydroelectricity-generating potential in north-east and south-west Cambodia. Other potential sites for hydroelectricity generation have been identified in areas south of Battambang and on the rivers flowing into the Great Lake, such as the Stung Chinit. However, most of these projects are primarily aimed at irrigation, with power generation given only a subsidiary role. A survey of future possibilities for hydroelectric power development shows that, among the three nations of Indo-China, conditions for development are least favourable in Cambodia.

Table 2.16 presents the ADB's 1995 estimates of the potential levels of hydroelectric power development in Cambodia. The mainstream Mekong projects referred to in this table are the Stung Treng and Sambor dams. The projected power output for the Stung Treng Dam is based on a plan which minimizes the area of probable inundation. Projects beyond the Mekong Basin, referred to in the table, include the scheduled developments on several smaller rivers which flow out of the Cardamom and Elephant mountain ranges of south-western Cambodia. These rivers, which terminate in the Gulf of Thailand, have a relatively steep incline and are suited to development. These projects include the Stung Mnam 2 Project (90,000 kW), the Stung Atay Diversion Project (110,000 kW) and the Stung Kamchay Project (85,000–125,000 kW). Among these projects, most progress has been made in the developmental survey of the Stung Kamchay Project. Projects within the Mekong Basin, referred to in the table, include the Prek Thnot Project near Phnom Penh (18,000 kW), the Stung Chinit Project (8,500 kW), and the Stung Battambang Project (36,000 kW). All of these are primarily aimed at the irrigation of the downstream plain areas and, with the exception of the project for the Prek Thnot River which flows into the Bassac River, all involve the development of river systems which terminate in the Great Lake. At one time, Japan was preparing to embark on the development of the Prek

Thnot Project, but the spread of fighting put a stop to these plans. Today, Japan is again considering this project.

Notes

1. Total rice production in Laos during fiscal 1993 amounted to 1.28×10^6 t. According to some sources, rice production in slash-and-burn fields in the mountainous regions accounted for 285,000 t (Boupha 1995).
2. These figures cannot be said to be accurate in comparison to the figures given in Table 2.1 but have nevertheless been included here (Mekong Secretariat 1988).
3. According to the 1992 report of the Vietnamese Trade Information Center, total land under cultivation in the Mekong Delta was 2.58×10^6 ha. Production of unhulled rice is given as 9.48×10^6 t, while per-unit yield is reported to be 3.67 t/ha. The comparable figures for rice production in the Red River Delta were 1.06×10^6 ha under cultivation, 3.62×10^6 t of unhulled rice harvested, and a per-unit yield of 3.42 t/ha. This comparison indicates the higher agricultural value of the Mekong Delta.
4. Anywhere between 5,000 and 30,000 ha of land in the Vietnamese portion of the Mekong Delta are used in the cultivation of long-stalked "floating" rice. Per-unit yield of unhulled rice is estimated at 1.5–2.0 t/ha.
5. The most extensive flood damage in recent years was experienced in October 1966. Total damage was estimated at US14×10^9 in Cambodia, 10.9×10^9 in Laos, 11.5×10^9 in northern and north-east Thailand, and 20.1×10^9 in the Vietnamese delta (Mekong Secretariat 1990). It is worth noting that, while this flood affected all regions of the lower basin, the most serious damage occurred in the northern areas around Vientiane.
6. Rubber is also extensively cultivated in the province of Hsishuangpanna at the southern end of the upper Mekong Basin. At first, rubber was successfully cultivated by Sen Hoshu, a Thai of Chinese descent. The Chinese government has continued full-scale rubber cultivation since 1953 with 80,000 ha under cultivation as of 1993.
7. As of 1995, approximately 1.3×10^6 people, equivalent to 30 per cent of the population of Laos, were involved in slash-and-burn agriculture (Boupha 1995).
8. The total area of forests stood at 16.5×10^6 ha in 1940, but had dropped to 11.2×10^6 ha by 1995 (Boupha 1995).
9. Thai newspapers reported in April 1995 that the necessary legal changes would be made during 1995 to allow passenger cars to drive directly over the bridge into Laos.
10. As mentioned in note 5 of chapter 6, another bridge is planned over the main Mekong at Kompong Cham in Cambodia.
11. According to material published by the ADB (ADB 1995), the waterway connecting Phnom Penh to the South China Sea is currently being used to transport 6×10^6 t of cargo per year. Estimates indicate that this will increase to 7.2×10^6 t by the year 2000.

BIBLIOGRAPHY

Asian Development Bank (ADB) (1994). *Economic Cooperation in the Greater Mekong Subregion Towards Implementation*. ADB, Hanoi, April.
——— (1995). *A Compendium of Project Profiles, Forum for Comprehensive Development of Indochina*. ADB, Tokyo, February.

Boupha, P. (1995). *Shifting Cultivation Extermination Program in Lao PDR*. Proceedings of Tokyo Symposium on Sustainable Agriculture and Rural Development, Tokyo, November.

Economist Intelligence Unit (1991). *Viet Nam, Country Profile 1990–91*. Vietnamese Government, London.

——— (1991). *Laos, Country Profile 1990–91*. Economist Intelligence Unit, London.

——— (1993). *Country Report, Indochina*. Vietnamese Government, London.

Finance Ministry of Laos (1990). *Basic Statistics (1975–1990)*. Finance Ministry of Laos, Vientiane.

——— (1992). *Policy Framework for Public Investment Program*. Finance Ministry of Laos, Vientiane, December.

Food and Agriculture Organization (FAO) (1980). *Forest Resources*. FAO, Rome.

Kanou, I. (1995). *On Mekong Power Express*. ECFA, Tokyo, December.

Mekong Secretariat (1970). *Report on Indicative Basin Plan*. Mekong Committee, Bangkok.

——— (1988). *Perspectives for Mekong Development*. Mekong Committee, Bangkok, April.

——— (1990). *Review of the Fishery Sector in the Lower Mekong Basin*. Mekong Committee, Bangkok, October.

Snowy Mountains Engineering Company (SMEC) (1990). *Nam Theun Hydroelectric Project Feasibility Report*. SMEC, Australia, November.

United Nations (UN) (1957). *Development of Water Resources in the Lower Mekong Basin*. Flood Control Series No. 12. United Nations, Bangkok.

World Bank (1992). *Country Economic Memorandum, Cambodia, Agenda for Rehabilitation and Reconstruction*. World Bank, Washington D.C.

3

Planning of dam development in the lower Mekong Basin: Chronology and international cooperation

Chapter 3 of the Japanese edition of this book presents a detailed chronological account of the planning of dam development in the lower Mekong Basin and touches on various issues of interest to specialists. However, in this English edition, the author has excluded the very lengthy section designated 3.2 in the Japanese edition and has instead included a summary and overview of the chapter as a section describing the chronology of planning of dam development and a brief history of international cooperation, for the benefit of the general reader.

Chronology of planning of dam development and a brief history of international cooperation

Development plans prior to the Japanese reconnaissance of the major tributaries in 1959–1961

The upper Mekong Basin in the territories of China and Myanmar forms a long and narrow corridor which accounts for only 20 per cent of the area of the entire Mekong Basin. The global political situation at the time of the United Nations' initial involvement in the Mekong development projects in the 1950s led the UN organizations to focus on the development of the Mekong only in the four countries of the lower basin.

When the United Nations Economic Commission for Asia and the Far

East (ECAFE, the predecessor to the present-day Economic and Social Commission for Asia and the Pacific, or ESCAP) first began to survey the lower Mekong Basin, its primary objective was the control of flooding in the Mekong Delta and the other leading river systems of the Asian continent. However, this narrow approach was soon amended. In the case of the Mekong, various other additional objectives were identified, including the promotion of irrigation in the lower basin, hydroelectric power generation, navigation, and the development of fisheries.

With the progress of its survey, ECAFE became aware of a very important point. Whereas the goal of ECAFE was to improve the welfare and standards of living of the people of this area through a well-balanced and rational development of the water resources of the lower basin, it was realized that it would be necessary for the governments of the four countries of the lower basin (Thailand, Laos, Cambodia, and Viet Nam) to give priority over the pursuit of national interests to the development and progress of the entire basin. This would require a fundamental unity of purpose among the four nations; in other words, special attention would have to be given to the "Mekong spirit." To foster this spirit and to ensure the success of the comprehensive development of the Mekong Basin, ECAFE set up the Committee for Coordination of Investigations of the Lower Mekong Basin, which was popularly referred to as the Mekong Committee.

The Mekong Committee was established in Bangkok in October 1957. Consisting of "the plenipotentiary representatives of the four countries of the lower basin," its primary objectives were defined as "the formulation, investigation, coordination, supervision, and control of water-resources development plans for the lower Mekong basin." In addition, the Committee was charged with the responsibility for "soliciting financial support for such projects on behalf of the member governments and receiving and controlling such funds."

Before the establishment of the Mekong Committee, ECAFE had already dispatched a five-member investigative team to the lower Mekong Basin in 1951 and a second investigative group of eight experts was dispatched in 1956. These studies led to a historic report entitled *Development of Water Resources in the Lower Mekong Basin (1957)* (UN 1957), which was distributed on the day of the establishment of the Mekong Committee. The report contained a proposal for the development of the lower basin based on the March 1956 recommendations of the US Bureau of Reclamation, which had identified five priority projects for the Mekong among seven major projects: these consisted of four mainstream dam projects and a large-scale weir project on the Tonle Sap river around the eastern mouth of the Great Lake. The dam projects consisted of the Pa Mong Dam Project immediately upstream from the Laotian capital of

Vientiane, the Khemarat Project (Laos) below the Pa Mong Dam, the Khone Falls Project (at the Laos–Cambodia border), and the Sambor project (Cambodia) (see figure 3.1).

In September 1957 (one month before the establishment of the Mekong Committee), the Preparatory Committee for the Mekong Committee dispatched an investigative team headed by Lieutenant-General R. A. Wheeler of the US Army Corps of Engineers, which resulted in the Wheeler Report published in February 1958. This report emphasized the need for developing topographical maps and observations of water levels and climatic conditions over adequate periods of time. It also stressed the importance of collecting information on such matters as fisheries, irrigation, mineral resources, demand for electric power, and flood damage (see pp. 104–105).

These two reports, released at the time of the formation of the Mekong Committee, were instrumental in defining the specific items of international assistance required. On this basis, assistance was solicited immediately. The first response was received from France, which contributed equipment for water-level observations. Following this, the United States offered to carry out these observations and New Zealand came forward with the vessels needed for conducting the observations. Various other countries made similar contributions and pledges. As for Japan (at this time, Japan had just recently been admitted to the United Nations and had yet to fully recover from the devastation of the war), it pledged to conduct what was certainly the least conspicuous of the investigative surveys which had been listed in the reports: this consisted of the reconnaissance of the major tributaries of the lower basin. Although this would require only a small outlay of funds, Japan thus became involved in the "investigation of dam construction plans for the entire lower basin," the most promising aspect of the whole development programme. These events occurred in late 1958.

During the 2-year period between January 1959 and September 1960, the Japanese government dispatched a total of three reconnaissance missions to investigate the development of the lower Mekong tributaries. The Japanese reconnaissance team used the newly developed Toyota Land Cruiser and chartered aeroplanes to travel through north-east Thailand and the interior of the three countries of Indo-China. The teams managed to develop flow-volume estimates for a total of 34 tributaries in the lower Mekong Basin. In addition, elevations were measured at various key points. (At the time, the only available map was a partial US military map drawn to a scale of 1:50,000.) On the basis of these observations, Japan was able to propose numerous dam projects for the major tributaries. Quite notably, the Japanese teams took the initiative of drawing up a proposal for the construction of a set of serial dams on the Mekong itself, in which the team incorporated various original ideas for development. The Japanese report containing these data and proposals was submitted to the Mekong Committee in September 1961.

The Japanese reconnaissance report made it possible for the first time to develop a concrete vision for the comprehensive development of the entire lower basin. (The Japanese teams marked the locations of a number of proposed dam projects, reservoirs, and irrigation systems on a map to a scale of 1:1,000,000. In addition to laying out a broad outline of the possible hydroelectric and irrigation projects in the lower basin, this report reviewed the possibilities for flood control in the Mekong Delta and navigation along the main stream and tributaries of the Mekong.

This Japanese reconnaissance report caused quite a sensation and the governments of the four countries of the lower basin were extremely happy with the massive scale of the vision which had been outlined for the development of the lower Mekong and its tributaries. The report had provided these countries with a very concrete idea of where their "Mekong spirit" could ultimately lead them.

The White Report

The expansive Japanese vision did much to build up the hopes of the people of the entire region. However, the Japanese and ECAFE proposals contained a certain flaw which was unavoidable at that time. The problem was that adequate consideration had not been given to such related matters as the socio-economic conditions in the lower basin and the relocation of people from inundated areas. Because the Japanese effort was no more than a reconnaissance report, it can be excused for neglecting to consider these human issues. Nevertheless, the report was heavily biased toward technical questions, a tendency that was reinforced by the massive scale of its proposals.

The countries that were involved in collecting basic data at this time, as well as ECAFE and the Mekong Committee itself, gradually came to recognize the importance of the socio-economic aspects of the proposed development programmes. Consequently, it was decided to invite Professor Gilbert F. White, the renowned geographer then at the University of Chicago, to study these issues. This led to the publication in January 1962 of the still-famous White Report entitled *Economic and Social Aspects of Lower Mekong Development*. This document effectively rectified the one-sided emphasis on technical questions which had until then characterized the development proposals.

Subsequent progress and studies

At about the same time as the submission of the Japanese development proposals, each of the four lower-basin riparian countries was beginning to undertake surveys and studies of the major development projects in the domestic portions of their Mekong tributaries. This work was being carried out through the support and assistance of the Mekong Committee and culminated in the construction of the Nam Pong Dam (completed in 1966 with German funds, with an initial power output of 16,600 kW and final power output of 25,000 kW, and the irrigation of some 53,000 ha) and the Nam Pung Dam (completed in 1965 with Japanese assistance, with a power output of 6,300 kW and the pumped irrigation of 8,000 ha). Located in north-east Thailand, both of these dams made a major contribution to the economic development of this region. In addition, work was started in 1968 on the construction of the Nam Ngum Dam near Vientiane in Laos (with funds from the World Bank, Japan, and others, with a first-phase power output of 30,000 kW). However, no progress was made in the various Cambodian and Vietnamese development projects which were scheduled to begin at the same time. The success of the development projects in Thailand and Laos can, in great part, be attributed to the financial assistance of the Japanese government and the heroic efforts of the engineering teams from Nippon Koei Co., Ltd. and the Electric Power Development Company (EPDC), Japan, whose work was supported by these assistance funds.

In other areas of the basin, the US Bureau of Reclamation was placed in charge of the feasibility study for the proposed dam at Pa Mong on the main stream of the Mekong, while the prefeasibility study for the Sambor Dam located far downstream from the Pa Mong Dam was assigned to the newly-formed Overseas Technical Cooperation Agency (OTCA, the predecessor of the Japan International Cooperation Agency, or JICA). The former study (4.6×10^6 kW power output, 1×10^6 ha of irrigation and flood control in the delta) was continued until 1972, while the latter study (power output of 875,000 kW for a single development) was completed in 1968. However, as the outcome of the war in Viet Nam became increasingly uncertain, both projects were abandoned without any actual construction being done.

Beginning in late 1964, the Secretariat of the Mekong Committee had embarked on formulating its Indicative Basin Plan report (Mekong Secretariat 1970). While the final plan was of a massive scale and covered the development of the lower Mekong and its major tributaries, this document was, in fact, based on the aforementioned Japanese reconnaissance report of 1961. To the Japanese proposals, the Secretariat had added additional details based on newly acquired data, expanded or elaborated the Japanese plans, or suggested alternative designs and approaches. Furthermore, the Secretariat had arrived at cost estimates based on the data provided by EPDC for dam development in Japan. Sections on socio-economic issues were added to this report, which was released in

1970. Unlike the earlier Japanese report, ample attention had been given to non-technical considerations and the report included a wealth of information based on the Committee's general investigations. Nevertheless, as previously mentioned, the framework of this document was based on the Japanese reconnaissance report.

The Nam Ngum Dam in Laos (first phase) was completed in 1972. During the same period, various projects for agricultural development were started in the lower basin. In addition, the formulation of the Delta Development Plan for the Mekong Delta was completed by 1974.

In 1975, the long war in Viet Nam came to an end with the defeat and retreat of the American forces. Subsequently, the three countries of Indo-China came under communist rule. As a result, all development projects in this region – with the exception of the projects in north-east Thailand – were suspended until the creation of the Interim Mekong Committee in January 1978.

The Mekong Committee was virtually unable to function after the early 1970s. This situation continued until January 1978 when the Interim Mekong Committee was formed without the participation of Cambodia. However, the Interim Committee was impeded by the non-participation of Cambodia and was unable to commission new studies for the development of the basin, let alone initiate new construction work. It was even hampered in its efforts to collect basic technical data. Nevertheless, the Interim Committee was able to make some valuable contributions to Thailand, Laos, and Viet Nam in such areas as increasing food production, facilitating the supply of energy resources, flood control, and improvement of river navigation.

In 1978, work began on the Nam Ngum Dam in Laos to increase its installed capacity from 30,000 to 110,000 kW. This advance was made possible by the contributions of the Japanese government. (Nippon Koei and various other Japanese companies played an important supporting role.) Japanese financial assistance was again provided in 1980 when the installed capacity of this dam was further increased to 150,000 kW. As much as 90 per cent of this power is exported to Thailand and continues to make a valuable contribution to the national finances of Laos.

While the three countries of Indo-China were bogged down in a period of stagnation, Thailand continued to make substantial progress toward development throughout the 1970s. Although north-east Thailand (located in the Mekong Basin) generally lagged behind in these advances, notable advances were made in the area of agricultural development with approximately 500,000 ha of cultivated land coming under irrigation by 1985. Furthermore, beginning in 1986, high-yielding rice cultivation was made possible in 10,000 ha of paddy fields as a result of pumped irrigation from two major Mekong tributaries, the Mun and Chi rivers.

Since the mid-1980s, Viet Nam has been actively involved in developing the Mekong Delta. For instance, Dutch and Australian assistance has been used to construct a series of weirs used for irrigation and in blocking the influx of sea water. These efforts have led to the irrigation of a total area of 350,000 ha.

During this period, various new studies were undertaken for the development of the entire lower basin. In 1978, Australia embarked on a review of the Pa Mong Dam Project (based on the plan proposed by the Japanese, i.e. the High Pa Mong Dam) on the main stream of the Mekong. Between 1982 and 1989, Britain carried out a water-balance study in the lower basin. In 1981, the United Nations Development Plan (UNDP) and the Swedish International Development Agency (SIDA) implemented a water-quality study in the lower basin. Various other studies, such as research on migratory fish, have also been commissioned. The commissioning of these studies indicates that, as in the rest of the world, environmental issues and problems of inundation and relocation have come to be viewed as crucial concerns in the development of new dams in the Mekong Basin.

Shortly after its formation, the Interim Committee decided to revise and draft the development plan which had been drawn up in 1970. Following this decision, consultants from Switzerland, Holland, and Thailand were commissioned to revise some parts of the previous plan at a cost of US$600,000.

The revised dam-development plan, which was entitled *Perspectives for Mekong Development*, was released in 1988 and generally confirmed the 1970 proposals regarding the construction of a series of dams on the Mekong. The single modification made was the reduction of the height of the High Pa Mong Dam. The revised plan identified certain international projects to be completed by the year 2000: these included the construction of the newly revised Low Pa Mong Dam and the construction of the Nam Theun 2 Dam on the Nam Theun River, a Laotian tributary of the Mekong. In addition, the revised plan called for the construction of a total of 26 minor dams in three lower-basin countries (excluding Cambodia) for the development of domestic water resources, also to be completed by the year 2000. The consultants estimated the total cost of all these projects as US$4 × 10^9.

One of the most important features of the revised plan was to lower the height of the water level of the Pa Mong Dam to 210 m (the original Japanese plan was based on a full water elevation of 250 m in the reservoir) and to build another 250 m dam immediately upstream from the Pa Mong Dam. The purpose of lowering the height of the High Pa Mong Dam was to reduce the number of people who would have to be relocated as a result of inundation. (It was estimated that the relocated population would be reduced from 250,000 to 40,000 by this lowering of the dam.) This revision clearly reflected the growing concern for human rights in the development of the lower Mekong Basin.

Unfortunately, no action had been taken on the implementation of the revised development plan of 1988 until 1996, when its revision was proposed by the Swedish government. A detailed account of the developments in the ensuing years is given in the first section of chapter 6.

The Mekong and European countries (before the establishment of the United Nations)

The previous section provided an overview of the evolution of water resources development plans from after the Second World War up to 1988. Beginning with the involvement of France in the nineteenth century, this section examines in more detail the efforts to develop the lower Mekong Basin, focusing on the contributions of ECAFE, Japan, and the Mekong Committee in planning its development. It is hoped that this will be a useful source of reference for those involved in the future development of the Mekong.

France's involvement

The first westerner to see the ancient ruins of the Angkor Wat Temple is said to have been a French Catholic missionary, Father Bouilleveaux, in 1850 (Osborne 1975). It seems that the ruins of the temple did not make a deep impression on this zealous missionary, for he did not make any report of it to his fellow countrymen.

Just over a decade later, in 1861, the French naturalist and explorer Henri Mouhot visited the ruins.[1] The detailed description of the ruins published by Mouhot caused a sensation in certain social circles in Europe. As a result of Mouhot's report, many French intellectuals became interested in the Mekong Basin, but the Mekong itself was not the focus of the French and British governments' interest in the East in the mid-nineteenth century: both governments were attracted much more to the potential markets of China, with its huge population. Their chief interest lay not in Burma, Thailand, and Cambodia themselves, but in whether these countries could be used as stepping-stones to the heart of China (Osborne 1975).

In 1837, long before Mouhot's journey to Angkor, a British army captain named McLeod succeeded in taking a boat up the Salween River and reported that the river could be used as a trade route to Yunnan. A few years after this successful expedition, Britain sent a special envoy to the

Burmese court and began negotiations for the use of the Salween River as a water route. After this, the French government developed an obsessive fear that Britain might monopolize the profits from trade with the Chinese hinterland (Osborne 1975).

In 1866, the French government dispatched a six-man expeditionary force led by Commander Doudart de Lagrée. Its aim was to explore the possibilities of using the Mekong as a trade route into the Chinese hinterland. The expeditionary force embarked upstream from Phnom Penh in accordance with the French government's plan. However, as described in the previous chapter, it had great difficulty in negotiating the rapids at Sambor soon after its departure and was obstructed further upstream by the mighty Khone Falls. Further up from the falls, the group was halted once again by the rapids at Khemarat and was forced to abandon the plan of using the river to go upstream (Osborne 1975); thus, France lost its battle with the great natural forces of the Mekong.

Prior to this expedition, at the end of the 1850s, France had occupied Da Nang and Saigon (now Ho Chi Minh City) in Viet Nam, claiming that Christians were being persecuted. Lieutenant Francis Garnier, a member of the Mekong expeditionary force, claimed that, although the Mekong could not be used, there was a good chance of reaching the Chinese hinterland via the Red River in northern Viet Nam. The French therefore decided on the military occupation of northern Viet Nam, and Hanoi was occupied for a while in 1873. However, after he was killed by local "rebels" (from the viewpoint of the French army), France abandoned this project for several years.

France continued to drew up various plans for the control of northern Viet Nam and eventually, in 1887, established French Indo-China, comprising the three countries of Viet Nam, Laos, and Cambodia. Now that France had control of Viet Nam, it could finally realize its aim of using the Red River as a trade route to Yunnan (Osborne 1975).[2]

Having established French Indo-China, the French built a railway bridge over the Red River in Hanoi and laid down a railway running through Viet Nam. They also built ports along the coast and constructed a north–south road along the main Mekong all the way from Vientiane in Laos to the Mekong Delta. In addition, they set up navigation aids at key points on the main stream to ensure the safety of navigation by boat between riparian regions. The French successfully developed agriculture throughout Indo-China. They operated rubber plantations in selected areas such as Chepp in eastern Cambodia and, discovering that the Boloveng Plateau in southern Laos was particularly suitable for agriculture, cultivated vegetables, coffee, and other products there. According to Charles Fenn's *Ho Chi Minh: A Biographical Introduction*, this development was mainly achieved through the forced labour of the local people.

The Mekong Delta came under French control from 1862 to 1863, some 15 years before the establishment of French Indo-China. At this time, the French government was already pursuing a policy of forced colonization, seizing large areas of land without landholders and dividing it among those Vietnamese who cooperated with them, free of charge. Resistance to this forced occupation of land broke out in various parts of the delta and, in order to suppress these outbreaks more effectively, inland navigation was improved by digging canals to facilitate the movement of troops.

By increasing the efficiency of paddy-field cultivation, however, the French succeeded in developing the Mekong Delta into one of the world's leading rice-growing regions. According to the documents of the time, the total paddy-field area of the Mekong Delta in 1866 was only 216,000 ha but by 1895 it had been increased to 1,026,000 ha. From the beginning of the twentieth century, the waterways initially built mainly for the suppression of rebellions came to play an important role in agricultural development as irrigation and drainage channels.

By 1930, the total length of the Mekong Delta canal system had reached 1,790 km and the total paddy-field area was more than 2,100,000 ha. The rice produced in the delta was picked up by collection and delivery agents, supplied to dealers in unhulled and polished rice via middlemen, and exported as "Saigon rice" to China, France and its colonies (including Africa), and Japan. In 1930, total annual rice exports from the Mekong Delta amounted to 1,370,000 t.

The French government's policy in Indo-China was not restricted to these improvements in infrastructure: it also promoted medical treatment, education, and the diffusion of French culture; it fostered bureaucrats; and it established various administrative structures. France's colonial policies in this region at the beginning of this century appeared to have been completely successful.

France's insatiable expansionist policy gave rise to several delicate problems concerning the border with Thailand, one of which was the determination of the border line on the Mekong itself. Through a revision of France's treaty with Thailand in 1925, all the small islands dotted along the main stream of the Mekong became French territory. However, Thais who were dissatisfied with this revision succeeded in overturning this decision through the Thai–French Non-Aggression Pact in 1939 and the Thai–French Peace Treaty in 1942. Clause 2 of the Thai–French Peace Treaty signed on 11 July 1942 stipulated that "the borderline between Thailand and French Indo-China shall be the centre line through the main water route along the Mekong (with the exception of Khone Island immediately upstream from the Khone Falls, which shall continue to be the territory of French Indo-China)." Quite recently, just before the opening ceremony for the Nong Khai Bridge between Thailand and Laos completed in April 1994 with aid from Australia, the Thai and Laotian governments agreed to make the centre of the bridge the national border. However, the fact that the negotiations leading up to this conclusion were by no means smooth shows that the disagreements of the past have not been forgotten.

Through pressure from Japan, the 1942 Thai–French Peace Treaty incorporated into Thailand the four provinces of Luang Prabang, Champasak, Siem Reap, and Battambang (all in the lower Mekong Basin); however, in 1946, one year after the end of the war, these provinces were restored to French Indo-China. Their possession was not a matter of vital importance for Thailand, which above all wanted to avoid military conflict with France, one of the victors in the war. When the three countries of French Indo-China won their independence, two of the four provinces – Luang Prabang and Champasak – were made part of Laos, while the other two – Siem Reap and Battambang – became Cambodian territory.

The contribution of the USA

Although France seemed to have made a great success of the colonization of Indo-China, the time eventually came for her withdrawal from the region. The confused political situation after Japan's occupation during the Second World War continued for several years after the war, but in November 1953 France eventually suffered an overwhelming defeat at the hands of Ho Chi Minh's army at Dien Bien Phu in the most northeastern part of the lower Mekong Basin. After the withdrawal of the French army, the United States advanced into the whole of the lower Mekong Basin.

In 1957, the United States, making use of its close relationship with the Thai Prime Minister, General Sarit Thanarat, who had close links with the village of Mukdahan in north-east Thailand, constructed the "Friendship Road," a highway from Bangkok to north-east Thailand built with the strategic aim of defending this region.[3] The construction of this highway helped to vitalize the hitherto sluggish development of the north-east. One result of this was the conversion of the relatively unproductive paddy fields around Sayabouri and Korat into cornfields. Within just eight years from the development of this road to the north-east, Thailand's total corn production, which had been only 27,000 t in 1950, had increased to over 10×10^6 t, making corn Thailand's principal export.

About five years earlier, from 1955 to 1956, the Bureau of Reclamation of the US Department of the Interior had conducted a preliminary study on the whole of the lower Mekong Basin. The results of this study, published in March 1956, are outlined below (p. 94). North-east Thailand was, of course, included in the study.

The Thai government always viewed the north-east as a region where communism might spread. When the government's First National Economic Development Plan was implemented in 1960, the north-east was therefore designated as a priority area for development.

Development studies

In fact, it was not the United States but the United Nations that led the way in implementing modern development studies of the lower Mekong Basin. In 1947, ECAFE was established in Shanghai.[4] In 1949, ECAFE set up the Flood Control Bureau, which proposed a study of the Mekong at the Flood Prevention Technology Conference in New Delhi in 1951. At the Seventh ECAFE Conference in Lahore the same year, the Flood Control Bureau proposed that a local study be conducted on flood control, irrigation, and development of hydroelectric power in the lower Mekong Basin, and this proposal was approved. As a result, a five-member investigative team, consisting of two members of the ECAFE Secretariat and three specialists from Thailand and Cambodia, was dispatched to the area (Menon 1966).

Two Chinese nationals played leading roles in persuading the United Nations to take this action: they were the top engineers Dr Shen-Yi and P. T. Tan, who had both joined ECAFE in Shanghai, having become disillusioned with China as it became increasingly dominated by communism after the Second World War. Until China was taken over by the Communist Party, Mr Tan had been seconded by the government to the University of Dresden in Germany to take part in an experimental project on the control of flooding of the Yellow River. Mr Tan once said that, since he had been unable to realize his dream of controlling flooding of the Yellow River, he set up the Flood Control Bureau in ECAFE with the even more ambitious aim of controlling floods throughout Asia. Turning his attention to the Mekong, he devoted the rest of his life to the promotion of development to prevent flooding in the lower Mekong Basin (P. T. Tan's family 1972).

Both Dr Shen-Yi and P. T. Tan firmly believed that, if dams and water channels were developed in the lower Mekong Basin, flood damage could be reduced, navigation and irrigation developed, and the river utilized for hydroelectric power. Underlying this conviction seems to have been the success in river development of the Tennessee Valley Authority (TVA) in the USA, which thrived under the Roosevelt New Deal programme from the 1930s to the 1950s. In spite of the differences in period and socio-economic background between the development of the lower Mekong Basin and the Tennessee River Basin, these two Chinese flood-control experts undoubtedly felt that, just as TVA had succeeded in overcoming the natural menace of floods (as well as developing electric power and navigation and promoting industry and agriculture), the successive development of a large number of dams on the main stream and tributaries of the Mekong could bring great wealth to the countries of the lower Mekong Basin.

The results of the ECAFE preliminary study conducted in 1951 were announced in May 1952 in a short 18-page report entitled "The Development of the International River Mekong – Technical Problems related to Flood Control and Water Resources Development." The report stated that the key to development of north-east Thailand was to build the Pa Mong Dam for irrigation and power generation. It also concluded that, if two more dams were constructed at the Khone Falls and Sambor, this would contribute to the improvement of power generation and navigation, and stressed the need for a full-scale study.

By the Geneva Agreement of July 1954, the three countries of French Indo-China finally achieved their dream of independence. These three countries (Cambodia, Viet Nam, and Laos), together with Thailand, unanimously hoped for US assistance with their economic development. After the publication of the ECAFE report, they asked the US government to conduct a study of the Mekong. This was the background to the study by the Bureau of Reclamation of the US Department of the Interior mentioned in the section on the contribution of the USA.

As we have seen, there were very few reliable basic data (such as hydrological data, meteorological conditions, topography, or socio-economic data) regarding the lower Mekong Basin in the 1950s. Nevertheless, the survey report submitted in March 1956 by the Bureau of Reclamation was fairly thorough and reliable. The report emphasized the importance of building several mainstream dams, such as the Pa Mong Dam, and the need to collect more reliable data for this purpose. As promising projects that should be promoted for the time being, it recommended the drawing-up of development projects on Mekong tributaries such as the Lam Pra Plerng Project in north-east Thailand and the Bovel Irrigation Project in the vicinity of Cambodia's second-largest city, Battambang. It also stressed the need to resume small-scale projects such as the Tiep Nhut and An Truong projects in the Vietnamese Delta and the Choeung-Prey Project in Cambodia, and to promote a tank (small reservoir) irrigation project in north-east Thailand. As for large-scale projects, the report recommended that a study be undertaken for the construction of a large weir to regulate the volume of water in the Great Lake in Cambodia in order to increase the fishing catch. Finally, it reiterated the need to collect and collate basic information concerning hydrological and meteorological conditions, topography, and geological features of the lower Mekong basin (McKay 1956).

Main aims of the ECAFE and US studies and the situation in the lower Mekong Basin

The 1952 ECAFE report and the 1956 United States report both went beyond the studies made during the period of French rule, which were based only on single objectives such as the improvement of navigation or flood control. Having collected basic data on a wide range of themes, both studies concluded that it was necessary to promote the effective use of land and water to improve residents' lives and that the riparian countries must cooperate to draw up a development plan based on the ideal of balanced comprehensive development of the lower Mekong Basin. These two reports played a significant role in opening the eyes of those responsible for development of the basin.

Let us briefly consider the lives and developmental needs of the people living in the lower Mekong Basin after the Second World War.

In the mid-1950s, the population of the basin was about 15×10^6 with an annual rate of increase of almost 3 per cent. The total population of the four riparian countries was 45×10^6, but in money terms the average annual income was less than US$100. About 80 per cent of the people of the four countries were local farmers and the proportion of farmers was even higher in the Mekong Basin. Most of these farmers relied on the rainy season for their harvest, which inevitably made their livelihood fairly unreliable.

At the time, the total arable land in the lower basin was about 5.7×10^6 ha; only 0.6 per cent of the water was being used, and the total area of irrigated land was just 153,000 ha. Of the cultivated land, 86 per cent was paddy fields, from which the unhulled rice harvest was on average only 2 t/ha on the flat flood plains (or about 0.8 t/ha of "floating" rice in areas where flooding was particularly heavy), 1.5 t in hilly areas, and 1 t in mountainous areas. It was, therefore, natural that the governments of the four riparian countries should consider increasing the harvest of agricultural products by making double-cropping possible by means of irrigation and the use of fertilizers. With regard to energy, too, the abundant water resources of the Mekong Basin had hardly been developed, making it necessary to pay large amounts in foreign currency for fuel such as diesel and crude oils for thermal power.

According to the assessment at the time, it was hoped that development of the lower basin would result in a total of 4.3×10^6 ha of irrigated land and 24×10^6 kW of hydroelectricity. However, there were (and still are) differences in the development requirements of the four riparian countries. In Cambodia and north-east Thailand, the supply of water for irrigation in order to ensure rice self-sufficiency was an urgent requirement, whereas Laos was just about able to achieve rice self-sufficiency but needed to develop hydroelectric power and acquire more foreign currency through its exports. Viet Nam's chief requirements were to reduce flood damage in the Mekong Delta, to increase food production by stabilizing the water supply in the dry season, and to increase overseas exports. Particularly in the cases of Laos and Cambodia, most of the land was in the lower Mekong Basin, which meant that development of the country and development of the lower basin were considered to be almost synonymous.[5]

The 1957 ECAFE Study Report

The above-mentioned first study report on the Lower Mekong Basin published by the ECAFE Secretariat in May 1952 stressed the necessity of conducting a fundamental study for future development as soon as possible. The Secretariat reiterated this necessity at the eleventh session in April 1955 and dispatched a second study team of seven experts (all now deceased) in April–May 1956. This team included P. T. Tan of the ECAFE Secretariat, Kanwar Sain from India (who later became Director of the Mekong Committee Secretariat), G. Duval of France, and Yutaka Kubota and Masanobu Sakaida of the Japanese company Nippon Koei Co., Ltd.[6] The team produced an excellent draft report entitled *Development of Water Resources in the Lower Mekong Basin* (UN 1957), which was presented and adopted at the thirteenth session of ECAFE in Bangkok which was held in March 1957. Since this report made a number of important proposals which became the basis of the thinking behind subsequent reports, we must now examine its contents in more detail.

The 1957 ECAFE Report (UN 1957) assessed the development potential of the water resources in the lower basin, based on available data on such factors as precipitation, river flow, and topography, and conducted a rough analysis of the current status of development and utilization and of future demand for hydroelectricity, navigation, irrigation, and flood control. The report outlined the possibilities and requirements for developing an operating weir at the entrance of the Great Lake (Lake Tonle Sap), in addition to the development of six mainstream dams at Luang Prabang, Pa Mong, Thakhek, Khemarat, Khone Falls, and Sambor (figs 3.1 and 3.2).

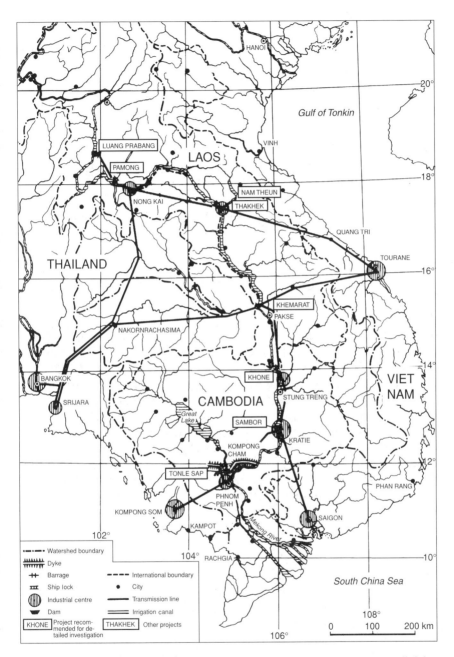

Figure 3.1. Development projects in the lower Mekong Basin recommended for detailed investigation
Source: UN (1957)

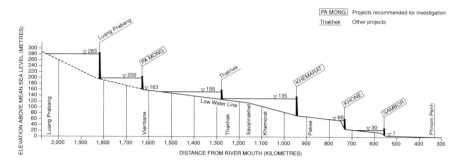

Figure 3.2. Profile of the Mekong showing dam projects recommended for detailed investigation
Source: UN (1957)

Planned mainstream dam projects and the Tonle Sap Weir

The Pa Mong

This development plan involves the construction of a dam with a surface water elevation of 200 m above mean sea-level (m.s.l) about 30 km upstream from the city of Vientiane on the main stream. The effective capacity of the reservoir behind the proposed dam would be 2×10^9 m^3 and the probable gross head and available river discharge of the power station attached to the dam would be 37 m and 200 m^3 (for 115 days), respectively. Its baseline output would be 19,000 kW (installed capacity: 250,000 kW) and the dam would provide 1,000 m^3/s of irrigation water, which could be transferred to both the Korat Plateau and the Vientiane Plain. In addition, if a ship lock were provided at the dam, this would permit navigation between the downstream and upstream areas of the dam and reservoir.

The Khemarat

If a dam with a full water elevation of 135 m (or 125 m) were built 14 km upstream from the confluence of the main stream and the Mun River (a major tributary), a reservoir with a dam 1,300 m long could be created with a baseline output of 585,00 kW (or 360,000 kW) with a gross head of 60 m. Assuming a drawdown of 10 m, the effective capacity of the reservoir would be as much as 5×10^9 m^3, which could be utilized to increase the flow in the dry season and to control flooding; a lock would also be provided for navigation. However, the construction of this dam would result in the inundation of a large area covering part of Savannakhet City and the districts extending downstream along another Mekong tributary, the Se Bang Hieng. It was therefore recommended that the full water elevation of the proposed reservoir should be examined further. The shortcomings of this project were that the proposed lock would be costly and its implementation would result in the extensive submergence of houses and land.

The Khone Falls

Khone Falls is a series of rapids extending over 20 km from Kinak to Voeun Khan with a total head of about 30 m. The longest fall is divided into two channels by a large island. The drop on the west side, known as the Chutes de Somphanit, has a high crest; it overflows only during the high-water period and dries up during the long low-water period. The bedrock between the main channel and the Chutes de Somphanit has a slight downward incline, forming a natural side spillway. The main channel, known as the Chutes de Phapheng, is situated on the east side and has a head of 18 m in the dry season. The mainstream flow passes through the Chutes de Phapheng incessantly. By constructing a dam 5 m high just above this fall and providing a low dam on the natural spillway near the Chutes de Somphanit, it should be possible to secure a head of 30 m by leading the flow to Voeun Khan below the fall, generating a baseline output of 265,000 kW (installed capacity: 320,000 kW). Alternative plans may be considered but, since this location constitutes the greatest obstacle to navigation in the lower basin, it was recommended that a lock (120 m long, 12 m wide, and 2.5 m deep) should be provided.

The Sambor

From its mouth as far as Sambor in Cambodia, the Mekong may be considered a good inland waterway, with a minimum depth of 6 m during the rainy season and 2 m during the dry season. Difficulties in navigation begin at the Sambor Rapids located 565 km from the mouth of the Mekong and stretching over a distance of 168 km to Voeun Khan, just below the Khone Falls. Along this stretch, the river bed is dotted with stone outcrops protruding above the surface at low water, making navigation during the dry season very difficult, if not impossible.

This difficulty could be overcome by heading up the flow with a dam at the lower end of the rapids, thus extending the perennial waterway to the border between Cambodia and Laos. A substantial amount of power could be generated from this dam, which would have the advantage of being situated very close to possible power-consumption centres in Cambodia. Furthermore, water up to 100 m^3/s could be diverted to irrigate the fertile plains extending some 100,000 ha along the Mekong below Kratie. In addition, when the dam is built, a baseline output of about 235,000 kW (producing an annual generated energy of 2,030 GWh) would be made available with a gross head of 23 m and a low-flow volume of 1,405 m^3/s. With the completion of both the Khemarat and the Pa Mong dams upstream, it would be possible to generate about 340,000 kW constantly; if all the proposed upstream dams were constructed, it was estimated that a baseline output of some 500,000 kW could be generated. It was also expected that a lock could be provided for navigation from downstream and that the low-flow volume downstream from the dam would be increased from 1,405 to 2,005 m^3/s.

The Tonle Sap

If a weir were constructed on the Tonle Sap river, which connects the Great Lake and the Mekong, it would be possible to keep the water of the lake at its lowest level by not admitting any flow from the Mekong during the rainy season. Water could be admitted into the lake by operating the weir gates only after the Mekong had reached a certain level, thus cutting off the flood peak. With the help of the weir, water stored in the lake could be released and used during the dry season for irrigation and the improvement of navigation.

After considering the alternatives, the report tentatively concluded that the most suitable weir site would be somewhere near Kompong Luong on the Tonle Sap river. The weir would ensure that the maximum water level of the Mekong immediately downstream from its confluence with the Tonle Sap would be about 9.70 m, just under the upper limit of the low flood level. As water would be released from the Great Lake through the control weir during the months of low-level flow from January to May, the flow volume during this 5-month period would be 4,500 m^3/s at the Mekong Delta, which is more than three times the present lowest rate of discharge of the rivers. The report thus predicted that navigation conditions would be considerably improved and that the current severe degree of damage caused by the influx of sea water into the areas furthest downstream would be drastically reduced. Moreover, with the proposed weir on the Tonle Sap and stored water supply in the Great Lake, it would be possible to provide gravity irrigation of at least 6×10^6 ha of the area on the west bank during the dry season (fig. 3.3).

The 1957 ECAFE Report not only outlined these advantages but also pointed out major problems that would be caused by the construction of a weir:

In spite of these merits, the proposed construction of a control weir on the Tonle Sap requires careful consideration. At present, the Great Lake is one of the richest fish-producing water bodies in Asia. Systematic investigation should be conducted by fishery biologists, technologists, and economists.

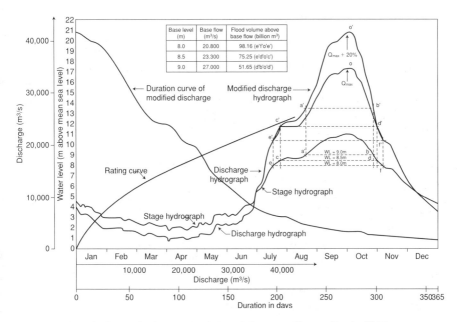

Figure 3.3. Discharge hydrograph of the Mekong at Phnom Penh, 1947
Source: UN (1957)

1957 ECAFE Report Annex: recommendations of the US Bureau of Reclamation

As an annex to the ECAFE Report (UN 1957), the US Bureau of Reclamation (USBR) appended a recommendation which included 29 articles. Most of these were related to the development of water resources in the lower Mekong Basin, but the USBR also recommended the implementation of overall studies of transportation, communication and the energy market covering the whole region. It also urged that public health be taken into account.

The annex also included results of a preliminary study comparing estimated costs of transportation by railway, highway, and river between major cities in the lower basin. These results showed that construction costs for river navigation are much lower than those for construction of railways (single line, metre gauge) and highways (6 m wide, asphalt), but in terms of the ton–kilometre cost per round trip, railways were found to be the cheapest means of transportation.

As the improvement of navigation was one of the major objectives of development of the Mekong Basin in the 1950s, this conclusion received considerable attention because it fundamentally questioned prevailing views concerning the utilization of waterways.

Emphasis on the Need for International Cooperation in the ECAFE Report

After outlining its development plan consisting of the five major projects described above, the 1957 ECAFE Report stressed that the realization of these projects would require international cooperation. First, an international agreement among the governments of the riparian countries would be required, after which funds would have to be made available through the developed nations.

The report also stressed that extensive basic data should be provided and that the collection of hydrological and hydraulic data, as well as topographical maps, must be performed in conformity with common standards. It also pointed out the need to carry out such data collection and recording on a long-term basis.

The ECAFE Report also noted that development projects in each tributary in the lower basin would often influence other riparian countries, particularly in the downstream areas. It therefore recommended that each country should supply sufficient information to the others, even when developing water resources within its own territory.

In its final section, the report emphasized that, since it would be necessary to take into account how each development project related to the comprehensive development of the entire river basin, all projects should be promoted in a cooperative atmosphere. For this purpose, an international organization should be set up so that all relevant information could be freely exchanged.

As can be seen from this outline of the ECAFE Report, there is not much difference between the problems already foreseen in the 1950s and those that are still being dealt with. Indeed, now that international and socio-economic conditions are so much more complicated, the burden of these problems related to water-resources development in the lower Mekong Basin has become heavier than ever.

Mekong Committee

Establishment (1957)

Although the 1957 ECAFE Report included a fairly precise description of the effects of development in the four proposed dam projects – Pa Mong, Khemarat, Khone Falls, and Sambor – as well as the Tonle Sap weir project, its main purpose was to insist on "the necessity of formulating a comprehensive development approach through close cooperation among the four riparian countries."

Most of the proposed mainstream dam projects either encompassed the border between two countries or, even if confined within one country, influenced a neighbouring country or countries. Moreover, some projects were expected to give rise to complex problems between countries in the upper and lower basins. It was therefore proposed that an international organization be established to coordinate studies and projects from a comprehensive viewpoint. Unfortunately, however, it was impossible (for political reasons) to involve the two upper-basin countries – China and Burma (now Myanmar) – in this international organization. At the time, China was not a member of the United Nations, while Burma did not show any interest in discussions on cooperative development and seemed almost indifferent to Mekong development, both politically and geographically.

Accordingly, soon after the thirteenth session of ECAFE in March 1957, at which the draft ECAFE Report was submitted and adopted, representatives of the four riparian nations of Cambodia, Laos, Thailand, and Viet Nam held a meeting in Bangkok and formally requested ECAFE to make legal arrangements for the establishment of a Mekong committee composed of these four nations. Several months later, the Committee for Coordination of Investigations of the Lower Mekong Basin, known as the Mekong Committee, was officially inaugurated as an autonomous organization for the achievement of common objectives.

Objectives

The main purposes of the Mekong Committee were to solve problems of poverty and political instability in the lower basin and to promote peace, progress, and prosperity through the effective utilization of the basin's resources. It was to implement, adjust, supervise, and control plans and studies of water-resources development programmes, to request governments to provide special financial and technical assistance on behalf of the member riparian countries, and to accept and manage this assistance.

The ideal of the Committee was to unite the four riparian governments in order to realize the following two aims through the cooperation of developed nations, the United Nations, and international lending agencies:
1. Optimization and unification of water-resources development in the entire lower Mekong basin;
2. Realization of development, particularly of large-scale multipurpose dams that would have a tremendous impact on the entire lower basin.

A Secretariat was appointed for the Committee. Its main duties were as follows:
1. Collection of basic data indispensable for planning;
2. Promotion of investigations related to land and water resources;

3. Planning of development projects for navigation, irrigation, hydro-electric power, and flood control;
4. Realization of planned projects in the lower basin by supplying the necessary funds.

In September 1957, the Mekong Committee Secretariat was set up in the office of the ECAFE Secretariat in Bangkok. Shortly thereafter, Dr Hart Schaaf arrived at the new office as the first Executive Agent of the Mekong Committee. The Committee was made up of minister-level representatives from the governments of the four riparian countries, each of which set up a national committee to deal with discussions and arrangements regarding domestic matters. As an autonomous body based on the official endorsement of the representatives of the four member countries, the Committee was to convene once a year and submit annual reports both to the four countries and to ECAFE.[7]

The Mekong Committee was thus inaugurated in the autumn of 1957 and continued its activities until 1977.

Recommendations of the Wheeler Mission (1958)

Immediately after the official publication of the ECAFE Report in October 1957, the UN Technical Assistance Administration dispatched to the lower basin a five-man mission headed by Lieutenant-General Raymond A. Wheeler of the US Army Corps of Engineers (who was a technical advisor of the World Bank at the time). The other four members of the mission were a Japanese (Yutaka Kubota), a Frenchman, a Canadian, and an Indian.

The mission studied the region from the end of 1957 to the spring of 1958. In its report, it concluded that during the next five years a basic investigation of the lower basin should be made with a budget of some US\$9.2 $\times$ 10^6. In its conclusion, the report also made the following recommendations:
1. To conduct topographical and aerial surveys;
2. To establish hydrological stations for monitoring river-water quality and the water level on the main stream;
3. To carry out geological studies and borings;
4. To conduct soil surveys;
5. To conduct geographical surveys;
6. To implement preliminary planning of promising reaches of the main stream;
7. To undertake preliminary planning of other reaches of the main stream and major tributaries;

8. To collect data on fishery, agriculture, forestry, mineral resources, transportation, irrigation, flood control, power markets, etc.;

9. To provide an advisory service.

On the basis of these recommendations, the United Nations proposed the first 5-year assistance programme (1959–1964). Accordingly, France offered US$120,000 to the Committee to provide hydrological and meteorological equipment. This was followed by an offer of US$2 × 10⁶ from the United States, as well as offers from New Zealand, Britain, Australia, the Philippines, and India, bringing the total amount to about US$5 × 10⁶. Also on the basis of these recommendations, an international advisory board was set up, which consisted of a top-level team of six water-resources development experts from the United States, Britain, France, and Canada. At first, this advisory service was limited to technical matters, but its scope was later extended to cover economic and financial matters as well.[8]

Comprehensive report on the major tributaries by the Japanese government reconnaissance team (1961)

As one of the developed countries, Japan proposed to undertake reconnaissance on the major tributaries of the lower basin – a relatively low-cost project requiring US$240,000, which reflected Japan's capacity at that time (the summer of 1958).[9] Prior to that, Japan had provided assistance with the studies and development of some infrastructure projects such as hydroelectric power projects in Burma and Viet Nam as part of its Second World War reparations. But now the way had been opened for the most desirable means of technical cooperation – the undertaking of international development cooperation to assist developing countries through the UN organization – and a brilliant future opened up for post-war Japan. The author himself participated in this reconnaissance from its preparatory stages and has never forgotten this valuable experience.

The author would like to stress that Japan's participation in this UN study on the Mekong Basin was the first step in the country's substantial contribution to overseas development in the post-war period.

Japanese reconnaissance team's historic survey

As soon as Japan's proposal to participate in the survey of the lower Mekong Basin had been accepted by the United Nations, the Japanese government set up a survey organization named the Overseas Electrical Industry Survey Institute in the former Tokyo head office building of the

Tokyo Electric Power Company. The role of the Institute was to organize a survey team composed of engineers of the Electric Power Development Company (EPDC) and from Nippon Koei. The team was also joined by several young engineers and researchers selected by various ministries involved in water-resources development, which included the Ministry of Construction, the Ministry of International Trade and Industry, and the Ministry of Agriculture.

The team carried out site surveys over three phases during a period of almost 2 years from January 1959 until the end of 1960. As a result, the Japanese government submitted to the United Nations a bulky report entitled "Comprehensive Reconnaissance Report on the Major Tributaries of the Lower Mekong Basin" in September 1961 (Government of Japan 1961).

The dynamic and positive attitude of the Japanese team was particularly noteworthy. The team not only achieved its main task of formulating preliminary plans for dam development on the 34 major tributaries of the lower basin but also drew up preliminary development plans for dams on the whole of the main stream. These plans for mainstream dams included several dynamic concepts that were completely novel.

When the Japanese team submitted its development plans for the tributaries and main stream of the Mekong to the United Nations, the developed countries had only just embarked on cooperative surveys of the basin and there were very few basic hydrological, meteorological, geological, or other data, all of which were vital for the formulation of development plans. In spite of such unfavourable conditions and the very limited budget of US$240,000, the team succeeded in compiling a comprehensive development plan which covered most major mainstream and tributary dams in the lower basin. The report was, of course, drafted by more than one person, but most of its contents were produced by the EPDC team headed by Takeshi Tokuno, an eminent dam planner, under the leadership of Yutaka Kubota, who was then President of Nippon Koei. As a member of this team, the author feels he should mention that, while the team was working on the report, the US Army Map Service kindly provided all the required sections of maps (scale 1:50,000) requested by the team, utilizing the aerial photographs available at the time. The author would like to take this opportunity to thank the US Army for this hitherto unacknowledged contribution to Mekong development.

Outline of the Japanese report

The Japanese team began by referring to the 1957 ECAFE Report, but it included in its report very original points of view based on information gathered during its site visits, as well all the other data it managed to collect.

The report consists of a total of 300 pages, including 120 pages of appendices, and was written in English, French, and Japanese. The main contents were as follows:

Chapter 1. General features of the basin and importance of major tributaries:
   (a) General features of the basin;
   (b) Importance of major tributaries.

Chapter 2. Reconnaissance of tributaries undertaken by the Japanese team:
   (a) 34 major tributaries;
   (b) Reconnaissance survey;
   (c) Studies in Tokyo.

Chapter 3. Potential of development of major tributaries:
   (a) Laos;
   (b) Thailand;
   (c) Cambodia;
   (d) Viet Nam.

Chapter 4. Selected projects of high priority for development:
   (a) Projects under investigation;
   (b) Projects expected for early development;
   (c) Other promising projects.

Chapter 5. Suggestions for mainstream projects, particularly those related to tributary development plans:
   (a) Projects upstream from Pa Mong;
   (b) Pa Mong Project;
   (c) Thakhek Project;
   (d) Khemarat Project;
   (e) Khone–Sambor Project;
   (f) Outline of reservoirs and development programmes on the main Mekong.

Chapter 6. Agriculture and irrigation:
   (a) Climate and agriculture;
   (b) Soil and agriculture;
   (c) Improvement of soil;
   (d) Farm products;
   (e) The basic principles of irrigation.

Chapter 7. General review of development of Mekong Basin water resources:
   (a) Flood control;
   (b) Navigation;

(c) Hydropower;

(d) Outline of transmission scheme.

Chapter 8. Further investigation of major tributaries and comprehensive recommendation:

(a) Further investigation of major tributaries;

(b) Comprehensive recommendation.

Appendices

1. Itinerary of the reconnaissance survey;

2. Data;

3. Related studies:

(a) After-effects of development of the main Mekong and tributaries on the Great Lake and the delta region;

(b) Geographical study of flooding in the basins of tributaries;

(c) On drafting development plans using a provisional survey map showing classification of flood-stricken areas.

The report described the general features of the lower basin. First, it briefly outlined the targeted 34 tributaries and their development potential for each country. Next, it selected a number of high-priority projects on each major tributary, again by country, explaining the possible dam and power-station projects and devoting considerable space to particularly promising projects. As mentioned above, the report outlined a development plan for building successive cascade-dam projects on the main stream in the lower basin from a completely new viewpoint, just as TVA had done in the United States. It also provided precise explanations on such terms as agriculture and irrigation, flood control, and navigation. Finally, it made various recommendations regarding projects that should be undertaken in the future on the major tributaries from a "comprehensive" viewpoint.

The report was mainly technical, with hardly any consideration of the influence on the natural and social environment, including the problem of inundation of land and housing. Nevertheless, its bold views and grand vision surprised all of those concerned at that time in the United Nations and the governments of the riparian countries. The Japanese government received unanimous and high praise for the report.

The Japanese development plan was to build a series of 10 large dams on the main stream in the lower basin and to construct hydroelectric power stations at all these dams, which would provide a total installed capacity of over 10×10^6 kW (total baseline output: 8×10^6 kW), and to build a number of other dams and hydroelectric power stations with a total installed capacity of 7×10^6 kW. A total of 17×10^6 kW (total baseline output: 15×10^6 kW) would therefore be generated in the lower basin. The plan was to interconnect the power by long transmission lines to supply power-consuming industries with a matched capacity that would be provided in key locations in the four riparian countries, from which various industrial products were to be exported. At the same time, a million hectares of farmland would be irrigated in both north-east Thailand and in Cambodia, resulting in the stabilization of agriculture as well as food self-sufficiency and exports of some of the surplus. In addition, the main stream in the lower basin would be made navigable by the installation of locks at all the dam sites from the estuary of the main Mekong to the upstream border between Thailand and Burma (Myanmar); furthermore, flooding at the delta would be drastically decreased. In short, the development plan was to generate a considerable amount of power in the four riparian countries by constructing a large number of dams and power plants on both the major tributaries and the main Mekong (fig. 3.4).

Since this was a seminal development plan which was to become the framework for future water-resources development plans in the lower basin, let us now examine it in more detail.

The Mainstream Dam Development Plan presented by the Japanese team

Dam projects upstream from the Pa Mong

The 1957 ECAFE Report proposed the development of a Luang Prabang Dam (full water elevation of reservoir: 280 m) which would hold a large volume of water between the city of Luang Prabang and its proposed downstream site of Pak Lay. After examining this proposal, the Japanese team decided on a different scheme of building the Pak Beng Dam (full water elevation: 380 m) at a site located 120 km upstream from Luang Prabang and to build a Luang Prabang Dam (full water elevation: 300 m) 10 km upstream from the city. The team also proposed that another dam, the Sayaboury Dam (full water elevation: 275 m) be built at the site proposed for the Luang Prabang Dam in the 1957 ECAFE Report.

The team thus concluded that, if the above three dams were constructed, some 2.6×10^6 kW of installed power (baseline output: 2.18×10^6 kW; annual generated energy: 19,000 GWh) would be made available and a total effective capacity of 26.5×10^9 m^3 of water would be stored in the three reservoirs.

Another important proposal made by the Japanese team was for a huge reservoir with an effective capacity of some 20×10^9 m^3 to be built immediately upstream from the borders of Myanmar, Laos, and Thailand, based on available maps drawn to a scale of 1:50,000. The report explained that, if this reservoir were built and it was also possible to construct dams on the Nam Ou (a major Mekong tributary which flows into the main stream immediately upstream from Luang Prabang) and at

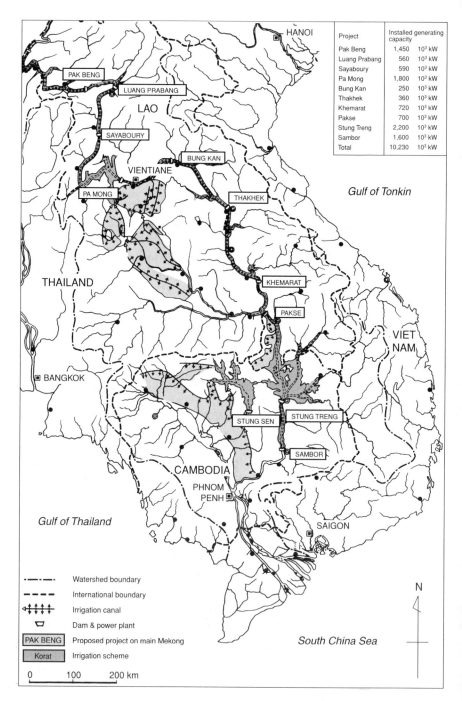

Project	Installed generating capacity	
Pak Beng	1,450	10³ kW
Luang Prabang	560	10³ kW
Sayaboury	590	10³ kW
Pa Mong	1,800	10³ kW
Bung Kan	250	10³ kW
Thakhek	360	10³ kW
Khemarat	720	10³ kW
Pakse	700	10³ kW
Stung Treng	2,200	10³ kW
Sambor	1,600	10³ kW
Total	10,230	10³ kW

Figure 3.4. Proposed development projects on the main stream of the lower Mekong
Source: Akira, 1993

110

other major tributaries, some 6.5×10^9 m^3 of water could be stored and the main stream flow at the Pa Mong Dam site could be adjusted to 4,000 m^3/s through the joint operation of all the dam reservoirs.

The High Pa Mong Dam Project

The 1957 ECAFE Report proposed the construction of a reservoir (full water elevation: 200 m) at the Pa Mong Dam site located upstream from Vientiane, but the Japanese team thought it would be preferable to build a far higher dam at the same site, raising the reservoir's full water elevation to 250 m.[10] The team pointed out that, by raising the water elevation, it would become possible to obtain an effective reservoir capacity of 38×10^9 m^3, producing a baseline output of water of some 2,500 m^3/s from the reservoir. By utilizing the effective head of 73 m, an installed capacity of 1.8×10^6 kW (baseline output: 1.53×10^6 kW) could be generated, producing annual generated energy of 13,400 GWh. A water output of 1,000 m^3/s could also be used in the dry season to irrigate farmlands in the Korat Plateau in north-east Thailand. In addition, the team foresaw that if some large-scale reservoirs were created upstream, an average water flow of about 4,000 m^3/s would become available immediately downstream from the High Pa Mong Dam site, making the downstream reaches more navigable.

Once the planned downstream dam(s), such as the Khone Falls Dam or the Stung Treng Dam, were completed, it would certainly be possible drastically to reduce flooding in the delta region. This idea of building the High Pa Mong Dam surprised those involved at ECAFE.

Dam projects downstream from Pa Mong (figs. 3.5 and 3.6)

The ECAFE Report proposed that the Thakhek Dam (full water elevation: 155 m) and the Khemarat Dam (full water elevation: 135 m) be built downstream from Pa Mong. The Japanese team proposed an alternative development plan: instead of the Thakhek Dam, the Bung Kan Dam (full water elevation: 155 m) should be constructed 200 km upstream from the planned dam site with the Thakhek Dam (full water elevation: 145 m) 10 km upstream from the proposed site. This would eliminate the need for the long dyke that might have to be built along the right (Thai) bank, which was considered to be one of the drawbacks of the ECAFE plan.

This plan would also avert the problem of extensive inundation of the Huey Tuey tributary basin. The installed capacity of the Bung Kan Dam Project would be 250,000 kW (baseline output: 212,000 kW; annual generated energy: 1,850 GWh) and the installed capacity of the Thakhek Dam Project would be 360,000 kW (baseline output: 310,000 kW; annual generated energy: 2,720 GWh) once the proposed four upstream dams were completed.

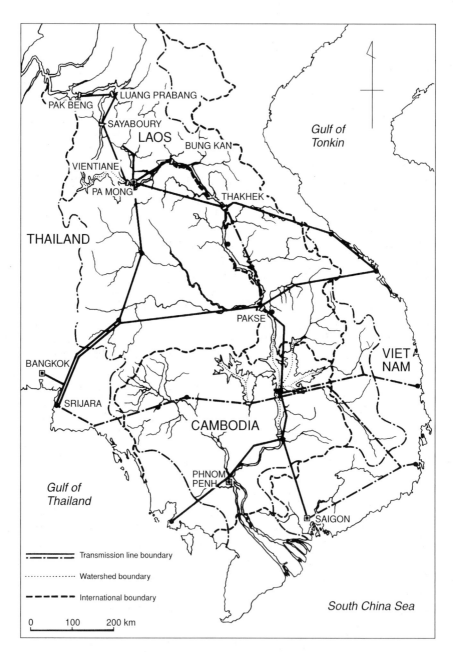

Figure 3.5. Proposed transmission-line network in the lower Mekong Basin
Source: Government of Japan (1961)

112

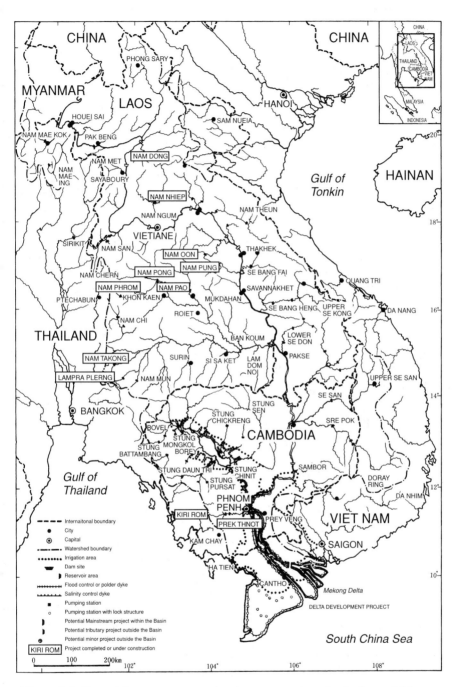

Figure 3.6. Potential water-resource development projects in the lower Mekong Basin
Source: Mekong Secretariat (1970)

113

The 1957 ECAFE Report included a plan to build the Khemarat Dam (full water elevation: 135 m) 14 km upstream from the confluence of the main Mekong and the Nam Mun. Instead of this, the Japanese team proposed two mainstream dams: these were the Khemarat Dam (full water elevation: 130 m) 25 km upstream from the confluence of the main Mekong and the Se Bang Hieng, and the Pakse Dam (full water elevation: 108 m) 7 km upstream from the confluence between the main stream and the Nam Mun. When all the dam projects upstream from the main stream were completed, the installed capacity of the Pakse Dam Project would be 70,000 kW (baseline output: 60,000 kW; annual generated energy: 5,250 GWh) (table 3.1).

Unless the full water elevation of the Khemarat Dam were reduced to 130 m, the low wet areas of the important city of Savannakhet, as well as most of the downstream area of the Se Bang Hieng River, would be submerged. If the full water elevation of the Khemarat Reservoir were 130 m, navigation between Savannakhet and upstream Thakhek, as well as on the Se Bang Hieng, would be much easier.

The full water elevation of the dam and reservoir at Pakse was set at 108 m in order to free the flood flow furthest downstream on the Nam Mun from all obstacles and to allow the water of the Nam Mun to be lifted for irrigation during the dry season. The installed capacity of the Khemarat power plant was to be 720,000 kW (baseline output: 620,000 kW; annual generated energy: 5,400 GWh) when all the main upstream dam projects were completed.

The Stung Treng Dam Project and Irrigation Plan for the area north of the Great Lake (Japan's grand vision)

The 1957 ECAFE Report included a plan to develop the Khone Falls, making use of a navigation lock. The Japanese team, on the other hand, proposed that a huge dam be developed at Stung Treng downstream from the falls. This multipurpose dam with a full water elevation of 88 m would submerge the whole Khone Falls area, providing an installed capacity of 2.2 kW (baseline output: 1.8×10^6 kW; annual generated energy: 16,350 GWh). The reservoir would have an effective capacity of some 31×10^9 m^3, which, as previously stated, would make it possible to bring about a drastic reduction in flood disasters in the delta, through joint operation with the upstream Pa Mong Dam Project. Moreover, if it were possible to divert part of the mainstream flow (say, at 1,000 m^3/s), which would be stored in the Stung Treng Reservoir through a long water channel to be built on the right bank of the Mekong to the Stung Sen (which flows into the Great Lake), a vast area of about a million hectares of farmland extending from the north of the Great Lake could be irrigated and energy generated.

The Japanese team conceived this far-reaching development plan and ascertained the technical possibilities of building the above-mentioned water channel by undertaking the daunting task of reconnoitring the steep, difficult route between the fringe of the proposed Stung Treng Reservoir and the Stung Sen in the dry season in 1960. The grand scale of the proposal again amazed ECAFE. If the idea were realized, flooding in the delta would be very much reduced, the river flow in the dry season would be considerably increased, and some two million kilowatts of electricity could be generated. If locks were attached to the dam, upstream navigation would become possible for the first time. The project would also make it possible to draw river water from the main Mekong to the Stung Sen River, which pours into the Great Lake through a relatively short water channel. Once a dam was built on the Stung Sen, other tributaries running into the Great Lake parallel with the Stung Sen were dammed up, and these dams were interconnected by water channels, the irrigation of the vast flat lands to the north of the Great Lake would no longer be a dream. This was the Japanese team's grand vision of development in Cambodia.

[After the submission of the Japanese report, its contents became one of the main concerns of the Mekong Committee and ECAFE. The author, who was responsible for basin-planning work in the Mekong Secretariat for a few years after the submission of the report, was asked to follow it up by carrying out further surveys and studies, which are described in the section on the Stung Treng Project (pp. 146–147).]

The improved plan to develop the Sambor Dam Project

When ECAFE made its proposal in 1957 for the development of the run-of-the-river Sambor Dam as the dam furthest downstream on the Mekong, the contours on the available topographical maps were, unfortunately, inaccurate. The Japanese team, which used newly obtained maps and took into account the probable elevation of the flow discharged from the proposed Stung Treng Dam, tentatively decided on a full water elevation of 40 m for the Sambor Reservoir.

The team therefore concluded that it would be possible to develop a hydroelectric power station with an installed capacity of 400,000 kW (with a baseline output of at least 340,000 kW) even as an isolated power plant. Once all the upstream dams were completed on the main Mekong, Sambor could be expanded to 1.45×10^6 kW or more (baseline output: 1.23×10^6 kW or more). In other words, the team estimated that the power generated at Sambor could be far greater than the amount projected by ECAFE. The team also expected that all the Sambor Rapids would be submerged by the construction of the Sambor Dam and that

Table 3.1 Projects on the main Mekong, including effects of tributary reservoirs

Project	Pak Beng	Luang Prabang	Sayaboury	Pa Mong	Bung Kan
Purpose[a]	P.N.F.	P.N.	P.N.	P.N.I.F.	P.N.
Drainage area (km^2)	227,000	260,000	279,200	305,000	341,600
Lowest natural run-off (m^3/s)	445	510	550	600	685
Increased discharge from upper reaches (m^3/s)	–	1,580	1,990	2,090	1,900
Reservoir:					
Surface area (km^2)	1,380	340	150	2,600	300
Effective storage (10^6 m^3)	20,000	(5,200)[b]	(1,300)	38,000	5,500
Regulated discharge (m^3/s)	1,580	(410)	(100)	1,345	(335)
Electric power project:					
Gross head (m)	80 = 380 – 300	25 = 300 – 275	25 = 275 – 250	80 = 250 – 170	10 = 155 – 145
Baseline discharge (m^3/s)	2,025	2,500	2,640	2,500	2,870
Installed capacity (kW)	1,450,000	560,000	59,000	1,800,000	251,000
Baseline power output (kW)	1,210,000	470,000	500,000	1,530,000	211,000
Annual energy output (10^6 kWh)	10,600	4,140	4,360	13,400	1,850
Irrigation:					
Discharge for irrigation (m^3/s)	–	–	–	1,500	50
Tributaries		Nam Ou Nam Seng	Nam Khan		Nam Ngum (Nam Lik)

Source: Government of Japan (1961).
a. P, power generation; N, navigation; I, irrigation; F, flood control.
b. Values in parentheses represent net storage and regulated discharge on tributaries.

boats could advance upstream after passing through locks attached to the Sambor Dam. Furthermore, it was estimated that about 30,000 ha of farmland on both banks downstream from Sambor could be irrigated.

Thakhek	Khemarat	Pakse	Stung Treng	Sambor	Total
P.N.F. 383,200	P.N. 427,700	P.N. 548,200	P.N.I.F. 640,000	P.N. 651,000	
780	890	1,170	1,380	1,410	
2,185	2,135	2,575	2,525	3,620	
900	360	220	6,100	800	
–	(11,340)	–	31,000	–	
–	(490)	–	1,995	–	
14 = 145 – 131 2,915	22 = 130 – 108 3,465	20 = 108 – 88 3,695	48 = 88 – 40 5,000	34 = 40 – 6 5,000	
360,000	720,000	700,000	2,200,000	1,600,000	10,230,000
310,000	620,000	600,000	1,860,000	1,350,000	8,662,000
2,720	5,400	5,250	16,350	11,850	75,920
50	50	50	1,000	30	
	Nam Theun Se Bang Fai Se Bang Hieng				

Some potential dam projects proposed by the Japanese team

The team which proposed this original plan of developing successive large-scale dams on the main Mekong also identified a total of 102 promising dam projects on the 34 major tributaries on the basis of 3 years' reconnaissance work (table 3.2). The following are some of the tributary projects found to be particularly promising.

Table 3.2 Dam-development sites identified by the Japanese team

Country	Tributary	Drainage area (km^2)	No. of sites
Laos	Nam Ou	25,600	6
	Nam Seng	6,780	1
	Nam Khan	7,410	4
	Nam Ngum	16,000	9
	Nam Nhiep	3,600	1
	Nam Sane	2,100	1
	Nam Theun	14,700	6
	Se Bang Fai	9,550	3
	Se Bang Hieng	19,800	3
	Se Done	7,790	4
	Nam Heung	4,360	2
	Nam Gam	3,440	2
	Nam Pong	15,900	1
	Lam Pao	7,600	1
	Huey Tuey (Nam Song Gram)	10,800	1
	Huey Bang Sai	1,350	1
	Nam Mong	2,390	1
Laos and Cambodia	Se Kong	25,600	16
Viet Nam and Cambodia	Se San	18,000	11
	Srepok	30,800	9
Cambodia	Prek Te	4,350	1
	Prek Chhlong	5,660	3
	Prek Thnot	5,050	3
	Stung Pursat	5,100	2
	Stung Dauntry	4,050	2
	Stung Battambang	4,520	4
	Stung Mongkol Borey	4,330	0
	Stung Sreng	6,270	1
	Stung Siem Reap	860	0
	Stung Stong	3,390	0
	Stung Sen	17,800	3
	Stung Chinit	6,820	0
Total			102

Source: Government of Japan (1961).

The Nam Ngum Dam Project

Before the Japanese team conducted its reconnaissance, France and Nippon Koei of Japan had already suggested the potential of the Nam Ngum Dam Project. The Japanese team proposed the construction of a dam 65 m high upstream from the confluence of the Nam Ngum River and its tributary, the Nam Lik, and the establishment of a hydroelectric power station with an estimated installed capacity of 90,000 kW (baseline output: 50,000 kW). It was also estimated that some 30,000 ha of farmland could be irrigated and flooding of the extensive Vientiane Plain could be considerably decreased.

The Yali Falls Development Project (Central Viet Nam)

There is a head of some 60 m at the Yali Falls upstream from the Se San, where the river flow volume is as much as 40 m^3/s even in the dry season. It was therefore estimated that a baseline output of 20,000 kW could be generated at this spot. If it were possible to build several dams upstream, the baseline inflow to the hydroelectric power station would be 160 m^3/s; furthermore, if a 7 km tunnel were constructed to draw this amount of water downstream from the dam at the foot of the falls to gain a head of some 250 m, a larger hydroelectric power station with an installed capacity of 400,000 kW (baseline output: 320,000 kW) would be possible.

The Stung Battambang Dam Project (Cambodia)

Long before the Japanese team conducted its reconnaissance, the government of Cambodia had, with the assistance of France, promoted the Bannan Irrigation Project for the irrigation of some 60,000 ha around the Battambang River, which flows through the Stung Sang Ker into the Great Lake at its north-west margin. When the team visited the project area, an irrigation network covering about 20,000 ha had already been completed;[11] however, this project did not include any plan to develop dams.

The Japanese team concluded that, if a dam were built in the middle reaches of the Battambang River, it could be expected to supply 60,000 m^3/s of water to irrigate 60,000 ha, as well as providing some hydroelectricity. The team also foresaw the possibility of creating two or three reservoirs in the river's most upstream reaches. However, before the team submitted its final report, the survey work for the project had been taken on by the UN Special Fund and a French engineering consultant had begun a survey for this dam project (for irrigation of 60,000 ha and power generation of 31,500 kW).

The Nam Pong Dam Project (Thailand)

Before the Japanese team conducted its reconnaissance, the Thai government had made a precise study of a plan to build a multi-purpose dam on the Nam Pong River, a tributary of the Nam Chi. The Thai plan was to build a dam 31 m high to generate 27,000 kW and irrigate 70,000–100,000 ha.

The Nam Theun Dam Project (Laos)

In Laos, a plan for developing a dam on the Nam Theun River had been drawn up by France at the beginning of the nineteenth century but had never been implemented. According to this plan, if a dam 85 m high were built in the middle reaches (elevation 523 m) of the Nam Theun River, a large reservoir of some 650 km^2 could be created. Precipitation is very high in this river basin, and if some 300 m^3/s of stored water were diverted to the Se Bang Fai, its neighbouring tributary, 1×10^6 kW (baseline output: 400,000 kW) could be generated with a head of about 330 m. This would result in the generation of about 5–6 $\times 10^9$ kWh of electricity.

A number of alternatives to this plan have been proposed and it is possible to build dams both upstream and downstream from this project site on the Nam Theun. Since the electricity obtained would be very cheap, the Thais had long been interested in the project. However, the Japanese team concluded that it would be necessary to install long transmission lines carrying some 400 kV to connect the power plant to Bangkok.

The Lam Don Noi Dam Project (Thailand)

If a dam 20 m high were constructed above a waterfall 5 m high situated near the lower reaches of the Lam Don Noi, about 34,000 ha of land could be irrigated by directly leading the water in the reservoir to the basin of the neighbouring Huai Kawn by means of a water channel; furthermore, hydroelectricity with an installed capacity of some 6,000 kW (baseline output: 2,600 kW) could be generated.

As this project site is located near the city of Ubon and is easily approachable, the project was considered quite attractive, even though its scale was modest.

The Prek Thnot Dam Project (Cambodia)

In the middle reaches and downstream areas of the Prek Thnot River in Cambodia, the local people till extensive rice fields; however, the region often suffers from water shortages in the dry season and from flooding in the wet season.

On the Prek Thnot River, a multi-purpose dam (30 m high) could be constructed for irrigation (about 60,000 ha), power generation (baseline output: 7,000 kW), and flood control some 20 km upstream from the town of Kompong Speu. The power generated by the project could be transmitted to both Kompong Speu and Phnom Penh.

The Nam Pung Dam Project (Thailand)

Upstream on the Nam Pung River, a tributary of the Nam Gam, there are several falls providing a combined head of 60 m. If a dam 33 m high were built immediately upstream from the falls, and the reservoir's water were drawn by a 720 m water channel to a downstream power plant, this would generate some 13,000 kW of power with a head of 88 m. If a weir were constructed 20 km downstream from the dam site, 23,000 ha of farmland on the left bank could be irrigated as well.

The Stung Sen Dam Project (Cambodia)

It is not clear whether the above-mentioned mainstream dam project at Stung Treng is practicable. However, if it were to be realized, and the water (maximum capacity: 1,400 m^3/s; minimum discharge: 600 m^3/s) were drawn from the Stung Treng Reservoir to the Stung Sen River on the north bank of the Great Lake through a water channel 30 km long and 50 m wide over the highest point at Chepp on the right bank of the main Mekong, one could expect to generate a baseline output of some 65,000 kW and to irrigate a vast area north of the Great Lake by building a dam 53 m high on the Stung Sen and constructing a long water channel from the Stung Sen Dam to the rivers to the west of it.

This idea was conceived by the Japanese team and, having later subjected it to a precise examination at the Mekong Secretariat in Bangkok, the author can attest to its technical feasibility.

Other promising tributary dam projects

Apart from the above, there are a number of other promising dam projects on Mekong tributaries such as the Nam Nhiep, Nam San, Se Dong, and the middle reaches of the Se Kong in Laos; the Nam Heung, Nam Man, Huey Tuey (or Nam Song Cram), Nam Gam, Huey Bang Sai, and furthest downstream on the Nam Mun (Falls) in north-east Thailand; the Se Kong, Se San, Srepok, Prek Thnot, Stung Pursat, Stung Dauntri, Mongkoi Borey, Stung Treng, Siem Reap, Stung Staun, and Stung Chinit in Cambodia; and the Upper Se San and Upper Srepok in Viet Nam. These projects were based mainly on the topographical maps available and some actual ground surveys, and desk plans for developing projects on the above rivers were outlined in the Japanese team's report.

Recommendations of the Japanese reconnaissance team and its follow-up study

Recommendation to continue tributary studies and follow-up study

The Japanese team carried out its reconnaissance in three stages. In the first report submitted to the Mekong Committee in May 1959, it recommended an immediate start on detailed studies on eight tributaries: these were the Nam Ngum, Se Bang Hieng, and Se Done in Laos; the Nam Pong in Thailand; the Stung Sen and Battambang in Cambodia; and the Upper Se San in Viet Nam. From these recommended eight tributaries, the Mekong Committee selected four rivers – the Nam Ngum (Laos),

Nam Pong (Thailand), Battambang (Cambodia), and Upper Se San (Viet Nam) – as the most promising, and embarked upon dam projects studies on these four rivers assisted by the UN Special Fund, as well as "pilot training projects" to provide equal opportunities for training on dam construction to each riparian country in preparation for future main-stream dam projects in the basin. The thinking behind this was that the development of large-scale dams such as those on the main stream should not be carried out without first improving the technical standards of each riparian country by gaining experience in the building of small-scale dams within their own borders, which would have an impact on each country.

In the second report, Japan recommended the initiation of detailed studies on seven tributaries: these were the Se Done and Nam Ngum in Laos; the Nam Pung in Thailand; the Prek Thnot and Stung Sen in Cambodia; and Upper Srepok in Viet Nam.

In the final report, the Japanese team mentioned that, in addition to the above-mentioned four tributaries on which studies had been started, there were promising dam sites on other rivers such as the Nam Lik, Se Done, Nam Theun, Se Bang Fai, and Se Bang Hieng in Laos; the Lam Don Noi, Chayapoom, and Nam Gam in Thailand; the Stung Sen, Stung Pursat, and Prek Thnot in Cambodia; and the Upper Srepok in Viet Nam. The team recommended that meteorological and mapping work be started at promising dam sites.

The report concluded that, among all these promising tributaries in the lower Mekong Basin, the cheapest and most abundant hydroelectric power could be made available by developing the Nam Theun River, and that a survey on this development should be commenced as soon as possible.

In response to the Japanese recommendation, the Mekong Committee requested the Food and Agriculture Organization (FAO), the World Health Organization (WHO), the International Labour Organization (ILO), and other UN organizations to provide assistance for studies for the implementation of small-scale dam projects. Among these four UN Special Fund projects, Japan was requested to carry out the Nam Ngum Dam Project (Laos) through Nippon Koei, while the Nam Pong Dam Project (Thailand) was to be conducted by a US consultant and the Battambang Dam Project (Cambodia) by a French consultant. Nippon Koei was also put in charge of the development of the Upper Se San (Viet Nam).

Meanwhile, the Japanese government decided to provide technical assistance for studies of the Prek Thnot (Cambodia), the Nam Pung (Thailand), and the Upper Srepok (Viet Nam). The Israeli government cooperated with Japan on the study for the Prek Thnot Project. While this work was proceeding, the developed countries continued with their collection of basic data, as they had promised in accordance with the recommendations made by the Wheeler mission.

Recommendation to initiate studies on mainstream dam projects and follow-up studies

In 1961, the United States agreed to conduct an overall study of the Pa Mong Dam Project, which had been recommended by both the Japanese team and ECAFE, with a budget of some US$2,500,000. In 1963, field studies were started in various sectors including agriculture, land utilization, forestry, hydrology, geology, sociology, and power demand. In addition, the Snowy Mountains Engineering Corporation (SMEC) of Australia conducted a geological survey at the proposed dam site between 1962 and 1964 and submitted a report. Taking the contents of this report into account, the US government submitted the first comprehensive study report in March 1966 and subsequently made its first feasibility report in 1970 and second feasibility report in 1971.

The US study was quite large scale and employed numerous people. It made a detailed examination based on the full-reservoir water elevation recommended by the Japanese team, i.e. a dam height of about 100 m. It concluded that the optimal installed capacity would be about 4.6×10^6 kW and the extent of irrigated land would be roughly a million hectares. The report foresaw the great advantages of flood control in the wet season and of irrigation in the dry season in the downstream delta.

The conclusions of the USA are outlined in the section entitled "Pa Mong Dam Project" in the report on the Indicative Basin Plan (Mekong Secretariat 1970), which is discussed later.

In October 1961, the Japanese government dispatched a provisional survey team to the Sambor Project site near Kratie in Cambodia for a short period. From January to September 1962, EPDC of Japan conducted a technical study at the proposed site and a provisional study report was submitted to the Mekong Committee in November 1962. On the basis of the report, further studies were carried out in the spring and winter of 1963 and from the autumn of 1963 to early 1964. As a result, a final report was submitted to the Mekong Committee in 1968.

Various other countries also cooperated in the studies for the Sambor Project: for example, Australia carried out geological studies at the proposed dam site and Canada provided mapping at the dam site (78 km^2) to a scale of 1:10,000, as well as of the reservoir area (3,200 km^2) to a scale of 1:2,000. The Philippines also carried out mapping at the dam site (100 km^2) and the reservoir area (1,400 km^2), and the Cambodian government did mapping to a scale of 1:20,000 at the proposed irrigation area (1×10^6 ha). It is noteworthy that the comprehensive study report was completed through this cooperative effort by several countries.

The conclusions of the Japanese team are outlined in the section entitled "Sambor Dam Project" in the report on the Indicative Basin Plan (Mekong Secretariat 1970).

The Mekong Committee organized three liaison meetings between the US Pa Mong team and the Japanese Sambor team at its Bangkok office. As a result of these efforts by the tripartite groups, the studies of both teams proceeded quite smoothly. The author attended these meetings as a member of the Committee and noticed differences in the style of presentation between the two teams: the Japanese team was rather too frank, revealing all its findings quite openly, whereas the US team appeared reluctant to disclose the data and information which it had obtained.

The White Report (1962)

The need for a socio-economic study

The reports on studies for development of the lower Mekong Basin conducted from the 1950s to the beginning of the 1960s concentrated on the technical aspects of development. In their forecasts of the possible effects of building dams at promising locations in the lower basin, these reports focused almost exclusively on matters such as the scale of hydroelectric power, irrigation of farmland, improvement of navigation, and flood control, giving little, if any, consideration to the socio-economic aspects of development. However, from the beginning of the 1960s, the feeling gradually spread among those involved in Mekong development that this type of technical development plan was not sufficient and that they should not embark on development without giving due consideration to social and economic questions.

When large-scale dam-development projects are implemented in developing regions, the government must prepare the required administrative structures beforehand and engineers have to be trained. It is also necessary to assess the "development needs," both in the region and outside it, by considering how the hydroelectricity generated by dam development will promote the region's industry, whether it will improve the standard of living of its residents, and what impact increased agricultural production will have on domestic consumption and exports. Indeed, consideration of the possible effects on the lives of the local people is a fundamental task which should be given priority over the technical investigation of the location to be developed. By making such forecasts concerning these various aspects, people engaged in development can gain a precise understanding of the social and economic situation of the region prior to development, and can make long-term predictions on the situation that will be likely several years – or even decades – after the project has been completed.

The Mekong Committee became increasingly aware of the need for this kind of extensive survey for social and economic development. In 1961, having obtained a subsidy from the Ford Group in the United States, the Committee commissioned a four-member study team led by Professor Gilbert F. White of Chicago University to prepare a report entitled *Economic and Social Aspects of Lower Mekong Development* (White 1962). Professor White and his team conducted studies in the lower Mekong Basin for two and a half months from September 1961 and submitted the report to the Mekong Committee in January 1962. The contents of the report are outlined in the following section.

Outline

The White Report consisted of about 100 pages with a few appended figures and tables. The report was organized in chapters, as follows:
1. Outline;
2. The various aspects of development;
3. River control as a means of national development;
4. Selection and timing of projects;
5. Manpower structures for development.
   It made the following recommendations:
 1. Strengthening of the Mekong Committee's executive organization;
 2. Extension of the work of the Mekong Committee;
 3. Cooperation with international organizations;
 4. Cooperative consideration of all measures for studies;
 5. Research into the size and scope of the Mekong River system after development;
 6. Research into the preparation of administrative structures to deal with construction and operation through international cooperation;
 7. Collection of materials on the necessary resources for development, their utilization, and related social structures;
 8. Studies based on aerial photographs;
 9. Research on agricultural development plans;
10. Comprehensive analysis of electric-power markets;
11. Flood prediction and mitigation of flood damage;
12. Fostering of rural village services and service personnel;
13. Promotion of pilot farm projects;
14. Experimental forest planting;
15. Schedule for implementation of recommendations.

Dividing the development of the lower Mekong Basin into three areas – future development needs, effects of river control, and selection and timing of river-basin development projects – the White Report provides a concrete summary of the view of American geographers. In conclusion, it lists the above-mentioned 15 recommendations to be taken into account by future developers. The contents of these recommendations may be summarized as follows: (1) a wide-ranging consideration of social and economic aspects is required for Mekong River development; (2) preparation of the necessary administrative structures by the governments of the four riparian countries is essential for effective international cooperation; (3) above all, the first step should be to collect and collate the various basic social and economic materials necessary for development.

Although the conclusions of the White Report have now become conventional wisdom, they had a great impact on those involved in development at the time. Let us now consider the report's contents in a little more detail.

After studying the living conditions of residents of the lower basin, Professor White stated that it was, first, vital to get a precise grasp of the overall situation regarding water resources and land. Next, it was essential to examine water-resources development projects closely in relation to national development plans, to select the most effective projects for the development of the region, and to promote their timely implementation. The report then stated that this selection and timely implementation of appropriate development projects would make it possible to assess accurately the various elements of river-development plans and to study different combinations of these elements and their mutual relationship, adding that it would be necessary for the four riparian countries to consider these projects carefully in the light of their economic situation and administrative capabilities.

The report indicated the production targets in the lower Mekong Basin from 1961 onwards. Assuming that the population of the lower basin would continue to increase at the present annual rate of 3 per cent, it stated that it would be necessary to set targets of average annual agricultural production increases of at least 3.5–4 per cent and industrial production increases of 6–7 per cent in the four riparian countries, in order to promote growth while keeping a balance between living standards and fiscal payments. It added that, to achieve this growth, improvements in agricultural and industrial facilities, transportation, and education were urgently required, and the advancement of commerce and technology was also essential. It strongly urged the formulation of comprehensive development plans including such factors as education, agricultural guidance, transportation, extension of credit, and market development, to ensure the success of development plans for irrigation, drainage, and flood control.

The report also stated that the implementation of large-scale hydroelectric power development should be based from the outset on the conviction that the unit cost of the power generated would be lower than the prevailing unit cost that had been imposed on large electric power-consuming industries aiming at world markets.

Finally, Professor White and his research team stressed that the immediate tasks of the Mekong Committee were to strengthen the executive organization of the Secretariat, to set up a division for socio-economic research, and to train personnel. Once these tasks had been completed, they recommended that the following aims be pursued as soon as possible in close collaboration with international organizations:

1. Compilation of marketing forecasts for agricultural products after development;
2. Securing of national revenue and promotion of economic plans on a major sector basis;
3. Standardization and quality control of products for export;
4. Clarification of aims for residents' living standards and lifestyles after development;
5. Studies on land use, forests, and residents' living conditions, and forecasts of future outlook;
6. Studies on the electricity market and electricity-consuming industries;
7. Studies on the current flood-damage situation and appropriate flood-control measures;
8. Establishment of 3,000–5,000 ha pilot farms;
9. Setting up of 500 ha experimental forest-stations.

Professor White stressed that, in the initial stages, those engaged in development should not embark on large-scale projects on the main stream of the Mekong but should concentrate on small-scale projects on the tributaries. The report stated that projects on the main stream should be begun only after the completion of studies on the amount of flow of the Mekong River system and the analysis of electricity and agricultural markets. The highest priority was to "collect basic data and conduct studies."

A general outline of development plans for the main stream and tributaries of the Mekong had already been provided by Japanese government reconnaissance teams, but this gave no indication as to the cost of development. Regarding costs, the White Report pointed out that, while the governments of the lower Mekong Basin countries were able to invest only small amounts, huge sums of capital would be necessary for the development of land, water resources, and hydroelectric power. The report stated that, judging from the plan outlined by the Japanese government reconnaissance teams, US$6–7 $\times$ 10^9 would be needed for the construction dams and hydroelectric power stations amounting to 20 $\times$ 10^6 kW in the lower basin. In addition to this direct investment, a total of US$14–

21×10^9 would be required to cover the costs related to agriculture, industry, transportation, telecommunications, and social expenses. Regarding the direct investment of US$6–7 $\times 10^9$ for dam and hydroelectric power-station construction on the main stream and tributaries, the report estimated that if about 65 per cent of this were to come from foreign capital (50 annual instalments), the total annual debt of the four riparian countries would come to about 20 per cent of their current annual revenue (at the beginning of the 1960s).

Previous examples of massive investment in construction were the US$2 $\times 10^9$ project of the Indian government for the construction of a canal in western Rajasthan to irrigate several thousand hectares of land and the investment of US$2 $\times 10^9$ by the US government for the upstream development of the Colorado River. Bearing these precedents in mind, the White Report emphasized the importance of initially limiting development to promising locations on the Mekong tributaries in order to demonstrate the local benefits.

During the 1950s, there was a worldwide craze for investment in ultra large-scale development projects. In many cases, this ultimately led to breakdown of the national economy, as in the case of the Volta River project in Ghana. Although the White Report did not specifically warn that development of the lower Mekong Basin might result in similar problems unless sufficient care was taken, it emphasized the need to gain experience gradually through small-scale projects, which would limit the socio-economic impact and provide useful practice for later large-scale mainstream development projects.

In addition to the above recommendations, Professor White and his research team made other very concrete proposals, such as detailed studies based on interpretation of aerial photographs. The report estimated that, including such studies, a complete socio-economic survey of the lower Mekong Basin would take at least 5 years and the total cost would be about US$100,000–250,000.

Follow-up to the White Report's recommendations

In its focus on the need for investigation of the social, economic, fiscal, and administrative aspects of development, the White Report differed radically from previous reports. In response to the report, the Mekong Committee immediately set up a division to deal with the socio-economic aspects of development. At the request of the Mekong Committee, France initiated studies on industrial development and the electricity market. In addition, the United States began a project to produce an atlas based on aerial photographs.

Shortcomings and merits of the White Report (author's opinion)

With hindsight, it appears that because the White Report viewed the lower Mekong Basin as a single enclosed "system," its analysis was, to some extent, divorced from reality. The White Report completely failed to anticipate the massive exodus of people from the lower basin, particularly north-east Thailand and the Viet Nam Delta, to large cities outside the region such as Bangkok and Saigon (now Ho Chi Minh City); this exodus, in fact, began not long after the report was submitted. In addition to the movement of population, goods could naturally be expected to flow into and out of the region; however, the report took no account of this, either. It seems that its target figures for increased agricultural and industrial production in the lower basin ignored the fact that goods were being freely distributed to and from the regions outside it.

Although the White Report dealt with the influence of development on the natural characteristics, economy, and society of the region, it did not consider its environmental impact. In this respect, it is certainly inadequate by today's standards. Nevertheless, one cannot praise too highly the splendid far-sightedness of the report in urging people engaged in Mekong development to reflect on the prevailing tendencies of the 1960s, which placed too much importance on technical aspects and virtually ignored the broad social and economic implications of development. However, when the report was submitted, all governments of the riparian countries were suffering from extreme shortages of foreign capital, basic natural and socio-economic data, and human resources; it was, therefore, far from easy for these countries to implement the various tributary development projects, let alone those of development of the main stream of the Mekong, and there was little they could do but await aid from the advanced countries.

The Mekong Committee's (1970) report on the Indicative Basin Plan

After almost 5 years of preparation from autumn 1964 to 1970 during which it utilized all the latest available data, materials, and information under the general guidance of the chief of the water-resources division of ECAFE, the Mekong Secretariat produced its report on a comprehensive development plan for the lower Mekong Basin.

Until it was completed, the report was referred to as the Amplified Basin Report within the Secretariat, but it was given the title *Report on Indicative Basin Plan* just before it was published (Mekong Secretariat 1970).

Outline

The report was composed of the main report and annexes, totalling about
600 pages. The contents were as follows:

Chapter 1. Summary.

Chapter 2. Description of the lower Mekong Basin:
    1. The basin today;
    2. Present economic conditions;
    3. Present status of data collection.

Chapter 3. Resources of the lower Mekong Basin:
    1. Water resources;
    2. Land resources;
    3. Mineral resources;
    4. Power and energy resources;
    5. Fishery resources;
    6. Human resources.

Chapter 4. Needs for development:
    1. The economic framework;
    2. Agriculture;
    3. Industry;
    4. Power;
    5. Transportation;
    6. Flood control, salinity control, and water-supply
       requirements;
    7. Education and manpower infrastructure.

Chapter 5. A plan to meet the needs:
    1. General considerations;
    2. Potential projects;
       (a) Mainstream projects and delta development
       (b) Tributary projects
    3. Plan formation;
    4. Indicative Basin Plan;
       (a) Short range 1971–1980
       (b) Long-range 1981–2000
       (c) Complementary programme
    5. Implementation aspects of Indicative Basin Plan;
       (a) Financial aspects of the Basin Plan
       (b) Legal and institutional framework.

Chapter 6. A programme of future investigation:
    1. Introduction;
    2. Central planning activities;
    3. Project investigations;
       (a) Mainstream projects

      (b) Electrical transmission network
      (c) Delta-development studies
      (d) Tributary projects
   4. Basinwide investigations;
      (a) Basic data collection
      (b) Sectoral programmes
      (c) Special investigation
      (d) Implementation.

Summary of chapters 2 and 3

The main points of chapters 2 and 3 of the Mekong Committee's *Report on Indicative Basin Plan* can be summarized as follows:
1. The waters of the Mekong flow down into the South China Sea without being sufficiently utilized, while droughts and floods occur frequently, causing distress to the people of the lower basin. The key issue in the future development of the lower Mekong Basin is to solve these two problems.
2. The lower Mekong Basin has so far succeeded in maintaining a surplus in agricultural production (mainly rice) and in exporting this surplus while supporting its increasing population. This has been possible because its inhabitants have enlarged their farmlands. However, because it will be difficult to expand farmlands in the future, it will not be possible to maintain a good standard of living unless floods are controlled and farmlands are irrigated.
3. At present (the late 1960s), the total installed capacity of electricity generated in the lower Mekong Basin is only 250,000 kW. Hardly anything is known about mineral resources in the lower basin except for limited information on the existence of small amounts of tin, copper, iron, and other minerals. Navigation has been developed locally (except in the delta) and there is no bridge on the main stream in the lower basin. Unless administrative systems are improved in each riparian country and international cooperation among these countries is considerably improved, it will not be possible to achieve significant development of the lower basin in the future.

Summary of chapter 4

The main points of chapter 4 are as follows:
1. It is necessary to expand and diversify agriculture and industry and to improve services in the four riparian countries. By doing so, it would be possible to double the average per capita income during the coming 30 years (from 1970 to 2000). The current populations of the four

countries and of the lower basin are 61×10^6 and 30×10^6, respectively. It is estimated that these figures will increase to 134×10^6 and 62×10^6 by the year 2000.

2. In order to cope with this population increase, it would be necessary to do everything possible to develop agriculture, primarily by increasing unhulled rice productivity from 12.7×10^6 t/year to 37.0×10^6 t/year. To achieve this aim, substantial investment would be necessary to promote research, improve stock-breeding, increase fishery products, reform the current single-crop system to diversify agricultural products, and improve institutions. There is a particular need to improve current land ownership, marketing systems, credit systems, communication and transportation networks, and agricultural unions. Human-resources development will also be essential.

3. The peak-time electricity requirement of the four countries is expected to increase from 1.99×10^6 kW in 1971 to more than 20×10^6 kW in 2000 as a result of the establishment of power-consuming industries using plants processing iron and steel, ferro-alloy, phosphoric acid, and aluminium in the lower basin.

4. It is necessary to improve navigation in anticipation of a growing demand for transportation of agricultural products, mineral products, timber, and related processed goods. At present, some 50,000 km^2 of land in the delta are flooded. It is essential to both implement a system that enables a timely response and take permanent technical counter-measures such as the construction of protective dykes and gates to prevent sea-water influx.

5. At present, water is supplied to less than 15 per cent of the population of the lower basin. By the year 2000, it will be necessary to supply some 3×10^9 m^3 of water per year for industrial use. Meanwhile, it is imperative to promote education and improve the level of public health.

Summary of chapter 5

The main points of chapter 5 are as follows:

1. In order to respond to development needs in the lower basin, it is necessary to consider a plan for the systematic development of water and related resources. The plan would have to be integrated with the socio-economic development programmes of the four riparian countries, increasing the agricultural productivity of each farmer and assisting the generation of hydroelectricity as the basis of industrial development. To this end, it is essential continuously to collect basic data, enlarge the scope of studies, and raise the level of accuracy of the development plan. The plan should be reviewed and improved from

time to time in accordance with the increase of available information and advancement of development in the region.

The division of the development plan into two parts – a short-range plan (1971–1980) and a long-range plan (1981–2000) – should be considered, and various measures, such as the provision of necessary funds, should be taken for the implementation of each plan.

2. Division of the development plan: development projects with high potential can be divided into two types – mainstream projects and tributary projects. The implementation of these projects is to be divided into two stages – a short-range (first) period and a long-range (latter) period.

(a) The short-range development plan (1971–1980):

- Tributary projects in each riparian country are to be promoted to meet each country's development target during the coming 10 years (1971–1980).

- In Cambodia, the plan could meet the additional irrigation requirement (62,500 ha) when both the Prek Thnot and Battambang projects are developed. It could also meet the country's power requirement (a baseline output of 50,000 kW and a peak requirement of 97,000 kW) when the Kompong Som thermal power plant (installed capacity 45,000 kW) is mobilized in addition to the above-mentioned two projects.

- In Laos, the plan could satisfy the additional irrigation requirement for the coming 10 years when, in addition to the Selabam Project on the lower Se Done, both the Nam Ngum and Nam Dong projects are completed. It could also meet the power requirement (a baseline output of 25,000 kW and a peak requirement of 51,000 kW) through the completion of these projects.

- In north-east Thailand, the additional irrigation requirement (144,000 ha) could be satisfied by developing the Nam Pong, Nam Pung, Lam Don Noi, Nam Prong, Lam Prapang, Lam Takong, and Nam Oon rivers. (All these projects were under way when the report was drawn up.) When the Pak Mun is added to these projects, the power requirement (a baseline output of 1.27×10^6 kW and a peak requirement of 2.4×10^6 kW) will be met.

- In Viet Nam, the irrigation requirement for the coming 10 years (additional irrigation of 65,000 ha) could be met when both the Upper Se San and Upper Srepok projects are completed together with the construction of river and coastal banks in the Viet Nam Delta. The power requirement (a baseline output of 64,000 kW and a peak requirement of 1.23×10^6 kW) would be satisfied by the completion of the thermal plants in Saigon (now

Ho Chi Minh City), Can Tho, Da Nang, and Da Nhim, and the Tian hydroelectric power project outside the Mekong Basin, in addition to the above projects.

(b) The long-range development plan (1981–2000):
  • The long-range development plan (1981–2000) was drawn up to satisfy the increased requirements of agriculture and industry expected after 1980. The basic concept was to implement development in stages over the 20 years up to 2000.
  • On the basis of an anticipated total power requirement in the four riparian countries of 12×10^6 kW (baseline output) and 21.3×10^6 kW (peak) by the year 2000, a study was conducted on the development of large dam projects, both on the main stream and on major tributaries, to generate some 20×10^6 kW in total.
  • Various alternative development plans were considered. One of these was a plan for first developing the Nam Theun Tributary Project and then the Luang Prabang Mainstream Project. Next, the Sambor Mainstream Project would be implemented, together with some thermal power plants. After that, the Pa Mong Mainstream Project and more thermal plants would be developed, to be combined later with the Stung Treng Mainstream Project.

  Thus, a plan to combine three development projects, all of which were based on the assumption that the Pa Mong Mainstream Project would be implemented first, was compared with a plan to combine three development projects based on the initial implementation of the Nam Theun Project. The conclusion drawn from this comparative study was that, since the two plans both had merits and demerits, neither was clearly superior.

  It was assumed that self-sufficiency in terms of provisions in the lower basin could be realized through one of these combination plans, since the total irrigated area of the four riparian countries would be increased by as much as 1.58×10^6 ha over the 20-year long-range period, a significantly greater area than could be irrigated by the end of the short-range period.

3. Realization of the development plan: for the successful implementation of both the short-range and long-range development plans, various measures are necessary. These include the following:
  • Providing an appropriate amount of funds;
  • Developing water-resources projects and providing ancillary facilities for producing, processing, transporting, and storing agricultural and industrial products, as well as the construction of power-consuming factories;

- Developing mineral resources related to water-resources development;
- Implementing resettlement programmes for people living in project areas;
- Training skilled labourers;
- Improving various "software" projects.

It was estimated that the realization of the short-range development plan would require more than US$1.8 × 10^9 ($1.4 × 10^9 for water-resources development and $0.4 × 10^9 for the ancillary projects) and that the long-range development plan would require some US$10 × 10^9 ($7.5 × 10^9 for water-resources development and $2.5 × 10^9 for the ancillary projects). A total of about US$12 × 10^9 would therefore be required.

In addition to this direct investment, it was also assumed that substantial indirect investment, such as investment for various industries and for the implementation of agricultural programmes, would also be needed. The total of direct and indirect costs would therefore come to US$36–40 × 10^9.

Appropriate management systems would be necessary to implement such a vast international cooperative enterprise. Furthermore, a fundamental agreement and territorial consensus among all the riparian countries concerning the international development projects would be essential.

Summary of chapter 6

In Chapter 6, the following future investigation programme is recommended:
1. Continuous review and revision of development targets in the lower basin in response to changes in conditions in the region. In order to do this, it is necessary to carry out the following:
   (a) Central planning;
   (b) Extensive project investigations;
   (c) Basinwide investigations;
   (d) Implementation.

The main aim of (a) – central planning – is to make continuous efforts to provide a number of alternative plans for the comprehensive development of the lower Mekong Basin and to examine these plans.

Extensive project investigations (b) are needed to provide accurate information for the revision of the various plans in both the short-range and long-range programmes proposed in the report on the Indicative Basin Plan.

Also required are (c) basinwide investigations, covering not only hydrometeorological aspects but also wide-ranging studies in the socio-economic, agricultural, industrial, power, transportation, and human-resources sectors. Ecological and socio-economic research such as studies on fishery in the Great Lake, forestry, resettlement, sedimentation, and urbanization should also be included in these investigations.

Implementation (d) means the steps to be taken to promote the systematic development and operation of projects. This covers not only the establishment of management systems for orderly development but also investigations concerning administration, operation, fund-raising, allocation of costs and benefits, decisions on charges for power and water supply, distribution of income, and so on.

2. It was estimated that the total cost of all these activities would be about US$50 × 10^6.

That completes the summary of the contents of the Mekong Secretariat's *Report on Indicative Basin Plan* (Mekong Secretariat 1970). The plan outlined a total of 17 mainstream projects and 87 tributary projects. When the report was completed, it was assumed that large-scale dams such as the Pa Mong, Sambor, and Stung Treng dams would be started before the year 2000.

Mainstream dam projects in the Indicative Basin Plan report

The intensification of the Viet Nam War and the outbreak of the Cambodian conflict made it impossible to carry out the planned projects outlined in the Indicative Basin Plan report. All the general circumstances, development requirements, and schedules envisaged in the short-range and long-range plans became unrealistic. Nevertheless, the skeleton of chapter 5 ("A Plan to Meet the Needs") has remained almost the same as the original Japanese development plan drawn up in the 1960s and is still being seriously discussed, albeit in a modified form and scale of development.

All the mainstream and tributary projects outlined in the Indicative Basin Plan report were formulated by the Mekong Secretariat on the basis of hydrometeorological, topographical, and geological data and information collected between 1957 and the end of the 1960s. When drawing up the mainstream development projects, the authors of course referred to the ECAFE Report (UN 1957) and the Japanese Reconnaissance Report (Government of Japan 1961).

A geological (geotechnical) study was conducted for the Stung Treng Dam Project (which was considered the most important project for the development of the lower basin at the time), and some supplementary reconnaissance work was done in the critical area – the divide between the main stream and the Great Lake – which was to be crossed by a long-distance diversion channel through which water stored in the Stung Treng Reservoir could be conveyed to the northern bank of the Great Lake.

However, since the Secretariat's budget was very limited, reconnaissance work was done only at five project sites – Pak Beng, Luang Prabang, Pa Mong, Sambor, and the Great Lake. Studies on all the other mainstream dam projects could be made only on paper, utilizing the available maps drawn to scales of 1:50,000 and 1:20,000.

In parallel with the work at the Secretariat's office in Bangkok, the US Bureau of Reclamation was working on a feasibility study for the Pa Mong Dam Project with help from the governments of Australia (geological research), Thailand, and Laos, The Secretariat had been provided with the results of this study before it completed its draft of the Indicative Basin Plan report. The Japanese government had completed a feasibility study for the Sambor Project with assistance from the governments of Australia (geological research), Canada (topographical mapping), and the Philippines (topographical mapping). The results of these feasibility studies were therefore incorporated in chapter 5 of the Indicative Basin Plan without any modification.

Among the 17 mainstream development projects proposed and examined by the Secretariat (fig. 3.7), several alternative plans were included. The mainstream cascade development programme in the lower basin was

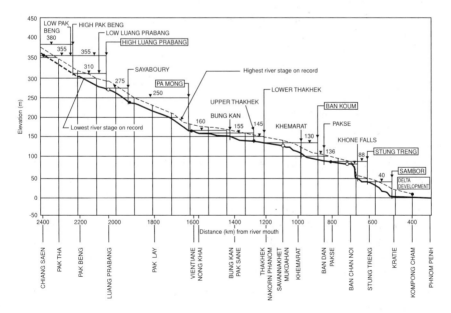

Figure 3.7. Profile of lower Mekong River showing possible mainstream projects
Source: Mekong Secretariat (1970)

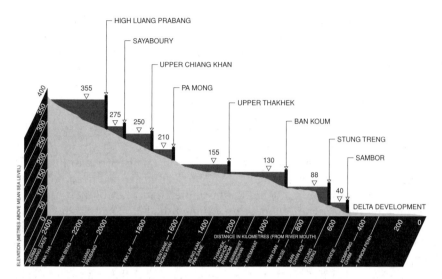

Figure 3.8. Profile of lower Mekong River showing cascade of 1987
Source: Mekong Secretariat (1970)

supposed to be implemented by selecting 9 or 10 projects (fig. 3.8) from among the tentatively proposed 17 projects.

In the following sections, the mainstream projects in the lower basin are outlined one by one in order from upstream to downstream.

High Pak Beng Dam Project

The High Pak Beng Dam Project is a modified version of the preliminary development project proposed by the Japanese reconnaissance team. The High Pak Beng Dam Project in the uppermost reaches of the main stream in the lower basin should be compared with the Low Pak Beng Dam Project described in the next section.

The plan is for the construction of a dam 90 m high with a full-reservoir water elevation of 380 m above m.s.l., which would provide about 25.900×10^9 m^3 of effective storage for power generation, flood control, and regulation of the flow in the upper reaches of the river. As an isolated project, the available flow and head would permit a power installation of 1.85×10^6 kW (baseline output: 1.1×10^6 kW). Storage of flood water and its subsequent release for power generation would reduce the flood peak just below the dam by about 9,000 m^3/s (from 15,000 to 6,000 m^3/s) during the rainy season in a wet year (such as 1939) and increase the low-water flow by about 1,400 m^3/s (from 600 to 2,000 m^3/s).

When the Mekong Secretariat made its reconnaissance survey, the survey team was unable to estimate how many people living in the town of Pak Beng would suffer from inundation as a result of the project proposed in the Japanese development plan. They therefore simply guessed that this number would be very high and the Secretariat decided to shift the proposed dam site 40 km upstream of Pak Beng to avoid the anticipated inundation.[12] However, the reservoir would extend up the river to just above the border between Burma (now Myanmar) and Thailand. Although the generated energy of 11,000 GWh would meet local needs for a considerable period while providing a large surplus to more distant centres, it would inundate about 37,000 ha of farmland in the Nam Mae Kok and Nam Mae Ing valleys and require the relocation of about 200,000 people living in the region of the reservoir. This was the problem foreseen at the time (the late 1960s) regarding this huge dam project.

Low Pak Beng Dam Project

In order to avoid this inundation problem, the Secretariat decided to present an alternative development plan, decreasing the proposed dam height by 25 m. This would reduce the effective storage from 25.9×10^9 to only 200×10^6 m^3, decreasing the baseline power output from 1.1×10^6 to 211,000 kW (installed capacity: 350,000 kW; annual generated energy: 2,850 GWh), which would serve to meet only local requirements for domestic and small-scale industrial use. However, navigation would be improved by the ponding of water upstream from the dam and the backwater of the reservoir would reach only the vicinity of the town of Chiang Saen.

High Luang Prabang Dam Project

Like the Japanese reconnaissance team, the Mekong Secretariat had planned to build a high dam 11 km upstream from the city of Luang Prabang; however, it decided to build a much higher dam, setting the full water elevation 55 m higher. The High Luang Prabang Dam Project could be viewed as an alternative to the above-mentioned Pak Beng Dam projects, since the full water elevation was set at 355 m, the same as that of the Low Pak Beng Dam.

The height of the dam was estimated at 93 m, producing an effective reservoir capacity of about 10×10^9 m^3 and a baseline output of 915,000 kW (installed capacity: 1.5×10^6 kW; annual generated energy: 10,800 GWh). The dam would serve for flood control, for the increase of the low flow in the dry season (the minimum flow could be increased from 700 to 1,350 m^3/s), as well as for the improvement of navigation upstream from the dam.

However, the Secretariat planned this project on paper using only a general map to a scale of 1:50,000 and a map of the main stream to a scale of 1:20,000. No field work (such as a geological survey) was done and no other basic data were collected or reviewed for the study.

Low Luang Dam Project

The Low Luang Dam Project is the alternative to the High Luang Prabang Project described above. The dam site was at the same location as (or about 1 km upstream from) the High Dam site. The dam height was reduced to 49 m and the full-reservoir water elevation was set at 310 m, 10 m higher than in the Japanese plan. The effective storage capacity would therefore be reduced from 10,000 m^3 (in the case of the High Dam) to 140×10^6 m^3, most of which would be needed to provide an operating head for power generation. This meant that flood control and increase of the low flow in the dry season could no longer be expected, so the only additional merit would be the improvement of navigation. The power output would be 350,000 kW and the annual generated energy 2,400 GWh; it was anticipated that this energy would be used mainly in Laos and northern Thailand. However, this project was formulated only on paper, utilizing the topographical maps available.

Sayaboury Dam Project

The Sayaboury mainstream-dam project site is located about 20 km above the confluence of the Mekong and Nam Met rivers. The dam is planned to be about 53 m high with a crest length of 700 m and a full-reservoir water elevation of 275 m. Through joint operation with the upstream High Luang Prabang Project, the power output would be 900,000 kW and the annual generated energy 5,335 GWh. This project was also planned by the Secretariat on paper only.

Pa Mong Dam Project

The study for the Pa Mong Dam Project, which occupies a key position in the development of the lower Mekong Basin, was conducted by the US Bureau of Reclamation. When the Mekong Secretariat drafted the Indicative Basin Plan report, this study had been almost completed.

The study was based on the full water elevation determined by the Japanese team, providing an installed capacity of 4.8×10^6 kW. The total irrigation area, including the Laotian and Thai sides, would reach some 13,000 ha in the first development stage and, at the end of the final stage 50 years later, 1.6×10^6 ha.

The full-reservoir water elevation of 250 m made it necessary to provide dams at the divides between the mainstream area and both the Nam Mong and Nam Lik tributary basins. The construction of these three dams would produce a huge reservoir with a gross volume of 107.4×10^9 m^3 and an effective capacity of 82×10^9 m^3, serving for power generation, irrigation, and flood control. It was estimated in 1970 that the number of houses inundated by the reservoir would be 25,000 on the Thai side and 20,000 on the Laotian side. On the other hand, the reservoir would reduce the largest flood in a 100-year period at the proposed dam site from 26,600 to 15,600 m^3/s. This meant that areas downstream from the dam site to the confluence of the main stream and the Mun River would be made safe against flood disasters, and the flooding of the Mekong Delta would be decreased. Furthermore, the low flow in the dry season, which increases to 840 m^3/s in the driest year, would be increased to 3,000 m^3/s after the completion of the Pa Mong Dam. This would contribute greatly to dry-season irrigation in the Mekong Delta as well as protecting the delta against sea water and improving navigation.

The dam height would be about 115 m and the crest length 1,360 m. The height of the dam was based on a comparative study of the economic benefits resulting from five hypothetical water elevations – 250, 240, 230, 220, and 210 m above m.s.l.

Part of the electricity generated by the Pa Mong hydroelectric power station is to be transmitted through 115 kV lines to the nearby city of Udone, but the rest is to be transmitted via 500 kV lines to Bangkok or via 230 kV lines to other cities or towns in Thailand. The power station will, of course, also supply electricity for the city of Vientiane through 115 kV lines.

Bung Kan Project

The Bung Kan Project is a essentially a run-of-the-river power project utilizing the available head between Vientiane and Thakhek, as proposed by the Japanese team. The Mekong Secretariat proposed a full water elevation of 160 m, 5 m higher than in the Japanese plan. However, the dam height is only 29 m, which would provide about 2×10^9 m^3 of effective capacity for power production and the improvement of navigation. The installed capacity would be 220,000 kW and the annual generated energy 1,170 GWh. The upstream end of the storage would extend a little further upstream of Vientiane, but 43 villages and 52 km of highway would be inundated. The city of Vientiane would not itself be affected, but, if the project were operated as an isolated power plant, it would not be possible to generate power in the wet season, as the water released would reduce the difference between the upstream storage-water level and the level of the downstream tailwater. However, with the completion of the Pa Mong Dam Project, when the Bung Kan Project would be operated as part of an integrated system utilizing water released from the Pa Mong Dam, this problem would be solved. Moreover, the economic value of the project would be enhanced, since it could receive 11 units of 50 MW each, totalling 550 MW of installed capacity, raising the annual generated energy to 3,180 GWh.

However, this project was drawn up only on paper, on the basis of the available topographical maps to a scale of 1:2,000, permitting a slight improvement to the original plan proposed by the Japanese team.

Upper Thakhek Project

The Upper Thakhek Project was drawn up as an alternative to the Bung Kan Project. The dam site is located about 60 km upstream from the city of Thakhek just above the confluence between the Nam Songkhram and the Mekong.

The normal full water elevation of the storage-dam project was tentatively set at 155 m and the effective storage capacity for power production was estimated at $3,400 \times 10^6$ m^3, providing an installed capacity of 400,000 kW and annual generated energy of 2,800 GWh. Once the upstream Pa Mong Project is completed, the installed capacity and annual generated energy would be increased to 1,125,000 kW and 5,920 GWh, respectively.

This project was drawn up on the basis of a photogrammetric map of the dam site drawn to a scale of 1:5,000 and a geological report based on surface studies.

Lower Thakhek Project

The dam site for the Lower Thakhek Project is located about one kilometre upstream from Thakhek, but since it is further downstream on the Mekong than the above-mentioned Upper Thakhek Project, it was named the Lower Thakhek Project. It was drawn up as an alternative to the Bung Kan and Upper Thakhek projects described above. Considerable inundation of the Nam Songkram Basin was to be avoided by lowering the normal full water elevation to 115 m, but 32,000 ha of arable farmlands in the lower reaches of the Nam Songkhram would be inundated.

As in the Bung Kan and Upper Thakhek projects, the Lower Thakhek Project could not operate independently from the Pa Mong during the wet season, owing to the high tailwater levels in a normal year. On its own, the project would provide an installed capacity of 250,000 kW and annual generated energy of 1,550 GWh. Combined operation with the Pa Mong Project would increase these figures to 700,000 kW and 4,100 GWh, respectively.

The project was drawn up on the basis of topographical maps to a scale of 1:2,000 and geological studies.

Khemarat Project

The dam site for this project is located about 24 km below the town of Khemarat, where the river channel of the Mekong is rather narrow. In order to restrict the inundated area of the town, the normal full water elevation was to be kept as low as 130 m, resulting in a small storage capacity. Owing to the small difference between the upstream and downstream water levels, power production would frequently be impossible during the wet season under independent operation. The project should therefore not be realized unless the High Pa Mong Dam becomes operational. Nevertheless, if it is developed as an independent project, the installed capacity and annual generated energy are expected to be 420,000 kW and 2,800 GWh, respectively, which would be increased to 1.05×10^6 kW and 7,550 GWh through combined operation with the Pa Mong. However, in both cases, the project may have to be shut down during the wet season.

This project was drawn up only on paper, using the topographical maps available drawn to scales of 1: 50,000 and 1: 20,000.

Bang Koum Dam Project

The proposed site of the Bang Koum Dam is located 14 km above the confluence between the Mekong and the Nam Mun. From the viewpoint of hydroelectricity generation, a project developed on the main stream here would be superior to any dam projects located between the Pa Mong Dam site and the town of Pakse.

This project was drawn up as an alternative to two development plans – the Khemarat Project upstream and the Pakse Project downstream. The purposes of the development of the Bang Koum are power generation and improvement of navigation. When the normal full water elevation of the reservoir is 130 m, its effective storage capacity would be 1.4×10^9 m^3 and power could be generated throughout the year, even if the dam operates independently. The output would be 900,000 kW (baseline output: 515,000 kW) and the annual generated energy 7,170 GWh. If this project were developed after completion of the Pa Mong Dam, the installed capacity would be 3.3×10^6 kW, baseline power output 1.96×10^6 kW, and annual generated energy 19,800 GWh.

The normal full water elevation was set at 130 m to avoid inundation of the important city of Savannakhet, located upstream from the dam site on the left bank of the Mekong. The project was drawn up on the basis of the maps available drawn to a scale of 1: 5,000, but the dam site was not surveyed since it was found to be unapproachable from the Laotian side. This project is compared with the Low Pa Mong Dam Project later in this chapter.

Pakse Dam Project

The Pakse Dam Project was conceived as a run-of-the-river power project located between the Bang Koum and Khone Falls projects. The dam was to be constructed in a narrow section of the Mekong main stream, some 6 km below the confluence with the Nam Mun. Its normal water level was set at 104 m. The project would result in the inundation of the low-lying land in the Nam Mun Valley, including the proposed Pak Mun Project, but since this area is regularly flooded to quite a depth, the loss in terms of agricultural production would not be great.

If this project were developed independently (of the Pa Mong Project), the power output would be considerably reduced by the rising tailwater level at the dam site during the wet season. In this case, the installed capacity would be 350,000 kW, baseline power output 122,000 kW, and annual generated energy 2,400 GWh. When combined with the Pa Mong, these figures would be 1.4×10^6 kW, 594,000 kW, and 8,400 GWh, respectively.

The project was drawn up on the basis of the topographical maps available to a scale of 1:2,000, covering the proposed dam site and the results of site reconnaissance.

Dam project immediately downstream from the Khone Falls

As previously mentioned, before the Japanese reconnaissance team carried out its study, the ECAFE Secretariat had already conducted its own study and formulated a plan to develop a relatively small-scale hydroelectric power station by constructing a dam 5 m high just above the Chutes de Phapheng in the Khone Falls. The Japanese reconnaissance team examined the site and proposed the construction of the Stung Treng Dam, a huge dam and hydroelectricity project far downstream, instead of a small weir or low dam at the falls. One of the drawbacks of this development plan was that it would inundate a vast area of farmland and housing when fully implemented.

In the Indicative Basin Plan report (Mekong Secretariat 1970), all such project plans were examined and the Mekong Secretariat decided to propose another Khone Falls Project, which could be viewed as a compromise between the ECAFE and Japanese plans. The main difference between this compromise plan and past proposals was that the planners thought it would be possible to move the dam site from upstream of the Khone Falls (proposed by ECAFE) to downstream on the main stream of the Mekong, where a favourable configuration could be formed utilizing two small islands. The spillway and the power station would be separated by the islands, which are located upstream from the town of Stung Treng.

Simple sounding work was carried out at the dam site on the main stream and a preliminary plan was drawn up on the basis of detailed topographical maps drawn to a scale of 1:2,000. The power station would have an installed capacity of 750,000 kW (baseline output: 460,000 kW; annual generated energy: 6,220 GWh) when operated independently;

with the construction of the Pa Mong Dam, the increase in the dry-stream flow would provide an installed capacity of 1.75×10^6 kW (baseline output: 1.01×10^6 kW; annual generated energy: 12,580 GWh) with upstream control.

In addition to the merit of power generation, the plan envisaged the improvement of navigation on both the main stream and tributaries, as well as the irrigation of a vast area of farmland, as in the Japanese development plan. The project would result in the falls being covered by water, thus removing a major barrier to navigation on the Mekong. If locks were installed, boats (or even ships) could pass upstream and downstream of the dam. Furthermore, the dam's storage would make it possible to irrigate low-lying areas downstream on both sides of the Mekong and to divert part of the total flow into the Stung Sen Valley to irrigate the vast area extending to the north of the Great Lake. It would therefore be possible, technically, to irrigate about 200,000 ha downstream of the dam and 700,000 ha north of the Great Lake.

Another, more remarkable, development plan had been proposed by the author himself before the above scheme was presented. While the Indicative Basin Plan report was being drawn up, he formulated a development plan at the same dam site downstream from the Khone Falls. The plan was to build two additional dams further downstream on the Se San and Srepok rivers in parallel with the mainstream Khone Falls Dam, thereby creating three reservoirs. If two weirs with gates were constructed at their divisions, these three reservoirs could be operated jointly. The size of this jointly operated reservoir would be about two-thirds of the Stung Treng Reservoir proposed by the Japanese team. The joint operation of these three dams and power stations would significantly diminish the flood peak, considerably reducing the damage frequently caused to the Mekong Delta downstream. In addition, a large amount of hydro-electricity would be produced and it would become possible to navigate from the downstream to the upstream reaches of the Se San, Se Kong, and Srepok, as well as to the upstream areas of the main Mekong. As a result, the areas along the three large reservoirs covering extensive areas of Laos and Cambodia could be developed in the future, and a large amount of water from the reservoirs could be used for irrigation – as in the Stung Treng Dam Project. In addition to these great benefits, the expected extent of inundation caused by this joint reservoir-operation project would be very much less than in the plan proposed by the Japanese government.

Unfortunately, however, this plan was not adopted and was forgotten by the Secretariat, since the author left the Secretariat soon after presenting the plan.

Stung Treng Project

In the case of the Stung Treng Project, the Mekong Secretariat followed the development plan proposed by the Japanese reconnaissance team. With the limited budget allocated for the study, the Secretariat carried out geological surveys at the proposed dam site, including core drilling down to 400 m and a seismic study over 20 km. A topographical map to a scale of 1:5,000 covering the dam site and a simple layout and design of the dam were provided, and all available hydrometeorological data related to the project were examined. As a result, the Stung Treng Project outlined below was formulated as one of the most important projects in the Indicative Basin Plan. However, this plan is essentially the same as that proposed by the Japanese team. In the Japanese report, the full-reservoir water elevation was set at only 88 m; the Secretariat, on the other hand, made a comparative study of four alternative heights above m.s.l. – 80, 85, 90, and 95 m – as well as an analysis based on three different reservoir drawdowns. The Secretariat's conclusion was that the Japanese team's plan, based on a full-reservoir water elevation of 88 m and a drawdown of 7 m, was the most appropriate.

The dam site is located just downstream from the confluence between the main stream and the Se San River, near the town of Stung Treng. With a dam 77 m high, the reservoir would cover an extensive area of about 8,000 km^2 bordering the main stream and tributaries. The reservoir would have a gross storage capacity of 110×10^9 m^3, of which 46.5×10^9 m^3 could be effectively used for flow regulation. Construction would result in the submergence of the town of Stung Treng and about 40,000 ha (estimated area of farmland as of 1985). In addition, 250 km of highways and 200 km of other roads would have to be built to replace several road connections in the area. At the time when the study was made, the population of Stung Treng was estimated at roughly 5,000, but the number of farmers living outside the town who might have to be relocated was not counted.[13]

If this huge reservoir with a full water elevation of 88 m and a drawdown of 8 m is realized and the project is operated in conjunction with the upstream Pa Mong Dam, it would be possible to reduce the largest flood in a 100-year period (estimated at 73,900 m^3/s) to 46,000 m^3/s. However, as it would be necessary to reduce the maximum flow volume in the Mekong Delta to 20,000 m^3/s in order to protect it completely, the reduction of the peak flood volume by this joint operation would not be sufficient to protect the inhabitants of the delta from flood disasters. A study by a French firm of consultants, SOGREAH, concluded that, if both the Pa Mong and Stung Treng dam projects were completed and additional dykes, control gates, and discharge pumps were installed at various critical points in the delta, it would be possible to give the people of the delta complete protection from flood disasters.

The minimum recorded dry-season flow would be increased by the Stung Treng project alone from 1,375 to 5,055 m^3/s. This increase could be used, as in the case of the Pa Mong Project, for developing dry-season irrigation in the delta, reducing salt-water intrusion, and improving navigation. Furthermore, it would be technically feasible to irrigate about 700,000 ha of land north of the Great Lake by means of a canal connecting the reservoir to the Stung Sen. However, a study on soils north of the Great Lake, carried out by the Secretariat with help from an agricultural expert, showed that the condition of the soil was, in general, not good.

Regarding the prospects for hydroelectricity generation, some 3.4×10^6 kW could be installed, even with independent operation, producing annual generated energy of as much as 24,500 GWh. Combined operation with the Pa Mong Project would increase the installed capacity and annual generated energy to 7.2×10^6 kW and 35,000 GWh, respectively.

The average height of the Stung Treng Dam in the river section (about 1,000 m wide) would be 70 m, although this height would be much lower outside the river. However, the exceptionally long crest length of about 27 km would make this a very major and expensive project.[14]

Sambor Project

As mentioned above, the feasibility study for the project furthest downstream, the Sambor Project, had been initiated by the Japanese government. Since the Mekong Secretariat was provided with all the results of this study by the end of the 1960s, it included them in the 1970 Indicative Basin Plan report.

The Sambor Project was essentially a run-of-the-river hydroelectric power project located 147 km below Stung Treng and 560 km from the sea. The dam would consist of a 26 km earth dam section, a 2.35 km rockfill dam section and a 1.47 km concrete section, with a maximum height of 54 m in the river section (rockfill dam). The gross storage capacity would be 10×10^9 m^3, of which 2.05×10^9 m^3 would be effective. The full water elevation would be 40 m, with a 2 m drawdown. The storage area extending upstream as far as Stung Treng would submerge some 16 km of highway, 2,400 ha of farmland, and 2,000 dwellings.

As an independent project, the installed capacity of the hydroelectric power station would be 875,000 kW (baseline power output: 532,000 kW; annual generated energy: 7,080 GWh). When combined with the Pa Mong Project, the installed capacity would be 3.0×10^6 kW (baseline output 2.07×10^6 kW; annual generated energy: 19,580 GWh). Furthermore, if it were combined with both the Pa Mong and Stung Treng projects, the installed capacity would be 3.25×10^6 kW (baseline output

2.27×10^6 kW; annual generated energy: 22,300 GWh). Initially, the project would include two transmission lines – a 345 kV single-circuit line from Sambor to Sihanoukville (Kompong Som), a seaport on the southern coast of Cambodia, over a distance of 350 km via Phnom Penh, and a 345 kV single-circuit line from Sambor to Saigon (Ho Chi Minh City) over a distance of 230 km.

According to the Japanese plan, the Sambor Project could irrigate, by means of gravity flow and pumping, about 34,000 ha on both sides of the river downstream from the dam site. This area could be extended further downstream along the Mekong and Bassac to 621,000 ha with the additional water released from Pa Mong and Stung Treng. In the tentative plan, provisions for the passage of river traffic were also included.

Tonle Sap Project and its influence

The Tonle Sap river, which connects the Great Lake with the Mekong at Phnom Penh, has a natural reversible flow. The aim of the Tonle Sap project is to control the flow into and out of the Great Lake by means of a barrage (dam) across the Tonle Sap at the outlet of the Great Lake. The operational strategy is to maintain a low water level in the lake until the flow of the Mekong reaches a critical stage at Phnom Penh, at which point the gates of the weir would be opened and water permitted to flow into the lake. The gates would be closed again towards the end of the rainy season, when floods in the river begin to recede. The water thus impounded in the lake would be released during the dry season. The water in the lake would be drawn down to its lowest level by the beginning of the rainy season, at which time the gates would be closed again.

In the 1957 ECAFE Report, it was clearly stated that this cycle of operation of the Tonle Sap barrage (dam) would ensure a measure of flood relief as well as augmenting the low flow volume, which would assist irrigation downstream in the dry season. The Japanese reconnaissance team later examined this plan and expressed its approval. Consequently, the Indian government studied the project site at Kompong Chhnang (located at the outlet of the Great Lake) and commenced design work on the weir and its control gates. The French government conducted a study on fish in the lake, and the Philippine government made a study of sedimentation using isotopes. Studies of the hydrological and hydraulic aspects of the projects using a mathematical delta model were made with assistance from the UNDP, UNESCO, and the French consulting firm SOGREAH. The Mekong Secretariat took charge of studies on the probable economic impact of flooding in the delta. In addition, the Development Resources Corporation utilized US funds to conduct studies on flood control, irrigation and drainage, and protection against salt-water influx, and submitted a report in 1969.

Before and during the course of these studies by various groups, including the UNDP/UNESCO/SOGREAH team, the Mekong Secretariat was optimistically considering the development of a large weir at Kompong Chhnang on the Tonle Sap river. The Secretariat expected that the operation of the weir would help to stabilize the water surface of the lake, increasing fish production and promoting the industrialization of areas around the lake. Furthermore, some hydroelectricity generation could be expected by utilizing the water released from the lake through the weir between January and April. However, SOGREAH's mathematical model revealed that this project would be less effective in reducing flood levels than anticipated: the resulting reduction in the depth of flooding in the delta in the rainy season might be only dozens of metres at most. The members of the Mekong Secretariat were disappointed by these findings and concluded that the Tonle Sap project would not play as important a role in flood reduction as the Stung Treng and Pa Mong projects, which the Secretariat now considered to be the only projects that could drastically reduce flooding in the delta.

Nevertheless, the Indian government reported that the proper operation of the weir gates would result in the following benefits:

1. The inevitable reduction in the fish catch in the Great Lake during the dry season could be considerably improved;
2. The problems experienced by large ships in navigating the Tonle Sap river during the driest season (usually 2 months) could be solved;
3. Pump irrigation during the dry season in both Cambodia and the Vietnamese delta could be improved;
4. The current problematic salt-water influx in the delta could be greatly mitigated by the expected increase in the low flow volume during the dry season, improving the irrigation of the extensive surrounding farmland;
5. Stable irrigation could be promoted on the land around the lake;
6. Some areas of the fringe of the lake could be converted into polders for fish breeding.

The proposed Kompong Chhnang weir was a large structure 678 m wide expected to pass 15,000 m^3/s of water, maintaining the greatest lake-water elevation at 12 m. It was to be equipped with two fish locks and one navigation lock with gates measuring 28 × 20 m. It was estimated that construction would take 5 years. Unfortunately (or not), the weir was not constructed, owing to the changed situation in the region after the mid-1970s.

Delta-Development Project and joint US–Vietnamese Study

The Mekong lowland flood plain is an extensive tract of land with an elevation of below 14 m extending from Kompong Cham to the sea. The flood plain covers an area of 7.40×10^6 ha, of which an area of about 510,000 ha is perennially under water and totally unproductive. Of the remaining area, 4.95×10^6 ha are subject to seasonal flooding and several varieties of flood-dependent rice are grown in this area during the rainy season. In the broad belt of land bordering the sea, a single crop of rain-fed rice is cultivated. The development of this lowland floodplain in order to increase agricultural production was referred to as the Delta-Development Project

As we have seen, the report on the Indicative Basin Plan (Mekong Secretariat 1970) concluded that, in order to protect the delta from flooding and increase agricultural production, it would be necessary to implement both the Pa Mong and Stung Treng projects, to construct low dykes along water channels in the delta, to build a number of small weirs, and to carry out drainage there. The report stressed that this was the only solution.

In fact, the gross storage capacity of the Pa Mong and Stung Treng reservoirs would be about 218×10^9 m^3 if they were realized as proposed in the report. If these two reservoirs were operated to provide some 171.9×10^9 m^3 during the flood period from August to October to accommodate the Mekong flood water, it would be possible to reduce the largest flood in a 100-year period from 73,900 to 46,000 m^3/s at Stung Treng. If, in addition to the above, low dykes were constructed along water channels in the delta, the reduced flood water would be held down within the water channels without spilling over the channel banks.

Looking at the current flood situation in the downstream region of the Mekong River Basin, when the Mekong River flow elevation exceeds 8.2 m at the city of Kompong Cham in Cambodia, the flow begins to spill over both river-banks. Similarly, when the flow elevation exceeds 2.5 m at Can Tho in the Vietnamese delta, flooding begins. It is therefore necessary to construct low dykes at various points along rivers all over the delta. As long as such dykes are constructed, the flooding situation in the delta would be moderately improved, even without the construction of major dams such as the Pa Mong and Stung Treng. In view of this, the Vietnamese government has started to build low dykes in the delta without any expectation that the upstream riparian countries will implement large-scale dam construction in the future. These efforts extend, of course, to the belt zone beside the sea. The government has also provided a number of control weirs on water channels at various locations in the delta to counteract the troublesome influx of salt water from the sea in the dry season.

During the period when the Mekong Secretariat was making its final efforts to complete the Indicative Basin development plan (1970), the US government cooperated with the government of South Viet Nam on a delta-development plan. According to the Indicative Basin Plan report, the joint study report drawn up by the US and Vietnamese governments proposed that some US$1 $\times$ 10^9 be invested over 20 years to control (partially) flooding in the delta by constructing dykes and control weirs at a number of sites and to develop 1.6 $\times$ 10^6 ha to produce a double crop of high-yielding varieties of rice which had just been developed. The plan was ultimately to develop 2.1 $\times$ 10^6 ha of irrigated land which would be fully protected from flood disasters within 30–40 years through the investment of US$2.5 $\times$ 10^9. This plan was called the Delta-Development Project.

Plans for dams on Mekong tributaries in the report on the Indicative Basin Plan (1970)

As we have seen, when the Japanese reconnaissance team started its investigation of the developmental possibilities of the 34 main tributaries of the lower Mekong Basin during the 2-year period from 1959, there were no maps available to a scale of 1:50,000 (apart from a map of Thailand) and only one map of the whole region to a scale of 1:100,000. The reconnaissance team requested the US Army Map Service in Oji, Tokyo, to assist them by providing maps (to a scale of 1:50,000) of the key locations from aerial photographs and used these to choose 102 candidate locations for development projects. However, most of the locations they selected were in mountainous, unpopulated, and inaccessible regions.

The Secretariat of the Mekong Committee set about formulating a development plan referring to the plan recommended by the reconnaissance teams. It made use of hydrological and meteorological data collected during the 1960s after the completion of the reconnaissance teams' report. The only maps used were a 1:50,000 scale map of the whole region and a 1:20,000 scale map of the main stream that were finally completed at the beginning of the 1960s. The Mekong Secretariat selected 87 promising locations for development and made rough estimates of the benefits to be obtained from hydroelectricity, irrigation, flood control, navigation, and fishery that would result from development at each location. The proposal for the development of the tributaries in the Indicative Basin Plan (1970) was thus only a paper plan but, when it was finally drawn up, development had already been completed in 4 locations and was in progress at 9 of the 87 locations.

As there is not sufficient space here to describe each of these tributary plans, the following summaries outline the most noteworthy development projects in each of the four riparian countries.

Cambodia

Cambodia can be broadly divided into three regions – the north-eastern region, the region around the Great Lake, and the southern mountainous region of the Elephant Range extending to the west of the Bassac River; however, only the first two regions are located in the lower Mekong Basin. On the basis of maps of the main tributaries (to scales of 1:250,000 and 1:50,000), the development report outlines the results of studies making use of hydrological and meteorological data.

Development plans for the Se Kong, Se San, and Srepok rivers in the North-East. The middle and lower reaches of the three major Mekong tributaries – the Se Kong, Se San, and Srepok rivers – are situated in Cambodia. In this region there are several promising locations for dam development for hydroelectric power and irrigation. If dams were developed in the lower and middle reaches of these tributaries near national borders, the reservoirs would extend into neighbouring Laos or Viet Nam, necessitating negotiations between these countries. If dams were developed upstream in Laos or Viet Nam, on the other hand, they would affect downstream regions in Cambodia. If a mainstream dam were later constructed at Stung Treng, the downstream parts of these three rivers would be completely submerged by the reservoir. Therefore, plans for the development of the Se Kong, Se San, and Srepok rivers required comprehensive forecasts covering the whole region, as well as international negotiations. At any rate, the Indicative Basin Plan report (1970) stated that the development of these tributaries in Cambodia would yield over 300,000 kW of electricity and irrigate more than 500,000 ha of land.

The Battambang and Stung Chinit rivers and other development plans in the vicinity of the Great Lake. Compared with the other two major regions of Cambodia, the region around the Great Lake has low rainfall and consists of relatively flat terrain sloping gently down to the lake. This region therefore has a certain potential for dam construction for irrigation.

As we have seen, of all the rivers flowing into the Great Lake, the Battambang is the most promising for development. It was estimated that it would yield more than 31,500 kW of hydroelectricity and irrigate an area of 68,000 ha, and a feasibility study for development was conducted. The Japanese government conducted a similar study on the Stung Chinit River, where development was considered mainly for the purpose of irrigation (25,400 ha).

There were other tributaries with development potential that was hard to ignore, such as the Stung Mongkorborey River (rainy-season irrigation 15,000 ha; dry-season irrigation 10,000 ha) and the Stung Pursat River (rainy-season irrigation 43,000 ha; dry-season irrigation 21,000 ha; hydro-electricity 9,700 kW).

Because it was the project located closest to Phnom Penh, particular attention was paid to the plan to build a dam on the Prek Thnot River flowing into the Bassac, which was expected to yield substantial benefits in terms of hydroelectricity, irrigation, and flood control. Japanese companies had been closely involved in this dam project before the outbreak of the Cambodian conflict. According to the 1970 plan report, the dam would be 28 m high with an effective reservoir capacity of 670×10^6 m^3, yielding 2,000 kW (baseline output) of hydroelectricity and irrigating a total area of 70,000 ha. However, the economic viability of this project was questionable.

Laos

Almost all the Mekong tributaries in Laos flow into the left bank of the main stream. Although most of them have hydroelectric power development potential, they are not all suitable for irrigation purposes.

Since the Indicative Basin Plan (1970) made use of a 1:50,000 scale map and relatively thorough hydrological and meteorological data that were not available when the Japanese reconnaissance team studied the region, it was possible to investigate many tributary development plans from various viewpoints. The 14 tributaries studied were, from the north, the Nam Ou, Nam Suang, Nam Khan, Nam Dong, Nam Met (on the right bank of the Mekong), Nam Ngum, Nam Lik, Nam Nhiep, Nam San, Nam Theun, Se Bang Fai, Se Bang Hieng, Se Dong, and Se Kong rivers. The candidate locations for development furthest downstream on these tributaries have been removed from the table of tributary dam projects because they are expected to be submerged when dams on the main stream are completed.

The studies on these tributaries showed that the two most promising tributaries for development were the Nam Ngum and the Nam Theun rivers.

The Nam Ngum Dam Project. The construction of the concrete Nam Ngum dam, 75 m high, had already started while the Indicative Basin Plan (1970) was being drawn up. According to the report, the Nam Ngum Dam was to have a first-phase installed capacity of 30,000 kW which would eventually be increased to 130,000 kW. It was also expected to irrigate an area of 5,000 ha (ultimately to be increased to 114,000 ha) in the vicinity of Vientiane. After electricity generation had started, the surplus was to be transmitted to Thailand.

The Nam Theun Dam Project. The project for the development of dams on the Nam Theun River was an improved version of an existing development plan. The project to construct a dam 240 m high furthest downstream on the Nam Theun River in a valley immediately upstream from the confluence with the main stream of the Mekong was named the Nam Theun 1 Dam Project. The dam reservoir was to have an effective capacity of 12×10^9 m^3 and a hydroelectric power station with an installed capacity of 1.5×10^6 kW (7,787 GWh/year) was to be built. The Nam Theun 1 Dam would completely control flooding of the Nam Theun River and increase the flow volume of the main stream of the Mekong during the dry season. Upstream from this Nam Theun 1 hydroelectric power station, the Nam Theun 2 Dam (85 m high with an effective reservoir capacity of 6.2×10^9 m^3) was to be constructed. Water from the reservoir of the Nam Theun 2 Dam would be diverted into the adjacent Se Bang Fai River, generating 2.5×10^6 kW (8,400 GWh/year) with a head of 350 m. Of course, a plan was formulated for the effective use of the water flowing from the Nam Theun River into the Se Bang Fai River.

Having examined the various promising rivers for dam development in Laos, the Indicative Basin Plan report (1970) concluded that it was possible to generate a total of 2.62×10^6 kW (baseline output) of electricity and to irrigate 250,000 ha of land through dam development.

Thailand

The Mekong tributaries in Thailand can be broadly divided into (a) the rivers in northern Thailand which flow into the main Mekong and (b) the tributaries and small branch rivers in north-east Thailand. In all, 16 projects on 15 rivers were studied in the Indicative Basin Plan (1970). The 15 rivers were the Nam Mae Kok and Nam Mae Ing rivers in northern Thailand and the Nam San, Nam Oon, Nam Songkhram, Nam Cheng, Nam Phrom, Lam Pao, Nam Chi, Lam Takong, Lam Pra Plerng, Lam Don Noi, Nam Pong, Nam Pung, and Nam Mun rivers in north-east Thailand. Because of the country's topographical features, the potential for development of hydroelectricity is low (a total baseline output of 87,000 kW), but irrigation over a total area of 320,000 ha could be expected.

Regarding these rivers, the Nam Pong, Nam Pung, Lam Pra Plerng, Lam Pao, and Lam Takong dam projects in north-east Thailand had already been completed and operation started while the Indicative Basin Plan (1970) was being prepared. Dam development was also in progress on the Lam Don Noi, Nam Oon, and Nam Phrom rivers. In addition, feasibility studies for the Nam Cheng, Pak Mun (the most downstream location on the Nam Mun River), and Nam San projects in north-east Thailand had been finished.

Since the small ponds (water tanks) that had long been used to deal with water shortages in the north-east were neither effective nor adequate, dam development was considered essential. However, it was pointed out that there was a danger that some of these projects might be submerged when the Pa Mong Dam on the main stream of the Mekong was completed.

Viet Nam

The upper reaches of the Se San and Srepok rivers in Viet Nam were considered very promising locations for development, owing to their high precipitation and steep slopes.

The Upper Se San River Development Project.   A study on the upper reaches of the Se San River was conducted while the Indicative Basin Plan (1970) was in preparation and, in 1966, Nippon Koei Co., Ltd produced a report outlining the development potential. According to this report, it would be possible to generate a total of 800,000 kW of electricity through seven hydroelectricity development projects: these would be at the Yali Falls; Prey Krong; the Upper Se San No. 1, No. 2, No. 3 and No. 4 projects; and the Dak Bra. In addition, a total area of 21,250 ha could be irrigated (naturally and by pumping).

The Upper Srepok River Development Project.   When the 1970 plan was being drawn up, development of the Upper Srepok River was being promoted by the Overseas Technical Cooperation Agency (OTCA, the predecessor of JICA) in Japan as part of the Krong Buk irrigation project. The Drayling small hydroelectric power station (500 kW) had already been developed in this region and Nippon Koei had a plan to enlarge it. On the basis of this plan, the Vietnamese government produced a report stating that a total of nearly 300,000 kW of energy could be generated at nine locations and that eight irrigation projects (covering a total of 75,500 ha) could be implemented. As a result of this development, it would eventually be possible to supply 12,000 kW of electricity to this region.

The Indicative Basin Plan report (1970) concluded that a total of 28 dam-construction projects could be considered in the upper reaches of the Se San and Srepok rivers.

Report on the Indicative Basin Plan (1970): summary

The aim of the Indicative Basin Plan was to create a blueprint and framework for development of the lower Mekong Basin during the 30-year period from 1971 to the end of the twentieth century. The common goal of those who drew up the plan was that the comprehensive development of the considerable water resources of the lower basin would stabilize and improve the lives of the people residing there, whose total population was expected to increase to 60×10^6 by the end of the century.

Making use of all the data accumulated by putting a total of US$60 \times 10^6 into pre-investment studies over a dozen or so years, this massive development plan aimed at the most rational and economical development of the water resources of the lower basin. The plan itself was put together by the Secretariat of the Mekong Committee, but the data on which it was based were collected by thousands of experts in 30 countries throughout the world – an operation that required a huge amount of time and money. The Indicative Basin Plan (1970) was the culmination of the tremendous efforts of all of those who had been involved in lower-basin development until then.

According to the report, there were more than 100 possible development projects on both the main stream and the tributaries of the Mekong. When the plan was drawn up in the second half of the 1960s, the lower basin as a whole was still at an almost primitive stage of development. No projects at all had been started on the main stream of the Mekong and, on the tributaries, only a few small or medium-sized dams had been completed (such as the Nam Ngum Dam in Laos) or were under construction. In spite of this (or, perhaps, because of it), this plan reflected the mood of the 1960s in its dream of improving the situation of the lower basin at a stroke by means of large-scale development.

Let us now consider the contents of the Indicative Basin Plan (1970) section by section. As far as electricity was concerned, the annual per capita electricity consumption of people in the lower basin in 1970 was only 10 kWh. As a result of the plan, demand for electricity in Thailand, Viet Nam, and the other two riparian countries suddenly shot up: by 1988, this demand had increased to between 2.5 and 4.0×10^6 kW in these countries. Furthermore, in view of the explosive industrial growth through overseas aid centring on large cities such as Bangkok and Saigon, it was forecast that the construction of a large dam would be necessary and, by the year 2000, power facilities generating $12–21.3 \times 10^6$ kW of electricity would have to be set up. On the basis of this forecast, the concept of short-range (up to 1980) and long-range (up to 2000) plans was formulated: the short-range development plan was to yield a total of 3.3×10^6 kW of electricity in the four countries, which would be increased to 20×10^6 kW through the long-range plan for large-dam development.

Regarding the development of irrigation and drainage facilities, it was estimated that, during the 10 years up to 1980, the irrigation of 700,000 ha would be necessary in the lower basin alone and that this would have to be increased by further development to 1.6×10^6 ha by 2000.

The plan report stated that, in addition to hydroelectric power and irrigation, dam development had various other advantages, such as the improvement of navigation and flood control. One of the design provisions for the mainstream dams in the long-range plan was the installation of lock gates. These gates would ensure that, by the year 2000, 5,000 t ocean-going vessels could pass upstream from the South China Sea to Phnom Penh in the rainy season; that 2,000 t large vessels could travel the same route, both upstream and downstream; and that 500 t ships in the rainy season and 100 t ships in the dry season could go as far as Kratie and back.

One of the most important aspects of the plan was flood control. The reduction of flood damage in places such as the lower reaches of the Nam Ngum and Se Bang Fai rivers, the area around the Great Lake, and the Mekong Delta was considered indispensable for social and economic development in the lower basin. The plan stated that, through the construction of the Pa Mong and Stung Treng mainstream dams and several large dams on the tributaries, flood-damage reduction equivalent on average to US$25 $\times$ 10^6/year could be expected. It also estimated that the increased agricultural production likely to result from reduced flooding would raise average agricultural profits to about US$200 $\times$ 10^9/year.

The plan stated that the total capital required for development over the whole 30-year period up to 2000 would amount to the vast sum of US$12 $\times$ 10^9.

Looking back on this plan from today's perspective, it seems extraordinarily optimistic. In the 1960s, however, most people believed that any large development project was possible as long as the money could be raised, and very few doubted the necessity and potential of this kind of development. In fact, if Viet Nam and Cambodia had not been ravaged by war, and the three countries of Indo-China had not come under communist rule in the 1970s, development would no doubt have proceeded (at least to some extent) as outlined in the plan. If this had happened, the lower Mekong Basin, for better or worse, would have differed greatly from what it is today.

Dam development during and after the formulation of the Indicative Basin Plan report (1970)

The fact that the Indicative Basin Plan report (1970) was formulated by the Secretariat of the Mekong Committee ensured that clear guidelines were laid down for the future development of the lower Mekong Basin for the first time. It also meant that all developmental effects of the Mekong Committee thereafter should be pursued within the framework of the plan.

Soon after the completion of the report, however, the situation in the lower Mekong Basin rapidly worsened and the development plan became impossible to implement. The Secretariat accordingly made efforts to improve and refine plans for relatively safe regions in the lower basin and had to content itself with modest activities, such as supplementing existing facilities or preparing a substitute development plan and comparing this with the original plan. Now that it was possible to continue only minor operations in areas not affected by the Viet Nam War and Cambodian conflict, the plan's aggressive and virile image of taming the mighty Mekong completely lost its force. Furthermore, the focus of attention on environmental problems that had begun at the end of the 1960s steadily grew stronger. Reflecting this growing environmental concern, several valid and thorough studies were conducted on the influence of small- and large-scale projects on the natural environment and socioeconomic conditions of the lower basin.

The following section outlines water-resources development in the lower basin while the 1970 plan was being drawn up.

Water-resources development in the lower basin during the preparation of the Indicative Basin Plan (1970)

During the 5-year period between the commencement of formulation of the plan at the end of 1964 to its completion in 1970, the construction of several dams for power generation and irrigation in the lower basin was started. Some of these dams had been completed and were already operating when the plan finally appeared.

Dams completed during the preparation of the Plan

All of the completed dams were hydroelectricity or irrigation projects in north-east Thailand:

- Nam Pung Dam: 6,300 kW; aid provided by Japan; completed in 1965; outlined below;
- Nam Pong Dam: 25,000 kW and 53,000 ha; aid provided by West Germany; completed in 1966; outlined below;
- Lam Pra Plerng Dam: 10,500 ha; aid provided by United States; completed in 1967;
- Lam Pao Dam: 54,000 ha; aid provided by United States; completed in 1968;
- Lam Takong Dam: 38,000 ha; completed in 1970.

In the south of Laos, a small-scale hydroelectricity project was completed in the lower reaches of the Se Dong River: this was the Saravan hydroelectric power-plant project, providing 2,040 kW and completed in 1968.

The Nam Pung Dam (Thailand). The Nam Pung Dam was the first project completed (October 1965) during the preparation of the Plan, and the ceremony for commencement of operation was attended by the King and Queen of Thailand. The dam was planned, designed, and constructed by EPDC with aid from Japan. This rockfill dam, with a height of 41 m and crest length of 286.5 m, was the first medium-scale lower-basin development project in north-east Thailand. The effective capacity of the reservoir is 122×10^6 m^3, the total installed capacity is 6,300 kW, and the annual generated energy is 32.4 GWh. Part of the output is being used to irrigate 8,000 ha of land near the dam.

The Nam Pong Dam (Thailand). Also known as the Ubolrathana Dam, the Nam Pong Dam was completed in north-east Thailand in March 1966 and operation was, again, commenced in the presence of the King and Queen of Thailand. The project was originally proposed by a Japanese reconnaissance team, but planning and construction were supervised by West German consultants and backed by West German aid. It is a rockfill dam with a height of 37 m and crest length of 800 m. The initial installed capacity of the facilities was 16,600 kW, and the estimated final installed capacity was 25,000 kW, with an annual generated energy of 104 GWh. In addition to power generation, the dam is used for irrigation, water supply (including water for industry), and flood control. An intake tower is located 25 km downstream from the dam for the irrigation of about 53,000 ha of land. The funding for this irrigation project was provided by the World Bank and the Asian Development Bank. The Nam Pong Reservoir was later to become well known through an environmental study.

Dam projects started during the preparation of the Plan

The most important project initiated while the Indicative Basin Plan (1970) was being drafted was the Nam Ngum Dam in Laos, which is discussed in detail later (pp. 160–161). The other main projects were the Lam Don Noi Dam (24,000 kW and 24,000 ha, aid from Japan, completed in 1971) and the Nam Phrom Dam (40,000 kW, completed in 1973) in north-east Thailand, and the Prek Thnot Dam (18,000 kW and 5,000 ha, project abandoned) in Cambodia.

Pilot farms completed during the preparation of the Plan

During the preparation period, pilot farms were opened in two locations. It was hoped that these farms would play an important role in the improvement of agriculture in the future.

Hat Dok Keo Pilot Farm in the vicinity of Vientiane in Laos. The Hat Dok Keo Pilot Farm consists of a 17 ha agricultural experimentation station and pilot farm of 300 ha. The farm was set up in 1962 with assistance from the UNDP and Israel prior to the implementation of an irrigation plan for the Vientiane Plain.

Kalashin Pilot Farm in northern Thailand. The Kalashin Pilot Farm consists of a 10 ha agricultural experimentation station and pilot farm of 300 ha. The farm was set up in 1965 with assistance from the UNDP and FAO.

Dam development from 1970 to 1978

Nam Ngum Dam Project

Among the tributary dams begun in the four riparian countries after the publication of the 1970 plan report, the most famous is the Nam Ngum Dam.

On 2 December 1971, the day of completion of the first phase of construction, the start button on the power station's generating panel was pressed by the kings of Laos and Thailand. This was followed by speeches by the prime ministers of the two countries and the Secretary-General of ECAFE, Mr U. Nyun. ECAFE and the Mekong Committee had played a leading role behind the scenes in promoting the construction of the hydroelectricity project and the transmission lines which crossed the Mekong between Laos and Thailand.

From its source on the Tran Ning Plateau at the Viet Nam–Laos border, the Nam Ngum River flows down through the Vientiane Plain to the east of Vientiane, where it joins the Mekong. The Nam Ngum is one of the Mekong's largest tributaries with a length of 420 km and total basin area of 16,640 km^2. The Nam Ngum 1 Dam was built immediately upstream from the confluence of the Nam Ngum and Nam Lik rivers. The basin area upstream from the dam is 8,280 km^2 with an average annual discharge of 328 m^3/s.

In January 1959, the Japanese government's Mekong Tributary Reconnaissance Team surveyed the dam site. In fact, engineers from Nippon Koei had already been conducting a survey focusing on the Nam Lik River since 1955. When Japan and Laos signed an agreement for economic and technical cooperation in May 1959, Nippon Koei immediately embarked on a further study of the Nam Ngum Basin. In 1960, Nippon Koei also conducted a development study of the lower reaches of the Nam Ngum River on the Vientiane Plain, using a special fund from the United Nations and, in 1962, drew up a comprehensive Nam Ngum development plan.

In 1964, the World Bank took over management of the funds for the construction of the dam hydroelectric power station. It was decided that 11 countries (the United States, Japan, Australia, Canada, Denmark, France, West Germany, India, Switzerland, Holland, and New Zealand) would provide foreign currency for construction, while Laos and Thailand would provide appropriate amounts of local currency.

In 1966, the World Bank chose the Acres Company from Canada and Nippon Koei as consultants for the project. Construction began in 1967 and, in December 1971, the first phase (30,000 kW) was completed. The developmental objectives were power generation, and irrigation and flood control on the Vientiane Plain.

The Nam Ngum 1 Dam is 75 m high with a crest length of 468 m. It is a concrete gravity-type dam with an overflow chute-type spillway. The reservoir has a full water elevation of 212 m, a reservoir area of 370 km^2, a total capacity of 7×10^9 m^3, and an effective capacity of 4.7×10^9 m^3. The head is 35 m, the average discharge 310 m^3/s, and the total installed capacity 150,000 kW. In 1992, almost 20 per cent of 882 GWh, the annual generated energy, was consumed by the city of Vientiane and the rest was transmitted to the transmission grid of the Electricity Generating Authority of Thailand (EGAT) via 115 kV lines passing across the Mekong to Thailand. Therefore, although the Nam Ngum 1 Dam was located in Laos, the transmission of energy to Thailand made it the first international project in the lower Mekong Basin.

For the second phase of construction, from 1975 to 1979, the ADB took over the management of funds, which amounted to US$47 $\times$ 10^6. In this phase, the generating facilities were enhanced to increase the total capacity to 110,000 kW by adding four gates and one penstock, bringing the number of penstocks to three, and the water-wheel generator was strengthened. In the third phase, from 1982 to 1985, with funding of US$2 $\times$ 10^6 from the International Development Association (IDA) and the Organization of Petroleum Exporting Countries (OPEC), another turbine was installed and five 150,000 kW generators were set up. Most of the electricity generated is still consumed in Thailand, providing a very important source of revenue for the Laotian government.

In 1970 – the year the Indicative Basin Plan report was completed – the Secretariat of the Mekong Committee discovered sites for two more promising dams upstream from the Nam Ngum Dam, which they named the Nam Ngum 2 and Nam Ngum 3 dam projects. The development of these two dams was later scrutinized in the 1988 development plan.

Other dams completed

In addition to the above-mentioned projects, the main projects in the four riparian countries among the those listed in the short-range development plan were as follows.

Laos. In Laos, the 1,250 kW Nam Dong hydroelectric power station was completed in 1972 on the outskirts of Luang Prabang. This project was the enhancement of an existing micro-hydroelectricity station. In addition, slight improvements were made to the micro-hydroelectricity station on the Boloveng Plateau and small-scale irrigation facilities on the Vientiane Plain were completed.

Thailand. In north-east Thailand, in addition to the above-mentioned Lam Don Noi Dam in 1971, two other dams were finished: the Nam Phrom Dam, completed in 1973, provided 40,000 kW; the Nam Oon Dam, completed in 1974 with aid from the United States, provided 32,500 kW.

In addition to the planned projects listed on pages 154–155, two irrigation dams on the Hoei Luang and Hoei Mong rivers were completed. However, the 100,000 kW Pak Mun hydroelectric power project planned in the lower reaches of the Nam Mun River had to be shelved for the time being, owing to the problems of submergence and resettlement.

Cambodia. In Cambodia, all construction, including the Prek Thnot Dam, was suspended in the first half of 1970.

Viet Nam. In Viet Nam, too, because of the war and shortages of funds, the planned construction for water-control projects, such as the Cai San (initial plan: 60,000 ha), Tiep Nhut (70,000 ha) and Go Cong (5,000 ha) projects, together with hydroelectricity and irrigation projects in central Viet Nam, such as the upstream projects on the Se San and Srepok rivers, were also suspended. However, the 16 ha Ehakmat pilot farm on the central plain was completed in 1972 and became operational the following year.

Although development could still proceed to some extent in north-east Thailand and Laos, this suspension of development in Cambodia and Viet Nam ended all hope of implementing the short-range plan as a whole for the 7-year period from 1971 to 1978.

During this period, the Secretariat of the Mekong Committee continued its efforts to prepare the ground for the implementation of the 20-year long-range plan from 1980 by conducting further research on plans for large international dams on the main stream and tributaries, and the development of the Mekong Delta.

Dark clouds over the Lower Mekong Basin Development Plan

As we have seen, a certain amount of progress was made in the early 1970s, including the completion of small- and medium-scale hydroelectric power-station projects such as the Nam Ngum 1 Dam, as well as the development of pilot farms. Nevertheless, the situation both inside and outside the lower Mekong Basin was rapidly deteriorating. For a start, the Secretariat of the Mekong Committee was losing some of its key personnel. After the completion of the 1970 plan, the main members who had made vital contributions since the establishment of the Mekong Committee in 1957 left the Secretariat one after another as they reached the mandatory retirement age. They included Dr Hart Schaaf, the Executive Agent who had been responsible for the overall direction of the Secretariat's activities, and Mr Kanwar Sain, the Director of the Engineering Division who played a vital role on the Secretariat's Advisory Committee. Dr Bunrod Binson, the Thai government representative who had so enthusiastically supported the Mekong Committee on behalf of the four riparian countries, stepped down at around the same time. Mr P. T. Tan, the Chief of the ECAFE Water Resources Development Division, who had directly led the Secretariat on technical matters, also retired and moved to Taiwan.

The same period saw dramatic changes in the political situation of the lower Mekong Basin. In March 1975, the pro-American Cambodian government collapsed and Saigon fell to the Vietnamese People's Army the following month. In December of the same year, the coalition government in Laos resigned and the Cambodian conflict intensified. Although the projects in north-east Thailand were not much affected by these events, it became impossible to continue development in Laos, Viet Nam, and Cambodia.

The continuation of transmission of energy to Thailand from the Nam Ngum hydroelectric power station provided a single ray of hope. In the year from 1 July 1976, a total of 101 GWh of electricity was transmitted to Thailand, for which Laos received US$1.8 × 10^6. After the output of the power station was increased from 30,000 to 110,000 kW, a total of 526 GWh was transmitted to Thailand in the year starting 1 July 1978, bringing Laos US$5.5 × 10^6.

However, funding for the Mekong Committee from the UNDP, which had amounted to US$5.6 × 10^6 in 1973, dried up almost completely during the next 4 years. (One year after the Interim Mekong Committee was established in 1978, funding from the UNDP – US$5.7 × 10^6 – finally returned to its original level.) The Mekong Committee was saved from collapse during this period by a generous pledge for a huge amount of funds (US$6.2 × 10^6 for studies and US$9.5 × 10^6 for construction) from Holland, which had supplied Dr Schaaf's successor as Executive Agent. However, Cambodia did not send any representatives to the meetings of the Secretariat after 1975 and, although the remaining members continued their work in Bangkok as before, the absence of a Cambodian representative meant that the Mekong Committee had, in effect, ceased to exist. This unfortunate situation continued until the establishment of the Interim Mekong Committee in 1978.

Meanwhile, the emergence of a quite different kind of problem dealt a further blow to the plans for development of the lower Mekong Basin: this was the increasing awareness of environmental problems and concern about human rights issues. The awakening in advanced countries at the end of the 1960s to the adverse social and environmental impact of development led to the reconsideration of, and opposition to, large-scale development plans. Among the riparian countries, criticism was particularly strong in Thailand, where the government had single-mindedly concentrated on development while achieving remarkable economic growth.

From the beginning of the 1970s, increasing fears were voiced in the other riparian countries concerning the environmental and socio-economic influence of development. In the heyday of the Mekong Committee up to the end of the 1960s, even when environmental studies were conducted, their conclusions tended to be ingratiating or optimistic, and close examination shows that some of them were even quite inaccurate. The studies made in the 1970s, however, were not only more meticulous but also included negative or critical assessments of certain aspects of development. Among the many environmental problems arising through this kind of dam development, particularly frequent mention was made of land submergence and the need to relocate local residents. The complaints whispered in the 1960s about the harsh treatment and inadequate compensation to the people displaced by the Nam Pong Dam in northeast Thailand and the Nam Ngum Dam in Laos grew to a crescendo in the Thai mass media, leading to criticism of the government. Against this background, large-scale dam development was becoming increasingly difficult.

This increasing concern about the influence of dam development on the natural environment and the question of submergence and resettlement was clearly reflected in the contents of reports published by the Mekong Committee. Most of the reports published by the Mekong Committee in the 1960s were explanations of development projects or surveys of the effectiveness or economic viability of development. From the end of the 1960s, however, research reports on the social, economic, and institutional aspects of development began to appear. These included *Socio-economic Research on Agricultural Development in the Lower Mekong Basin* (1966), *Social Feasibility of the Pa Mong Irrigation Plan* (1968), and *Importance of the Human Factor in Lower Mekong Basin Development* (1968). Environmental reports focusing on how development caused water pollution, or its impact on the health of residents, also began to appear, such as *The Influence of Lower Mekong Basin Water Resources Development on Pollution, Public Hygiene and Nutrition* (1969).

In the 1970s, research reports exploring the actual problems accompanying water-resources development started to appear in addition to the above-mentioned reports. These included research on local development such as *Local Development Planning and Policy in the Mekong Basin* (1974), studies on agricultural product processing like *The Agricultural Product Processing Industry in Lower Basin Countries – a Macro-Economic Approach* (1975), and research on human resources development such as *Human Resources Development to be Promoted together with Water Resources Development* (1973). In addition, theses appeared on the application of new technologies that were attracting attention, including remote sensing using satellites and computer processing of measurement data.

Together with these research reports, there was a marked increase in studies on environmental issues and the problem of submergence and resettlement. The large number of these reports makes it impossible to list them all, so the following are selected examples. The year 1973 saw the publication of reports such as *An Ecological Study of the Nam Ngum Reservoir*; *Study on Ancient Remains in the Vicinity of the Planned Pa Mong Reservoir*, and *Study on Fish in the Lower Basin*. Reports published in 1974 and 1975 included *Insect Life around the Nam Pong and Nam Phrom Dams*; *Malaria in the Lower Mekong Basin*; *Schistosomiasis on Khone Island, Drinking Water in Local Society*; *Submergence and Relocation Resulting from the Nam Ngum Dam and other Dams*; *Comparative Study to Determine the Most Appropriate Size of the Pa Mong Dam*; and *Possibilities of River Bank Erosion on the Fringes of the (Planned) Pa Mong Reservoir and Downstream from the Dam*. In addition, not only technical studies related to satellite technology but also many non-technical theses on socio-economic issues and human-resource development were published. In 1976, studies on the ecology and environmental management of the Nam Pong Reservoir were begun (completed in 1979), and Michigan University published the results of a full-scale survey of submergence and resettlement resulting from the construction of the Nam Pong Dam. From 1977 to 1978, reports were published on studies of water quality on the main stream and tributaries of the Mekong in Thailand and Laos, the overall environmental impact of Mekong river development, wildlife in the lower basin, the concept of turning land into national parks, and water quality in the Nam Ngum Reservoir.

Establishment of the three-country Interim Mekong Committee (1978)

In April 1977, agreement was at last reached among the three riparian countries of Thailand, Laos, and Viet Nam for the establishment of an interim Mekong committee. Accordingly, the Interim Mekong Committee, which excluded the government of Cambodia, was set up in January 1978.

Without the cooperation of Cambodia, however, it was impossible to implement plans for development of the lower basin as a whole, or even to collect basic data. Having been forced into existence by an extraordinary situation, the Interim Mekong Committee could do little more than assist in increasing food production and the energy supply in the three member countries and make a certain contribution to improvements to water-resources infrastructure in the areas of flood control and navigation.

The fragile edifice of the blueprint for development on a grand scale laid down in the Indicative Basin Plan (1970) had completely crumbled, and all that could be hoped for now was the accomplishment of small but valuable and realistic tasks.

After the establishment of the Interim Mekong Committee in 1978, the situation in the lower basin, with the exception of Thailand, was hopeless. Whereas Thailand actually benefited from the Viet Nam War and Cambodian conflict to achieve even faster development than expected, Laos was in no position to make a new start. At the beginning of the 1980s, the situation in Laos finally improved and rice self-sufficiency became possible, but the state of the economy and standard of living remained at an extremely low level.

When Viet Nam invaded Cambodia at the end of 1978, the Western nations cut off its financial support. In addition to the continuing civil war, the Cambodians now had to fight the Vietnamese, resulting in further disruption to the country and the lives of its citizens. (In July 1982, Prince Sihanouk formed the Coalition Government of Democratic Kampuchea, but Vietnamese troops remained in Cambodia until 1989.) Pol Pot's party tried to improve irrigation in Cambodia, using forced labour that resulted in the deaths of large numbers of citizens. This irrational and slipshod "development plan" actually made the situation worse and was finally abandoned.

Water-resources development in the lower basin from 1978 to 1987

In this situation, the initial activities of the Interim Mekong Committee could be only local and limited. It concentrated mainly on strengthening of the Nam Ngum Dam in Laos, continuing relatively small-scale development projects in the three riparian countries apart from Cambodia, improving existing facilities, and conducting new studies. Nevertheless,

these efforts were starting to bear fruit in each region by the beginning of
the 1980s.

Hydroelectric power and irrigation

Laos

The dark side of the success of the Nam Ngum Dam was the tragic displacement of people who had been living in the area. However, various
efforts were made to improve the situation, such as the construction of
the village of Pak Chen (600 houses) with Dutch aid. With the help of
funding from the Japanese government and the efforts of the companies
involved, the installed capacity of the hydroelectric power station was
expanded to 110,000 kW in 1978. By the mid-1980s, this was further
increased by 40,000 kW for a total output of 150,000 kW. Almost 90 per
cent of the energy generated was transmitted to Thailand, making a great
contribution to the increase of Laos's national revenue. At the beginning
of 1987, a study was commenced for the transmission of electricity from
Nam Ngum to Luang Prabang; all these activities were financed entirely
by Japan. In the same year, the study on the development plan for the Nam
Theun 2 Dam, which was to have several times the energy-generation
capacity of the Nam Ngum Dam, was begun in earnest.

In southern Laos, a study funded by Norway was commenced in the
1980s for the generation of 45,000 kW of hydroelectricity using the head
available at the Se Set Falls in the middle reaches of the Se Set River,
which flows into the Se Dong River – a tributary of the Mekong. Efforts
were also made to reduce flood damage from the Se Bang Hieng River.
However, the main concern of the government was, naturally, the development of the Vientiane Plain in central Laos. In 1979, an irrigation
project was started to supplement the water supply in paddy fields in the
rainy season and provide water required for irrigation in the dry season,
making use of energy from the Nam Ngum Project and water from the
main stream of the Mekong and lower reaches of the Nam Ngum River.

Among the projects to irrigate the Vientiane Plain, the most successful
was the irrigation of the Cao Rieo region backed by Dutch aid. This
project supplied water to 3,000 ha of paddy fields during the dry season
and diversified the region's agricultural products through the planting of
cash crops. About 70 km north of Vientiane, irrigation was also provided
for 1,500 ha of existing paddy fields near the upper reaches of the Nam
Ngum River, resulting in a considerable increase in the rice harvest. In
addition, a project funded by the European Economic Community (EEC,
now the EC) was started in 1980 to protect the wetlands in the vicinity of
Vientiane from flooding and to drain them. As a result of these efforts,
the total area of irrigated land in Laos was increased from 18,000 ha (2.2

per cent of arable land) in 1970 to 118,000 ha (17 per cent of arable land) in 1982, enabling Laos to achieve self-sufficiency in rice.

As the result mainly of these steady efforts to extend electricity and irrigation in Laos, the country's per capita GDP improved from less than US$100 in the 1960s to US$140 in the 1980s.

Thailand

While its neighbouring countries struggled, Thailand continued to enjoy rapid economic growth during this period. By the mid-1980s it was generally acknowledged to be a semi-advanced nation: in 1985 the per capita GNP was US$735 on a nationwide basis and had even reached US$440 in the relatively backward north-east.

By 1986, the total installed capacity of electric power facilities in Thailand was 2×10^6 kW, but hydroelectricity in the north-east accounted for only a small part of this. On the other hand, irrigation in north-east Thailand had increased significantly: whereas only 1.9×10^6 ha in the whole of Thailand were irrigated in 1970, this area had risen to 3.5×10^6 ha 15 years later, in 1984. Of this irrigated area, as many as 500,000 ha were in north-east Thailand, where irrigation had been particularly poor. During the period 1970–1984, whereas the rate of increase of irrigated land was 5 per cent for the whole of Thailand, it was 9 per cent in north-east Thailand, where 68 per cent of cultivated land was yielding rice by 1984.

Continuing with this rapid progress, 8,700 ha of lowlands in the Hoei Mong River Basin near Nong Khai in north-east Thailand were reclaimed by drainage and pumped irrigation facilities set up in 1985. The following year, a project was completed for the cultivation of high-yielding rice by pumped irrigation of 10,000 ha of paddy fields on the banks of the Mun and Chi rivers. The former project was funded by the EEC and the latter by Holland.

Viet Nam

The Viet Nam War was brought to an end by the dramatic invasion and occupation of Saigon by the North Vietnamese army in April 1975. The economy, however, showed no sign of recovery until the mid-1980s. In 1985, the per capita GDP was still only US$180, the average figure for the Vietnamese delta being somewhat higher at US$240.

The development of the delta gathered pace from the mid-1980s, when Holland provided funds to formulate a framework for the delta-development programme. As a result, the total area of irrigation projects included in the programme exceeded 350,000 ha. The planned projects were Tan An and Go Cong (irrigation and drainage; 81,000 ha), Kien Hoa (salinity control, irrigation; 58,000 ha), Huong My (salinity control, irrigation; 40,000 ha), Thau Lon (salinity control, irrigation; 83,000 ha), Tiep Nhut (salinity control, irrigation; 42,000 ha), Cai San (flood control, irrigation; 49,000 ha) and Tam Phong (salinity control, high-yielding rice irrigation to increase rice harvest).

In the lower Mekong Basin region outside the delta, small-scale hydro-electric power development was continued. For example, a study was commenced for the extension of the generating facilities (from 480 kW to 24,000 kW) at the Drayling hydroelectric power station on the central plateau, and surveys to re-examine the Yali Falls Project and to study irrigation projects in the Darlac Plateau on the Upper Srepok River were carried out.

List of lower basin hydropower and irrigation projects

In 1984, the Secretariat of the Interim Mekong Committee published a list of all the development projects on the major tributaries of the lower Mekong Basin, including Cambodia. Based on the latest hydrological data and maps, this list was essentially a reassessment and extension of the results of studies on the development potential of lower Mekong tributaries in the 1970 plan (including candidate locations for development in Cambodia outside the Mekong Basin).

Navigation

During this period (1978–1987), continuous efforts were made to improve navigation in Laos and the Vietnamese region of the Mekong Delta.

Laos

In Laos, improvements were made to water channels to secure the safety of passage of ferries on the routes between Vientiane and Nong Khai and between Savannakhet and Mukdahan. Small dredgers were donated by Australia for this purpose.

At a notoriously difficult section of the Mekong to navigate, 24 km north of Savannakhet at Keng Kabao, the increase of cargo being transported to Thailand from the Vietnamese port of Da Lang across the Annanese Mountains via Laos made it necessary to dredge the river bottom and set up a transit port where boats could unload and reload their cargo. A small river-port was completed with Dutch aid in 1984. Backed by funding from Japan, Pacific Consultants (PCI) made improvements to the port of Laksi near Vientiane, and Holland provided cargo-handling facilities. As a result of these improvements, the amount of cargo handled by the port increased from 15,000 to 50,000 t/year. Australia also contributed to the improvement of two river-ports further downstream – Tha Dua and Pata Khone.

For about four years up to 1978, the Dutch assisted in the setting-up of concrete navigation aids (markers) at key points on the main stream from Vientiane to Luang Prabang. The Interim Mekong Committee attempted to standardize the types of boats suitable for navigating the Mekong. According to these standards, two 100 t barges pulled by a 400-horsepower tugboat was considered the most appropriate combination in Laos. In 1980, the shipbuilding training centre set up by Britain in Nong Khai on the bank opposite Tha Dua in Laos was refurbished with West German aid and many technicians received training there in the 1980s. All of these achievements can be attributed to the efforts of the Interim Mekong Committee.

In connection with the improvement of navigation, the Secretariat of the Interim Mekong Committee also promoted the prevention of erosion of the banks of the main stream and tributaries of the Mekong. River-bank protection works were promoted along the banks of the main stream where it flows through major cities such as Vientiane, Thakhek, and Savannakhet and on tributaries such as the Nam Ngum and Nam Ou.

Viet Nam

From 1973 onwards, the amount of cargo carried by boats in the Vietnamese region of the Mekong Delta increased by about 20 per cent every year. In response to this growth, the Interim Mekong Committee formulated a plan for the improvement of navigation in the delta. The Committee provided assistance for the implementation of the plan, which included the establishment of an inland navigation school, setting-up of navigation markers, measurement of water channels, river-bank protection, and erosion prevention. The Secretariat also assisted with a study to determine the situation of cargo transportation by boat in the delta. The report revealed that nearly 1.5×10^6 t of cargo were handled by the port at Ho Chi Minh City in one year. In response, the Vietnamese government determined measures for the construction of a new port to alleviate the situation.

Few surveys had been made on water channels of the Mekong since the 1960s, but in 1987 the Interim Mekong Committee began preparations for drawing up a chart of water channels by conducting a detailed survey of the routes in the Mekong Delta, as well as from Savannakhet to the border with Myanmar. This was one of the extensive projects covering the whole of the lower basin, described in the next section.

Extensive surveys and growing environmental concern in the 1980s

The projects outlined above all brought immediate benefits to their respective regions and contributed to the development of the riparian countries of the lower basin. However, the Interim Mekong Committee had not forgotten its original purpose of conducting studies on development of the main stream in the lower Mekong Basin and other extensive surveys. Not only the countries providing support for the Secretariat but also the riparian countries understood that development surveys would be meaningless unless they were conducted on this scale. Once things had settled down after the establishment of the Interim Committee, mainstream development plans were reviewed and a substantial amount of basic research, including environmental studies, was conducted for the development of the lower basin as a whole.

The extensive surveys that constituted the central mission of the Interim Committee included *Summary on Engineering and Economic Studies for Development of the Pa Mong Dam* (1979) and *Organization and Financing for Project Implementation for Pa Mong Dam Development* (1982) conducted by Australia; *Study on Mainstream Development and Related Problems* (1983) and *Establishment of a Lower Basin Information System* (1984) by the Secretariat; and *Survey on Water Balance in the Lower Basin* (1982–1989) backed by Britain. On the other hand, there was a steady increase in interest in environmental issues and the problem of submergence. In December 1982, an international conference on environmental management was held in Stockholm. The Secretariat of the Interim Mekong Committee sent a representative to the conference, which gave strong impetus to its interest in environmental problems.

The following is a list of the main studies conducted by the Committee on environment-related themes and submergence and resettlement in the 1980s:

- *Submergence and Resettlement related to the Nam Ngum Reservoir* (1982);
- *Environmental Management of the Nam Pong Reservoir* (1982);
- *Sulphate Soil Management in the Vietnamese Delta* (1982);
- *Fishing Industry Management in the Mekong Basin* (1983);
- *Migratory Fish of the Mekong Basin in Thailand and Laos* (1983);
- *Properties and Management of Saline Soil in North-east Thailand* (1983);
- *Environmental Impact at the Pa Mong Dam and Downstream* (1984);
- *Waterborne Diseases in Laos and Viet Nam* (1984);
- *Environmental Impact of Developing Water and Land Resources in the Mekong Delta* (1985);
- *Quality and Problems of Ground Water in Thailand* (1985);
- *Management of Tha Gong Fishery* (1986);
- *Riverbank Erosion, Sedimentation, and Flash Flood Damage* (1986);
- *Water Quality Management in the Lower Basin* (1987).

The author also recalls another noteworthy study conducted during this period. Regrettably, the year of publication of the report cannot be ascertained, but this was a very interesting British-funded 1-year study on the influence of agricultural development and forest destruction on the river flow volume in the lower Mekong Basin. Unfortunately, because of the limited funds, the period of the study was short and no clear conclusions were reached; however, it did make the significant point that no changes in the low-water discharge of the main stream of the Mekong had been observed in the last 20–30 years. The author found it extremely interesting that this type of report favourable to the position of the developers, which had been quite common in the 1960s, was published in the 1980s. This is certainly a theme that should be studied in greater depth.

Towards the revision of the Indicative Basin Plan (1970)

In October 1980, only 2 years after its establishment, the Interim Mekong Committee decided to revise the Indicative Basin Plan (1970). In the annual report of that year, the Committee outlined the following reasons for this decision.

First and foremost, the long-range development plan that had been the most important part of the 1970 plan had completely broken down and the programme aimed at completing a dam on the main stream of the Mekong by the beginning of 1980 had not even begun. Secondly, because of the considerable economic and political changes, not only in the riparian countries but throughout the world, the forecasts of demand for energy and food in and outside the region were no longer applicable. Thirdly, as mentioned above, various studies had been conducted and a large amount of data accumulated on the region as a whole in the period between the announcement of the plan and the formation of the Interim Mekong Committee. At a time of heightened awareness of the environment and human rights, attention was drawn to the problems of submergence and resettlement and the need to conduct research on environmental management. Furthermore, with the rapid progress of computer technology in the 1970s, it became possible to use computers to process and analyse the data collected through this research: for example, hydrological data could be fed into a computer to simulate the operation of a series of projected reservoirs on the main stream. This considerably facilitated comparative studies of dam sizes and reservoir operation, and the Secretariat performed various calculations of this sort from the beginning of the 1970s.

At the end of 1968, when the Indicative Basin Plan (1970) was in its final stages of preparation, several young technicians returned to the Mekong Committee after being seconded through the Committee to the

United States by the governments of Thailand and Laos for a short period of training in computer technology at the Columbia River Basin Development Headquarters. One of these trainees served as a representative on the Secretariat until August 1995. However, it took time to apply the technologies acquired to the Mekong data and the analysis was not completed in time to be included in the plan report. In the end, the figures for the operation of dams for various purposes on the main stream (excluding the Pa Mong and Sambor dams) were based on primitive manual calculations by the authors. It was, therefore, natural that the Secretariat should apply computer calculations in order to reassess the effects of dam development on both the main stream and tributaries in the 1970s.

For the above reasons, and in view of the political developments in the region since 1970, the Interim Mekong Committee decided to use its new data and technology to reformulate a plan to meet demand for food and energy in the three riparian countries by the year 2000. However, in spite of the decision to revise the plan, the situation both in and outside the lower Mekong Basin from 1980 onwards made it extremely difficult to procure funds for this reformulation, and very little progress was made.

Upon the expiration of the term of the Chief of the Secretariat of the Interim Mekong Committee in the fall of 1983, an Egyptian, Mr Galal Magdi, was appointed to the post on the recommendation of the UNDP in accordance with the normal practice at that time. Up to the fundamental reorganization of the Mekong Committee in 1995, no other Asians or Africans were chosen for this post either before or after Mr Magdi's appointment. All the others were Caucasian and, with the exception of the first chief, Dr Schaaf from the United States, none was particularly positive or enthusiastic. But Galal Magdi was different: encouraged by the author, he decided to commission Japan to revise the 1970 plan. In the spring of 1984, with the assistance of the author, he came to Japan and made approaches to the government, bureaucrats, and the financial world. Although he did not get the response he had hoped for, we continued to cooperate until the following fall. The author drew up a rough plan estimating that revision would cost at least US$1 $\times$ 10^6 and would take a year to 18 months. With Mr Magdi's approval, this plan was submitted to the Ministry of Foreign Affairs and other related organizations in Japan, requesting assistance from them. However, in the end it proved impossible to create interest in funding the Interim Mekong Committee, even though Japan was in the process of providing quite a large amount of capital for official development assistance (ODA). (The UNDP stated that it would be able to make up to about US$150,000 available, but this was nowhere near enough.)

The Japanese government and private sector could not summon up any enthusiasm for revising the plan for the lower Mekong Basin (where the three main countries had been taken over by communist regimes after the withdrawal of the United States from the region), and showed no interest in the various proposals submitted up to the end of Mr Magdi's term as chief of the Secretariat.

In 1987, the UNDP eventually earmarked US$600,000 for a revision programme in the Secretariat's budget for the following year. Accordingly, the Netherlands Engineering Consultants (Nedeco), Electrowatt Engineering Services of Switzerland, and Asian Engineering Consultants (AEC) of Thailand formed a consortium which took a year to complete a revised development plan (*Perspectives for Mekong Development*, Mekong Secretariat 1988). After being approved at a special meeting at Bangkok in April 1988, the plan received official approval at the 27th session held at Vientiane in June of the same year.

The 1988 development plan

Aims

The aims of the 1988 development plan drawn up by the Interim Committee Secretariat differed from those of the 1970 plan. In the new plan, the Secretariat indicated guidelines on the development of international and domestic water-resources projects for the governments of the three riparian countries, excluding Cambodia. Made up of specific proposals for construction programmes in the three countries, the report was no more than a collection of piecemeal objectives adapted to the current situation, and certainly could not be called a comprehensive development plan.

On the basis of these limited aims, the first task the Secretariat assigned to the consortium of consultants was to draw up a list of the development plans of the governments of Thailand, Laos, and Viet Nam. Of the 62 water-resources development projects listed, 14 had already been started. Applying the same criteria to all the projects, they selected 30 priority development projects and formulated implementation schedules up to the year 2000. Most of these 30 projects were national in scope, but some required international cooperation. The report stated that the following should be investigated in connection with the implementation of the plan:

1. Investment plans necessary for implementation;
2. Long-term perspective of projects particularly likely to affect the international situation;
3. Reform of the organization of the Interim Mekong Committee in order to carry out development;
4. Related studies and research.

Outline of the plan

The 1988 development plan, drawn up through the procedure described above, was divided into two parts – the main text of 160 pages and a 28-page outline. Entitled *Perspectives for Mekong Development – Revised Indicative Plan for the Development of Land, Water, and Related Resources in the Lower Mekong Basin*, it was distributed by the Secretariat to the countries involved in April 1988.

The contents of the report were as follows:

Preface.
Part I: Setting the scene:
1. Background;
2. Physical features;
3. National economic and social features;
4. Sectoral potentials and demand.
Part II: Development of the basin's international potential:
1. Long-term perspective;
2. Priority international projects for the shorter term.
Part III: The basin's national potential:
1. Power development in the national context;
2. Irrigation projects.
Part IV: The programme of action:
1. Studies and investigations;
2. Institutional strengthening;
3. Investment plan and funding needs.
Appendices.
1. References;
2. Approach to cost benefit estimation.

Part I of the report describes how the plan came to be drawn up, outlining the historical development of the lower basin; the system of organization; physical features of the region; socio-economic features of the three riparian countries; and the demand and potential of each development sector such as agriculture, irrigation, and power.

Part II of the report deals with water-resources development projects to be promoted in more than one country, from which the benefits and profits should be shared among the countries involved. Section 1 of part II examines the long-term perspectives of a cascade of major dams proposed on the main stream of the Mekong. Table 3.3 and figure 3.9 show the salient features of mainstream development projects in eight locations from Luang Prabang to Sambor. Section 2 focuses on three projects with international scope: the Low Pa Mong Dam, which was to have a lower elevation than the Pa Mong Dam originally planned on the mainstream; the Nam Theun 2 Dam on the major Mekong tributary of the Nam Theun River in Laos; and the Nam Ngum 2 Dam on another major tributary, the

Table 3.3 Main characteristics of the Mekong cascade

Name of project	FSL[a] (m)	LWL[a] (m)	Net storage volume (10^6 m^3)	Head (m)	Installed capacity (MW)
High Luang Prabang	355	335	10,700	74	2,750
Sayaboury	275	270	100	18	900
Upper Chiang Khan[b]	250	230	5,150	39	1,500
Low Pa Mong[b]	210	192.5	7,310	41	2,250
Upper Thakhek	155	150	3,650	13	1,100
Ban Koum	130	128	1,400	34	3,300
Stung Treng	88	81	46,500	44	7,200
Sambor	40	38	2,050	27	3,250

Source: I. Kyama: *Mekong Delta – Present Situation and Its Development*, JICA, March, 1978
a. FSL, full surface water level; LWL, low-water line.
b. Pa Mong 250, as proposed in 1970 in place of these schemes, had LWL at 220 m, $48,340 \times 10^6$ m^3 net storage volume, 77 m head, and 4,800 MW installed capacity.

Nam Ngum River. (These projects are discussed in more detail below, on pages 176–181.) The report proposed that the development of these three projects be initiated within a short period (1988–2000).

Part III of the report deals with hydropower and irrigation projects that can be handled independently by each of the three riparian countries (see pp. 181–184).

Section 1 of part IV outlines the studies and investigations that have to be conducted prior to investment in the above-mentioned projects (see p. 184). Section 2 deals with the institutional strengthening of the Mekong Secretariat and the necessary costs, while section 3 outlines the investment plan for development up to the year 2000. In its outline of the investment plan, the report points out that, during the period from the establishment of the Mekong Committee in 1957 to 1986, a total of US$$56.34 \times 10^6$ in funding had been procured from each donor country and the various international organizations involved. Part of these funds had been used for pre-investment studies and the rest had financed actual construction.

Development proposals for the three international projects

Low Pa Mong Dam Project

One of the features of the 1988 plan was the great efforts made to reduce the number of people expected to be displaced by the submergence of land resulting from the construction of the Pa Mong Dam. This made it

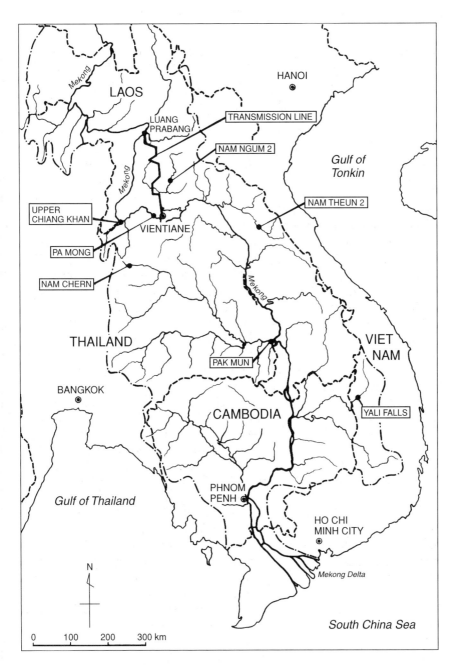

Figure 3.9. Power and multi-purpose projects in the lower Mekong Basin
Source: Mekong Secretariat 1988

necessary to reduce drastically the full water elevation of the reservoir, which had been projected at 250 m in the 1970 plan, following the proposal made by the Japanese reconnaissance team in 1961. The report proposed that, after the Low Pa Mong Dam (with an elevation of 210 m) had been built, the Chiang Khan Dam (with an elevation of 250 m) should be constructed immediately upstream.

On the basis of the potential construction of these two dams, a schedule was drawn up for eight dams on the main stream through the whole of the lower Mekong Basin. Upstream from the Low Pa Mong and Chiang Khan dams would be the Upper Luang Prabang and Sayaboury dams, while four dams – the Thakhek, Ban Koum, Stung Treng, and Sambor dams – would be developed downstream.

The flood-control effect downstream from the Low Pa Mong Dam around the river-banks in Thailand and Laos and in the Mekong Delta would clearly be lower than the 250 m High Pa Mong Dam. However, the report stated that, even if all the dams proposed in the 1970 plan could be realized, flooding in the delta could not be completely controlled, so the difference in the effects of the two proposals was only a matter of degree. It should have been possible to show this difference in the report, but this was not done. The report completely ignored the fact that the concept of development of the lower Mekong Basin had originated from the aim of controlling flooding, stressing only the advantage that a dam on the main stream would make it possible to meet the demand for water in the downstream delta in the dry season.

The report concluded that, even if construction of the Low Pa Mong Dam could be started immediately, it could not commence operation until around 2000. After that, the next stage would be the building of the Chiang Khan Dam directly upstream. This would be followed by either the Sambor Dam or the High Luang Prabang dam.

On the basis of a comparative study (1987), it was concluded that another international project, the Nam Theun 2 Dam, was more promising than the Nam Ngum 2 Dam Project, and a feasibility study was planned. One of the reasons why the Nam Theun 2 Dam Project was considered the better option was that it would directly increase the downstream low-water discharge on the main stream of the Mekong.

Let us now examine these proposals for the Nam Theun 2 Dam and Nam Ngum 2 Dam projects in a little more detail (table 3.4).

Nam Theun 2 Dam Project

The Nam Theun 2 Dam Project has already been described in the sections concerning the report of the Japanese reconnaissance team and the 1970 plan. However, the development concept in the 1988 plan differed somewhat from these.

Table 3.4 International projects: salient technical features

Feature	Units	Pa Mong 210	Nam Theun 2	Nam Ngum 2
Average net head	m	44	352	134
Mean annual inflow	10^6 m^3	135,600	6,500	6,938
Reservoir net storage	10^6 m^3	7,310	4,000	3,600
Ratio storage/ inflow		0.054	0.616	0.490
Installed capacity	MW	2,250	600	400
Energy output				
Baseline	GWh	4,250	4,590	1,580
Secondary	GWh	6,410	60	600
Total	GWh	10,660	4,650	2,180
Resettlement	No. of people	40,000	2,000	1,000
Project costs				
Construction	US$10^6	1,866	605	520
Resettlement	US$10^6	200	10	5
Total	US$10^6	2,066	615	525
Further investigations, study and design	US$10^6	7.9	6.5	3.6
Duration up to tendering	Months	47	44	28
Construction period	Years	11	5	5

Source: Mekong Secretariat (1988).

The Nam Theun River drains an area of 14,650 km^2 before it reaches the Mekong and has an average annual discharge of 900 m^3/s. Because of the great difference in altitude between the Nam Theun River and adjoining rivers to the south, it had long been viewed as a promising river for hydroelectric power development. The dam-development site (catchment area: 3950 km^2) is located on a plateau at an altitude of over 500 m, a little further upstream than originally proposed in the 1970 plan. The main dam would be an earthfill dam with a height of 40 m, a crest length of 330 m, and a total capacity of 2.1 × 10^6 m^3. The reservoir would have an elevation of 530 m at full supply level, would provide a total gross storage of 6.7 × 10^9 m^3 and an effective storage of 4.0 × 10^9 m^3, and would cover an area of 530 km^2 when full. The 1970 plan had proposed that the water from this reservoir should be channelled into the neighbouring Se Bang Fai River, but the 1988 plan recommended that the outflow should be into the Nam Hin Boun River, which joins the Mekong 120 km downstream from the Nam Theun–Mekong confluence. It was estimated that a head of 352 m would make it possible to construct a hydroelectric power station with an installed capacity of 600,000 kW, and an annual generated energy of 4,650 GWh. Since the estimated installed capacity had been 2.5×10^6 kW and the annual generated energy 8,400 GWh in the 1970 plan, the scale of this development was considerably reduced.

The construction period was estimated to be 5 years, with total project costs of US$615 $\times 10^6$ (1987 estimate). In addition, a further US$100 $\times 10^6$ would be necessary to fund the installation of an extra-high-voltage power-transmission line 374 km long to the city of Khon Kaen in north-east Thailand. Almost all this electricity was expected to be transmitted across the Mekong to Thailand.

It was estimated that the Nam Theun 2 Dam would increase river flows during the dry season to about 170 m^3/s. Like the Pa Mong Dam on the main stream, this would bring considerable benefits to the delta region. However, because the development site was located in a densely forested area, the project was expected to result in a considerable loss of natural forest. The number of people who would have to be resettled, on the other hand, was only about 2,000 (see table 3.5).

Nam Ngum 2 Dam Project

The proposal for the Nam Ngum 2 Dam was conceived by the Mekong Secretariat after the formulation of the 1970 plan. The idea was to build a gravity dam upstream from the existing Nam Ngum Reservoir, with a height of 165 m, with an annual generated energy of 2,060 GWh, and an installed capacity of 400,000 kW. The 1988 plan report outlines the development concept and the results of a prefeasibility study conducted by a Swiss firm of consultants, Motor Columbus. The catchment of the dam site was to be 5,750 km^2 with a mean annual discharge of 7×10^9 m^3, or 220 m^3/s. The reservoir would have an elevation of 360 m when full, and would provide a total gross storage of 5.4×10^9 m^3 and an effective storage of 3.6×10^9 m^3, covering an area of 119 km^2 when full. The dam would be of the concrete gravity type with a height of 172 m, a crest length of 425 m, and a total concrete volume of 1.6×10^6 m^3. The hydro-electric power station would be incorporated in the foot of the dam and have a mean net head of 134 m. As in the original proposal, the installed capacity would be 400,000 kW with an annual generated energy of about 2,180 GWh. As in the Nam Theun 2 Dam Project, much of the electricity generated was to be transmitted to Khon Kaen in north-east Thailand. It was estimated that about 1,000 people would be displaced and that the total project cost, including construction and resettlement expenses, would amount to US$525 $\times 10^6$ (1987). One of the attractions of this project was that it would increase the energy output of the existing Nam Ngum 1 Dam downstream.

A third project upstream from the Nam Ngum 2 Dam was also considered (dam height 235 m; installed capacity 800,000 kW; annual generated energy 3,000 GWh), but it was of little economic merit.

In the spring of 1987, the Secretariat conducted a comparative study of the economic merits of the Nam Theun 2 and Nam Ngum 2 dams in terms of such factors as energy output and fishing profits. The results of this study and salient features of these two dams together with the Low Pa Mong dam are shown in tables 3.4 and 3.5. Table 3.4 compares the salient technical features of the Low Pa Mong, Nam Theun 2, and Nam Ngum dams (it was estimated that the combined construction costs of the Low Pa Mong and Nam Theun 2 projects would be US$2,681 $\times$ 10^6). Table 3.5 shows the advantages and disadvantages of the three dams. On the basis of the study, it was concluded that construction should commence with the Low Pa Mong Dam, to be followed by the Nam Theun 2 Dam. It was thus decided that development of the Nam Ngum 2 Dam should be postponed until after the turn of the century.

Proposals for national projects in the riparian countries

Power-development projects

The 1988 plan report compares the technical and economic indicators of the following national power and hydroelectric power projects: the power transmission-line project between the Nam Ngum hydroelectric power station and the city of Luang Prabang in the north; the Nam Cheng Dam Project on the Nam Chi River in north-east Thailand; the Pak Mun Dam Project on the Nam Mun River (upstream from the river's confluence with the main stream of the Mekong); and the Yali Falls large and small projects in the upper reaches of the Se San River in Viet Nam. All of these projects were to be funded up to the year 2000.

Irrigation projects

The 22 irrigation projects to be funded up to 2000 were as follows: four projects in Laos – the Nam Ngum Pump Irrigation Project, the Nam Cheng Project, and two small-scale irrigation projects; seven projects in north-east Thailand – the Nam Cheng, Pak Mun, Nam Loei, Upper Chi, Nam Songkhram, and Nam Suai Stage I and Stage II projects (the Nam Cheng and Pak Mun projects were also hydroelectric power projects); and 13 projects in the Vietnamese delta including the Cai San, Tiep Nhut, Phu Tan, Than Nong, Tan Thank, and Go Cong projects.

Table 3.5 International projects: advantages (+) and disadvantages (−)

Aspect	Pa Mong 210		Nam Theun 2		Nam Ngum 2	
Location	On border	(−)	Not on border	(+)	Not on border	(+)
Access	Good	(+)	Difficult	(−)	Good	(+)
Construction site	Compact	(+)	Different for dam and power station	(+)	Compact	(+)
Resettlement	40,000 people	(−)	2,000 people	(+)	1,000 people	(+)
Sedimentation rate[a]	High	(−)	Low	(+)	Low	(+)
High/low investment	High	(−)	Low	(+)	Low	(+)
Potential for satisfying water demands of north-eastern Thailand	Fair	(+)	Limited	(+)	None	(−)
River flow increase in the dry season	Fair	(+)	Limited	(+)	None	(−)
Number of pending issues to be resolved before construction could start	Many	(−)	Many	(−)	Few	(+)
Long/short implementation time	Long	(−)	Short	(+)	Short	(+)

Source: Mekong Secretariat (1988).
a. Compared with storage capacity.

The following three sections (pp. 183–184) outline briefly some of the projects that were considered particularly important.

Pak Mun Project. The Mun River is the largest tributary of the Mekong in north-east Thailand, draining an area of 117,000 km^2 with a mean annual flow of 24×10^9 m^3, equivalent to a mean discharge of 760 m^3/s. The possibility of building the Pak Mun Dam (a rockfill dam 19 m high with an installed capacity of 136,000 kW) immediately upstream from the confluence of the Mun with the main stream of the Mekong had already attracted the attention of the Japanese reconnaissance team. In the 1988 plan, the project site was to be about 5 km upstream from the confluence with the Mekong. The dam would not only generate hydroelectricity but also promote agricultural production by irrigating an area of 33,000 ha around the Pak Mun Reservoir, of which about 7,160 ha were selected for development of a pilot pump-irrigated area near Sisaket.

Since the French consulting firm SOGREAH had conducted a study of this project in 1985 and concluded that it could be sufficiently guaranteed, both technically and economically, the Thai government had adopted the Pak Mun Project as part of its long-term investment plan. As we will see, this project was completed only recently after weathering a strong opposition movement by local residents because of the considerable resettlement that would result from its implementation.

Nam Cheng Project. The Nam Cheng pumped-storage hydroelectricity and irrigation project is located in the headwaters of the Nam Pong Basin, also in north-east Thailand. The plan was for a pumped-storage hydroelectric power scheme of 400,000 kW installed capacity and the expansion of the irrigation area to 8,000 ha of paddy fields and farms. The proposed pumped-storage scheme uses a head of 380 m between the upper and lower pondage, which are 3.2 km apart. This was approved as a promising project as a result of a feasibility study in 1986, and some of the irrigation had already commenced by the time that the 1988 plan was drawn up.

Yali Falls projects. Both the Japanese reconnaissance team and the 1970 plan report had pointed out that the Yali Falls in the upper reaches of the Se San River in central Viet Nam had good potential as a site for hydroelectric power development. Prior to that, Nippon Koei had conducted a thorough study from the early 1960s.

The dam was to be a rockfill dam 45 m high with a crest length of 1.3 km. The water-conveyance system would consist of a 4.5 km pressure tunnel and three penstocks each 415 m long. The effective head would be 200 m, generating an installed capacity of 480,000 kW. This project was named the Yali Falls Large Project.

Another proposal for the same site was the Yali Falls Small Project. This was to be a run-of-the-river scheme consisting not of a large dam but of a regulating pond (with 1.6×10^6 m³ of effective storage) that would be located immediately upstream from the Yali Falls. The waterway would consist of a 290 m headrace channel and 30 m penstocks to draw the water to a power station installed at the foot of the falls and generate power (installed capacity: 24,000 kW). From 1986 to 1987, three possible development plans for the Yali Falls were discussed, but a conclusion had still not been reached when the 1988 plan was published. The three plans were (1) to implement either the Yali Falls Large or Yali Falls Small Project, (2) to combine the two plans, or (3) to develop the Yali Falls Small Project first and then to construct a large reservoir upstream and switch to the large project later on.

Funding requirements for national projects

As outlined above, the 1988 plan proposed a total of 27 national-scale hydroelectricity and irrigation projects for development in the three riparian countries. The total funding required for these projects in all three countries was estimated at US$$1,320.9 \times 10^6$ – US$$40.7 \times 10^6$ in Laos, US$$318.9 \times 10^6$ in Thailand, and US$$961.3 \times 10^6$ in Viet Nam.

Necessity for, and funding of, studies for international projects

The 1988 plan made proposals for various studies for promotion of the two large-scale international projects (Low Pa Mong and Nam Theun 2 dams) and for the development of other dams on the main stream of the Mekong. For example, the report pointed out the need for studies on the influence of dam development on the fishing industry, soil fertility in the delta region, downstream discharge, and changes in water quality. It also stressed the need for research on the diversification of agriculture in the lower basin and proposed that a research institute for agricultural diversification be set up in north-east Thailand.

Moreover, in view of the shortage of water in the dry season, the flood disasters in the rainy season, the soil surface deterioration, and the lack of room for reservoir construction in north-east Thailand, the report pointed out that a master plan concerning land and water resources was urgently needed and indicated the necessary funds and schedule for the formulation of such a plan. It also stated that, among the themes for studies covering the whole of the lower basin, the most important problems were reservoir sedimentation and soil erosion.

Regarding dams on the main stream, the 1988 report stressed that the optimal size and operation of each dam should be thoroughly examined, and that feasibility studies would have to be conducted on the Low Pa

Mong and Nam Theun 2 dams. It also pointed out the importance of formulating a master plan for the Vietnamese delta. Towards the end, the report stated that it would be essential to strengthen the Secretariat of the Interim Mekong Committee[15] and that funding would have to be provided for the necessary reorganization and legislation.

Finally, the report concluded that the total amount needed to fund all of these studies from 1988 to 1996 would amount to US$94.2 $\times 10^6$.

Funding of studies for national projects

For development in Laos, it was estimated that US$1.1 $\times 10^6$ would be needed for studies on themes such as agricultural development potential and the productivity of pilot farms currently receiving technological assistance for irrigation. In Thailand, US$1 $\times 10^6$ would be needed for studies on the development of paddy fields in the north-east. In Viet Nam, it was estimated that US$1.9 $\times 10^6$ would be required to fund studies on such topics as the fishing industry and navigation. Total funding of about US$4 $\times 10^6$ would therefore have to be provided for studies on national projects in the three riparian countries up to 1996.

As far as funding for hydroelectric power projects was concerned, the 1988 report concluded that altogether US$27 $\times 10^6$ would be needed for the costs of studies and planning for projects such the Low Pa Mong and Nam Theun 2 dams, the power-transmission line from the Nam Theun 1 Dam to Luang Prabang, the Nam Cheng and Pak Mun dams in Thailand, and the Yali Falls projects in Viet Nam.

The report concluded that US$13.5 $\times 10^6$ would be required for irrigation projects in Laos and Thailand. In Laos, US$3.1 $\times 10^6$ were needed up to the end of 1991 to cover studies and planning costs for irrigation projects in Laos such as the Nam Ngum pumped irrigation project and the Nam Cheng Project. In Thailand, US$10.4 $\times 10^6$ would have to be provided up to the end of 1996 to fund studies on the Nam Suai, Upper Chi, Nam Songkhram, and Nam Loei irrigation projects.

Totalling the costs of all the studies in the above-mentioned four categories, the report estimated that, between 1988 and the end of 1996, a total of at least US$57.2 $\times 10^6$ would be needed.

Total funding for development and studies

As a sum of US$2,681 $\times 10^6$ was needed to implement the two international projects (Low Pa Mong and Nam Theun 2 dams) and US$1,320.9 $\times 10^6$ for the national projects (in the three riparian countries), a total of US$4,001.9 $\times 10^6$ would be required to fund development in the lower Mekong Basin by the year 2000.

Adding together the total estimated costs of the above-mentioned international and national projects and related studies, the 1988 plan report concluded that the total funding required would come to US$4,140.6 × 10^6 (Hori 1990).

Author's evaluation of the plan and subsequent developments

When a draft of the 1988 plan report had been completed, the Executive Agent of the Mekong Secretariat, Mr Kamp, asked the author to review it. After spending a month at the office of the Secretariat studying the draft report, the author made detailed comments focusing on the following two main points:

1. The consultants who drew up this plan have concentrated too much on trying to lower the height of the Pa Mong Dam in order to reduce as much as possible the number of people (over 40,000) who would have to be resettled as a result of dam construction, and have forgotten the need to compare the Low Pa Mong Dam plan with dam-construction projects on the main stream. The Ban Koum Dam Project planned further downstream from the Pa Mong Dam not only succeeds in significantly reducing the extent of submergence and resettlement (about 2,000 people) as a result of reservoir construction but is also superior to the Low Pa Mong Dam Project both in terms of energy (installed capacity and generated energy) and costs. In view of the considerable opposition expected to the construction of the Pa Mong Dam, it might be better to switch from this to the Ban Koum Dam Project.

2. The major worry with implementing the Nam Theun 2 Dam Project is that water may leak from the reservoir to the neighbouring Se Ban Fai River and other parts of the basin. Studies and investigation of this project should be preceded by a careful geological study of the ground on and around the proposed development site.

The Executive Agent of the Secretariat agreed with this second point made by the author, who later learned that this type of study had been implemented for the Nam Theun 2 Dam Project. Regarding the first point, however, since studies for the Pa Mong Dam Project had already begun several decades earlier, whereas the Ban Koum Dam Project had been studied only on paper, it proved impossible to follow the author's recommendation. (Soon after that, however, the Mekong Secretariat commissioned an Australian specialist to study the geological conditions at the Ban Koum Dam site. Although the conclusion was that there were no particular problems, no study on the Ban Koum Dam has been conducted to date.)

In spite of the enthusiasm of the drafters of the 1988 development plan, hardly any progress in development was made after 1988, owing to various factors in the riparian countries (the completion of the Nong Khai "Friendship Bridge" being a notable exception). This situation was to continue until the formation of the New Mekong River Commission in 1995.

Notes

1. Henri Mouhot's grave is located in the outskirts of Luang Prabang in Laos by the furthest downstream section of the Nam Khan River, a tributary of the Mekong.
2. In 1873, 14 years before the establishment of French Indo-China, a trader (Jean Dupuis) had succeeded in opening a trade route via the Red River, but it was abolished soon afterwards by the Vietnamese government.
3. The Bangkok–Sayaboury–Korat–Nong Khai north-east trunk road.
4. The ECAFE head office later moved to Bangkok.
5. The situation is still the same today.
6. During the Second World War, Japan dispatched a survey team to the Mekong Basin with the intention of examining the possibility of building a railway between Hanoi and (the then) Saigon along the left bank of the main stream of the Mekong, but this idea was never realized.
7. In 1965, the Mekong Committee planned to change its official name from the "Committee for Coordination of Investigations of the Lower Mekong Basin" to the "Committee for the Comprehensive Development and Investigation of the Lower Mekong Basin." However, it was unable to make this alteration because the Cambodian government did not sign the document authorizing it (Mekong Secretariat 1989).
8. One of the former senior staff of the Japanese Finance Ministry became a member of the advisory group to take charge of the financial aspect of the activities of the Mekong Committee.
9. According to Menon (1966), the official letter of request submitted by ECAFE to the Japanese government to implement a reconnaissance survey of the major tributaries of the lower Mekong Basin was made as a result of the special courtesy and expectations of the staff of the ECAFE office involved in Mekong development, as Japan was considered at that time to be the only country in the East that could contribute funds and technology for the development of the Mekong Basin.
10. In the Japanese reconnaissance team's report, the full-reservoir water depth was given as 240 m with a note that this would have to be corrected to 250 m if the results of the Canadian topographical survey were found to be accurate, as they were later found to be. Consequently, the figure proposed by the Japanese team as the full-reservoir water depth was 250 m, not 240 m.
11. When the Japanese reconnaissance team examined the old irrigation project, which had been constructed by the French and left in a devastated condition, the author realized that it was not enough for a donor country to provide hardware facilities: in order for irrigation systems to take root among indigenous farmers in a recipient country, it is necessary to provide education and training for local people so that they become fully aware of the requirements of irrigation and wish to initiate and implement development by themselves.
12. Upstream from the town of Pak Beng.
13. Using the available topographical map drawn to a scale of 1:50,000, a rough estimate was made of the probable area of farmlands and rice fields which would be submerged by the construction of the Stung Treng Dam Project.
14. The author and others who worked on the Indicative Basin Plan (Mekong Secretariat 1970) fully recognized the importance and difficulties of the problems of relocation, but did not pay much attention to other environmental influences of dam construction.

15. By employing more economists and specialists in agriculture and water-resources development.

BIBLIOGRAPHY

Government of Japan (1961). *Comprehensive Reconnaissance Report on the Major Tributaries of the Lower Mekong Basin.* Government of Japan, Tokyo.

Hori, H. (1986). *Planned Interbasin Water Transfer.* IWRA, Beijing.

———— (1988). *Economies versus Environment, Pa Mong in Mekong River.* World Water, UK.

———— (1990). *Economics of the Mekong River.* IWRA, China.

———— (1993). *Development of the Mekong River Basin, Its Problems and Future Prospects.* IWRA, USA.

Kawai, T. (1984). *Aspects of Mekong Basin Planning.* Japanese Society of Irrigation, Drainage and Reclamation Engineering, Tokyo.

McKay, M. (1956). *ICA Reconnaissance Report on Lower Mekong River Basin.* US Bureau of Reclamation, USA.

Mekong Secretariat (1970). *Report on Indicative Basin Plan.* Mekong Committee, Bangkok.

———— (1988). *Perspectives for Mekong Development.* Mekong Committee, Bangkok.

———— (1989). *The Mekong Committee, a Historical Account (1957–89).* Mekong Committee, Bangkok.

Menon, P. K. (1966). "The Mekong Project." *Indian Journal of Power and River Valley Development,* Vol. XVI.

Osborne, M. (1975). *River Road to China: the Mekong River Expedition 1866–1873.* Allen and Unwin, London.

P. T. Tan's family (1972). *A Memorial Album of the Late Mr. P. T. Tan.* Taipei.

Schaaf, H. et al. *(1963). The Lower Mekong, Challenge to Cooperation in Southeast Asia.* New Jersey.

United Nations (UN) (1957). *Development of Water Resources in the Lower Mekong Basin.* Flood Control Series No. 12. United Nations, Bangkok.

———— (1958). *Program of Studies and Investigation for Comprehensive Development of Lower Mekong Basin.* UN Survey Mission, United Nations, Bangkok.

United Nations Economic and Social Commission for Asia and the Pacific (UNESCAP) (1974). "Alternative Choices in Development Strategy and Tactics: The Mekong River Project as a Case Study." *Water Resources Journal,* Bangkok.

United Nations University (UNU) (1955). *Asian Water Forum Summary Report.* United Nations University, Bangkok.

White, G. F. (1962). *Economic and Social Aspects of Lower Mekong Development.* Mekong Committee, Bangkok.

4

Dam-development projects in the upper basin and the Lancang River

Various dam-development projects have already been started on the main stream of the Lancang river in the upper Mekong Basin, all of which are intended exclusively for hydroelectricity generation. The Manwan Dam Hydroelectric Power Project currently (as of December 1994) under construction constitutes the first of such projects being undertaken in south-eastern China by the Ministry of Electric Power of the Chinese government. The Manwan Dam, which will have an installed capacity of 1.5×10^6 kW when fully completed, has been producing 1×10^6 kW since the end of 1994 when the first phase of its development was finished. This output is being supplied to the city of Kunming, the capital of Yunnan Province, using 150 kV power lines. The Manwan Dam will be followed by the construction of the Dachaoshan Dam $(1.35 \times 10^6$ kW) located immediately downstream from the Manwan Dam. Following this, China is expected to embark on the construction of the Xiaowan Dam $(4.2 \times 10^6$ kW). Located immediately upstream from the Manwan Dam and boasting a very large reservoir lake, this third dam will constitute the key project in the overall development of the Lancang River. Further downstream from this series of dams is yet another development project located near the crucial junction at Jinghong. This Jinghong Dam, which is currently in the planning stage, will be located immediately upstream from the city of Jinghong and is being promoted by private funds from Thailand.

When completed, the Lancang will have a total of 14 dams with an

189

aggregate output of more than 20×10^6 kW which will be used to meet the growing power demands of the Yunnan, Sichuan, and Guizhou provinces. Additional power from these serial dams will be transmitted to demand areas throughout China, and also to Thailand and Indo-China.

Needless to say, the development of the Lancang River will have an important bearing on the development plans for the Mekong Basin. In view of this crucial connection, some of the issues covered in the first part of chapter 1 (pp. 1–10) are reviewed here, together with some basic facts about the Lancang River (see fig. 1.1) and the sources of the Mekong, before a detailed overview of the current development plans.

The Lancang River

Sources of the Lancang River

There is very little information about the sources of the Mekong. The author's own research has led him to a few records written by European explorers of the nineteenth and twentieth centuries. However, a very significant step forward was reported by an American newspaper in April 1995.[1] Under the heading "Source of the Mekong Finally Found in Central Asia," the paper reported that a joint party of four French and British explorers had reached a site known as the Rup-sa Pass on the evening of 17 September 1994 and had determined that the swamps of this area constituted the source of the Mekong. However, the article provided no information on the location of the Rup-sa Pass and the geographic details of this area.

Using a Chinese atlas published in 1991, it can be estimated that the source of the Mekong is located in the approximate vicinities of the intersection of latitude 33°N and longitude 94°E. Turning to a map of Japan, the thirty-third parallel cuts across the southern part of the Kii Peninsula. Following the thirty-third parallel westward, this latitudinal line eventually brings us to the highlands of Tibet (Xizang), where it passes through the northern face of the eastern end of the Tanggula Mountains and the southern portion of Qinghai Province: this is the location of the source of the Mekong. In other words, the Mekong originates in the swamplands of the Rup-sa Pass located to the west of the village of Mugxung (latitude 33°N, longitude 95°E). (According to the exploration group, the Rup-sa Pass is situated at an elevation of 16,300 feet or 4,968 m.) The village of Mugxung is itself located west of Zadoi in the autonomous region of Yushuzangzu in the Chinese province of Qinghai. The river takes a south-easterly course as it flows out from these swamps. The uppermost reaches of the Yangtze River and the basin of the Tongtian He River are located directly on the other side of the source of the Mekong. On the first leg of its long seaward journey, the river is referred to as the Za Qu River as it flows through the Xizang Plateau to Quamdo, the capital city of the Xizang Autonomous Region.

As it flows southward, the Za Qu River is joined by the Zi Qu River on its left bank. Further downstream, it meets the Ngom Qu on its right bank in the vicinity of the large city of Chengdu in the Xizang Autonomous Region and flows further south to meet the Zhi Qu River, again on its right bank, at Zhagyab. Thereafter, the river takes a south-easterly course through the quiet mountains of this region. The river takes the name of the Lancang after it leaves the city of Quamdo.

The Lancang River

As the Lancang River flows southwards, it directly faces the upper reaches of the Yangtze River (i.e. the Jinsha River) on the east. Towards the west, the Lancang faces the Nu River, the headwaters of the Salween River. Immediately after it enters Yunnan Province, the Lancang passes beside the beautiful snow-capped Degin Meidi Mountain (6,740 m) situated on its right bank. A further 250 km downstream, the river moves away from the Jinsha River and assumes a parallel course to the Nu River as it maintains a southward course. Before arriving at the southern extremity of Yunnan Province, the Lancang absorbs the Yangbi River, a major tributary of this area which flows down from Lake Erhai in the vicinity of the city of Dali.[2] The southern extremity of Yunnan Province marks the end of the Lancang River. Thereafter, the river takes the name of the Mekong as it takes a south-westerly course marking the border between eastern Myanmar and northern Laos, eventually arriving at the area known as the Golden Triangle, where the borders of Myanmar, Laos, and Thailand meet.

Length and slope of the river

The distance travelled by the river from its source to the southern tip of Yunnan Province (where the Chinese border meets the eastern end of northern Laos and the north-western tip of Myanmar) is approximately 2,000 km, as estimated from maps. Of this total distance, the length of the Za Qu River from the headwaters to the confluence of the Zhi Qu river near Chengdu is roughly 500 km. Next, the length of the river from this confluence to the beginning of the Lancang River at the northern border of Yunnan Province is roughly 250 km. The length of the river from its entry into Yunnan Province to the end point of the Lancang has been accurately measured by the Ministry of Electric Power of the Chinese government as being 1,250 km; thus, the total length of the Lancang section of the river is 1,500 km. Adding the 500 km of the Za Qu river, the entire length of the river comes to approximately 2,000 km.

The height of the river above sea level falls continuously as the river flows southward from its source (4,968 m) to the northern border of Yunnan Province (2,300 m) and on to the southern end of the province (400 m). This yields an average slope of 1:290 between the source of the river and the northern border of Yunnan Province. In comparison, the river assumes an average slope of 1:660 as it traverses the province; at this stage, the slope of the river is similar to that of an average Japanese river.

According to the Ministry of Electric Power, the area of the river basin between the headwaters and the southern end of Yunnan Province (that is, the area of the basin covering the territories of China and Myanmar) comes to 174,000 km^2.[3] (In addition to this figure, the Ministry of Electric Power states that the elevation of the source of the river is 5,500 m and that the total length of the river in Chinese territory is 2,000 km.)

In formulating its development plans for the lower Mekong Basin, the United Nations has separated the upper and lower basins at the region of the Golden Triangle where the borders of Thailand, Laos, and Myanmar come together. After leaving Chinese territory, the Mekong defines the border between Myanmar and Laos over a distance of 220 km. While this section of the river obviously cannot be included in the Chinese development plans, the United Nations has also excluded this stretch from its plans. However, the 1961 report of the reconnaissance mission of the Japanese government does mention this region and points to the possibility of constructing a reservoir dam which might store water with a capacity of 20×10^9 m^3 immediately below the three-country border (at an elevation of 380 m). Although the report does not specify the height of the dam required to maintain full capacity, it is very probable that the construction of such a reservoir would have a direct impact on the development plans for the Lancang River in the southern part of Yunnan Province (410 m elevation at the southern end). In any case, such a project would affect all development projects for the Mekong below this point by influencing the flow volume of the Mekong throughout its entire length.

Rainfall and flow volume

The source of the Mekong is in the snow-covered mountains of Tibet. At first, the river is fed and nourished by the melting snows of these slopes. After its entry into Yunnan Province, it takes on greater volumes of water from rainfall over the forests of this region. By this point, rainfall is the greatest source of its waters.

Average annual precipitation at Manwan amounts to 1,027.6 mm with an annual average of 154 days of rain. During the summer, trade winds carry moisture into this region, while the mountains to the north shut out the cold winds of winter. This creates a high-precipitation subtropical environment marked by temperate-zone forests which are home to the laurel forest culture. This region is noted for its production of rice and tea (black tea, green tea, and puer tea).

The flow volume at the southern end of the Lancang River at the three-country border stands at an annual average of 2,020 m^3/s. The flow volume at Manwan (with a basin area of 114,500 km^2) is relatively stable with maximum, average and minimum flow rates of 9,150, 1,230, and 273 m^3/s, respectively.

The Lancang River takes on a light brown colour between the Jing-hong Bridge and the planned site of the Gunianba Dam. The bridge at Jinghong, the capital city of the Thai Autonomous Region of Xishuang-banna at the southern extremity of Yunnan Province, was completed in 1964 (a concrete bridge with a span of 160 m and a width of 14 m). As observed in the month of December, this section of the river has a width of roughly 100 m, a depth of 5 m, and a flow velocity of 2 m/s. Both banks are sparsely cultivated and very few houses can be seen outside Jinghong. The gradient of the river is quite steep at this point and outcrops and mounds of gravel can be seen here and there.

Electricity demand and facilities in Yunnan Province and future development plans

Electricity demand and facilities in China

In 1995, China stood at the final year of its eighth Five-Year Plan (1990–1995). Total installed capacity for the entire nation at this point, including generation for private use, amounted to 183×10^6 kW. The generating facilities of the Ministry of Electric Power accounted for roughly 150×10^6 kW. Of this amount, approximately 25 per cent or 38×10^6 kW, was derived from hydroelectric power generation; nuclear power provided 1×10^6 kW, while the remaining 110×10^6 kW was derived from coal-fired power plants.

As a result of the rapid increase in electricity demand, the Ministry of Electric Power estimates that total demand will top the 300×10^6 kW mark by the year 2000. Current development plans are geared towards meeting this accelerated increase in electricity use. Because coal-fired plants will cause additional environmental pollution, the Chinese authorities have indicated a strong interest in markedly increasing hydroelectric power generation in the future. Fortunately, China has a theoretical development potential for 400×10^6 kW of hydroelectric power generation. Currently, slightly less than 10 per cent of this potential has been tapped and the national policy calls for greater emphasis on such development.

China's electricity network currently consists of five component regions: these are the north-east (the region of former Manchuria centred around Changchun), Huabei (Beijing, Tianjin, and inner Mongolia), the northwest (Xian, Ningxia, Gansu, Qinghai Province), Huadong (centred around Shanghai), and Huazhong (centred around Wuhan). The power networks of Huazhong and Huabei are linked by 500 kV power lines. In addition to these power grids, the provinces of Yunnan and Shandong have their own independent power networks.

With the completion of the Three Gorges Dam (started on 15 December 1995 and scheduled to be completed in 2009 with an installed capacity of 18.2×10^6 kW), the entire nation will be connected by a radial array of 500 kV power transmission lines.

Electricity demand in Yunnan Province and the development of the Lancang River

As mentioned above, the province of Yunnan operates its own independent power network, with a total installed capacity of 4.6×10^6 kW. Of this amount, 15 generating facilities operated by the Ministry of Electric Power provide 3.32×10^6 kW (as of July 1994). By type of generation, hydroelectricity accounts for 62 per cent (eight power plants) and coal-fired power for 38 per cent (seven power plants). The hydroelectric power stations are concentrated in the western sections of Yunnan Province, whereas the thermal plants tend to be located in the eastern regions. Total energy output in 1993 amounted to 12.78×10^9 kWh/year, rising to an estimated 14.7×10^9 kWh/year in 1994. The peak load registered during 1994 came to 2.21×10^6 kW (excluding facilities not operated by the Ministry of Electric Power). As in previous years, this peak load was recorded during the month of November when industrial production picks up and household demand for heating increases.

The breakdown in power demand is 40 per cent for agricultural use, 40 per cent for industrial use, and 20 per cent for commercial, service-industry, and household use. Current projections indicate that agricultural demand will account for 35 per cent of total use in the year 2000, while the share of industrial use will increase to 41 per cent and other commercial and household demand will rise to 24 per cent. Thus, the relative share of agricultural use is expected to decline while household and commercial demand increases rapidly.

As a result of the accelerated growth in demand (demand for electricity has been growing at an average annual rate of 10 per cent since 1990), Yunnan Province will require an installed capacity of approximately 8×10^6 kW by the year 2000 (assuming an average annual growth rate in demand of 11–12 per cent between 1995 and 2000). To meet this demand, the generating facilities of the Ministry of Electric Power will have to be raised to 6.3×10^6 kW. Of this total, the Ministry expects to derive 3.5×10^6 kW from coal-fired plants and 2.8×10^6 kW from hydroelectric power stations.

The mountainous Yunnan Province boasts the second-largest potential for hydroelectricity generation in all China. In fact, China's first hydroelectric facility, the Shilongba hydroelectric power station, was completed on the western outskirts of the provincial capital of Kunming in 1912 and remains in operation to this day. In addition, four power stations with a total output of 250,000 kW are in operation on a tributary of the Yangbi River which flows out from the south-western corner of Lake Erhai near Dali.

More recently (December 1988), the Lubuge Hydroelectricity Project was completed on the Huangni River, a tributary of the Nanpan River (headwaters of the Hongshu River). This project, located 150 km east of Kunming, features a dam-fed underground generating plant with a current output of 600,000 kW to be increased to 1.5×10^6 kW when the entire project is completed. (The dam is a rockfill dam 103.8 m high and 217.7 m long with a reservoir capacity of 110×10^6 m^3 and with four generators and turbines of 153,000 kW each.) Following closely on this project, the Manwan Dam, the first hydroelectricity-development project on the Lancang River, was near completion by the end of 1994. This hydroelectric project will have an initial output of 1×10^6 kW to be increased to 1.5×10^6 kW in the final stage of development.

Other projects are currently being pursued to meet the projected power demand in the year 2000. Work continues on the Dachaoshan Hydroelectricity Project located immediately downstream from the Manwan hydroelectric power station. (When completed, this plant will have an installed capacity of 1.35×10^6 kW. Construction of the temporary diversion canal was started in December 1994.) In addition, the strategy of the Ministry of Electric Power is to develop numerous small-scale hydroelectric power stations on the tributaries of the Lancang Basin. As for matching the growth in demand for electricity after the year 2000, both the Ministry of Electric Power and the authorities of Yunnan Province intend to exploit the outstanding hydroelectric potential of the lower sections of the main stream of the Lancang.

It is estimated that Yunnan Province has a total hydroelectric potential of 100×10^6 kW. According to the Ministry of Electric Power, it is technically possible to develop no less than 71×10^6 kW of this potential. Although it is unclear what proportion of this amount is to be derived from the Lancang and its tributaries, the theoretical hydroelectric potential of the upper Mekong Basin (including areas other than Yunnan Province) is estimated to be 76×10^6 kW, as indicated in table 2.9.

According to the current estimates of the Ministry of Electric Power, the technical and economically feasible hydroelectric potential of the main stream of the Lancang River is 22.11×10^6 kW. Adding the hydroelectric potential of the tributaries and upper reaches of the Lancang, the entire Lancang River system should have a total installed capacity of approximately 30×10^6 kW. As indicated by these figures, the hydroelectric potential of the upper Mekong Basin, which has now begun with the completion of the Manwan Dam, is vast.

A few comments should be made on the power transmission-line systems of Yunnan Province. In the area to the west of Kunming, a single 500 kV line covers the distance (230 km) between Kunming and the Manwan hydroelectric power station. Work is now being done to add a second cable to this section. This western section of Yunnan Province is covered by other power transmission lines of 200 kV or less, which cover a distance of roughly 2,000 km. The area to the east of Kunming is served by a single 500 kV line which connects the Lubuge hydroelectric power station to Guangdong.

Lancang River Serial Dam Development Project

Hydroelectric development potential

Near its source, the Za Qu River (headwaters of the Lancang) is nourished by the melting snows of the Tibetan mountains, but as it flows through the Xizang Autonomous Region it is replenished by the ample rains which fall in this area between the months of June and September. As the Za Qu River drops from an elevation of 5,000 to 2,500 m, it can be assumed that it has great potential for hydroelectricity generation. However, the author has not found any information on the feasibility of developing the sections of the river above Yunnan Province, so cannot comment further on this matter.

Developing the Lancang River within Yunnan Province and the TVA Project of the United States

For sections of the Lancang River situated within Yunnan Province, the Yunnan Provincial Electric Power Bureau of the Chinese government has identified 14 sites for dam construction (table 4.1) with a total installed capacity of 22.11×10^6 kW (baseline output of roughly 11×10^6 kW). In its estimates of July 1994, the Bureau states that these facilities would be capable of generating 109.5×10^9 kWh annually.

These 14 facilities would take the form of a set of serial dams and power plants. The uppermost dam would be the Liutongsiang (installed capacity of 550,000 kW) with a normal reservoir elevation of 2,174 m; the lowermost dam would be the Mongsong (installed capacity of 600,000 kW) with a normal reservoir elevation of 519 m, yielding an elevation differential (head) of 1,655 m.

If the Lancang River development projects proceed as planned, the completed project will far eclipse the well-known Tennessee Valley Authority (TVA) system in the United States. The TVA consists of nine serial dams beginning with the Fort Loudoun Dam (installed capacity of 128,000 kW and a normal reservoir elevation of 247 m) and ending with the Kentucky Dam (installed capacity of 160,000 kW and a normal reservoir elevation of 114 m). In total, the TVA system has an installed capacity of 1,654,000 kW. By comparison, the Lancang River projects would yield an installed capacity of 20×10^6 kW, 13 times higher than that of the TVA. Similarly, the elevation differential of the Lancang River projects would be 1,500 m, more than 12 times higher than that of the TVA. These comparisons clearly indicate the monumental scale of the Lancang River projects.

Comparison of the Lancang River serial dams and the Three Gorges Dam

It is interesting to compare the scale of the Lancang River serial dams project and the Three Gorges Dam, construction of which officially began on 15 December 1994. The final installed capacity of the Three Gorges Dam is to be 18.2×10^6 kW with an annual generated energy of 84.6×10^9 kWh; obviously, the total installed capacity of the 14 Lancang River dams will exceed this figure.

According to the *China Daily* (15 December 1994 issue), the total cost of the Three Gorges Dam will be 95.4×10^9 yuan (approximately $¥1,145 \times 10^9$). As opposed to this, in January 1991 the Ministry of Electric Power estimated the total cost of the Lancang River projects to be 40.8×10^9 yuan. For the Three Gorges Dam, the actual dam-construction cost will account for 52 per cent of the total cost, while the power-transmission system and resettlement of the population affected by inundation will account for 17.4 per cent and 30.6 per cent of the total cost, respectively. (Inundation will affect a population of more than a million people in two cities and eleven towns and villages. Some 1,600 factories and 28,000 ha of agricultural land will be inundated.) By comparison, the

Table 4.1 Hydroelectric potential of the Lancang River

Dam number		1	2	3	4	5	6	7
Name of dam		Liutongsiang	Jiabi	WuengLong	Tuoba	Huangdeng	Tiemenkan	Gongguoqio
Drainage area	(10^4 km^2)	8.3	8.4	8.55	8.8	9.2	9.34	9.73
Average annual flow	(m^3/s)	698	720	754	809	898	929	985
Reservoir's full water level	(m)	2,174	2,054	1,964	1,820	1,640	1,472	1,319
Drawdown	(m)	120	90	144	180	168	153	77
Average head	(m)	108.75	81.35	134.65	154.52	147.94	136.92	72.05
Gross storage volume	(10^8 m)	5.0	3.2	9.8	51.5	22.9	21.5	5.1
Baseline output	(10^8 kW)	33.64	26.52	48.94	76.23	84.96	82.71	46.74
Installed capacity	(10^4 kW)	55	43	80	164	186	178	90 (75)
Annual generated energy	(10^8 kWh)	33.62	26.52	48.78	80.79	89.20	85.94	47.11
Total investment	(10^8 yuan)	11.20	8.18	14.48	28.27	30.04	29.19	19.47
Unit investment cost	(yuan/kW)					1,615	1,640	2,163
Construction period	(year)	6	6	7	9	9	9	7
Type of dam					rockfill	concrete	gravity	
Height of dam	(m)							130
Farm land (to be inundated)	(6.7 a)							5147
Paddy field (to be inundated)	(6.7 a)							3146
Population (to be relocated)	(person)							4596

Dam number		8	9	10	11	12	13	14	Total
Name of dam		Xiaowan	Manwan	Dachaoshan	Nuozhado	Jinghong	Gunlanba	Mengsong	
Drainage area	$(10^4$ km$^2)$	11.33	11.45	12.10	14.47	14.91	15.18	16.00	
Average annual flow	(m^3/s)	1,220	1,230	1,230	1,750	1,840	1,880	2,020	
Reservoir's full water level	(m)	1,240	994	899	812 (807)	602	533	519	
Drawdown	(m)	252	99	84	205	67	11	33	
Average head	(m)	237.97	92.67	75.72	178.42	59.87	7.00	24.00	1,501.67
Gross storage volume	$(10^8$ m$)$	153.4	9.2	8.84	246.71	10.4	/	/	543.84
Baseline output	$(10^8$ m$)$	180.33	78.70	70.93	226.71	84.74	10.08	37.39	1,108.43
Installed capacity	$(10^4$ kW$)$	420	150	135	500	135	15	60	2,211.0
Annual generated energy	$(10^8$ kWh$)$	191.7	76.67	68.77	226.55	76.06	8.99	33.97	1,094.58
Total investment	$(10^8$ yuan$)$	86.93	20.00	19.98	99.44	25.64	3.09	11.32	407.83
Unit investment cost	(yuan/kW)	2,070	1,333	1,586	1,989	1,899	2,060	1,887	1,839.29
Construction period	(year)	10.5	8	7	12	8	4	6	
Type of dam		conc. arch	conc.	gravity	rockfill	conc. grav.	lock	gravity	
Height of dam	(m)	284.5	132	120.5	260.5	118			
Farm land (to be inundated)	(6.7 a)	35,460	4,670	9,800	36,000	6,000	180	875	
Paddy field (to be inundated)	(6.7 a)	26,021	2,918	3,740	17,519	2,268	/	32	
Population (to be relocated)	(person)	28,748	3,042	5,200	14,800	1,700	58	230	

Source: Yunnan Provincial Government experts, 1995.

costs incurred by inundation will be relatively low in the case of the Lancang River dams. According to a 1994 estimate, the eight dams of the middle and lower sections of the river, bounded by the Gongguoqio Dam and the Mengsong Dam, will displace a total population of 58,374 people and the total area of inundation would come to 98,132 mo (6,542 ha). However, it must be noted that the cost estimates for the two dam systems cannot be readily compared because of China's high rate of inflation. That is to say, the nominal cost of the 14 Lancang River dams estimated in 1991 must be corrected for the rate of inflation in the subsequent period of 1991–1994.

One of the 14 Lancang River dams is the Xiaowan Dam. According to an estimate made in January 1991, this dam will entail total costs of 8.7×10^9 yuan (including the resettlement of 228,787 people and the inundation of 35,460 mo [2,364 ha]). However, the cost of this same dam was estimated at approximately US$$3.5 \times 10^9$, or 29×10^9 yuan, according to information issued by UPI on 4 January 1994. This cost estimate is 3.3 times higher than the figure announced by the authorities of Yunnan Province in January 1991. If this multiple can be used to arrive at a very rough approximation of the cost of the 14 Lancang River dams in current monetary terms, the 40.8×10^9 yuan estimated in 1991 would escalate to over 130×10^9 yuan.

Comparison of the three possible reservoirs of the Lancang River

The Chinese government has embarked on the development of the Lancang River with only one purpose in mind – to develop the hydroelectric potential of this river. The Manwan Project located in the middle section of the Lancang boasts an average annual flow rate of 1,230 m^3/s, a total annual flow volume of 38.5×10^9 m^3, and an elevation differential of 1,600 m. In addition, the river has three potential dam sites where relatively large reservoirs could be created at its upper, middle, and lower reaches.

The projected site of the Touba Dam, the fourth dam on the Lancang counting from the entry of the river into Yunnan Province, would yield a total reservoir capacity of 5.15×10^9 m^3 (effective capacity of 3.40×10^9 m^3). This figure is exceeded by the 15.3×10^9 m^3 (effective capacity of 10.3×10^9 m^3) reservoir capacity of the Xiaowan Dam, the eighth dam in the series, which is located in the middle reach of the river. The third projected site of the Nuozhado Dam, the eleventh dam in the series, has the potential to hold 24.3×10^9 m^3 (effective capacity of 12.4×10^9 m^3); thus, the third dam site at Nuozhado could potentially be the largest. However, in considering the downstream flow augmentation effect of each of the 14 dams, it is obvious that a reservoir at the upper end of the river would have a greater positive impact. A comparison of Xiaowan in the middle reaches of the river with the Nuozhado Dam, situated near the lowermost reaches of the river, indicates that a reservoir at the Xiaowan Dam would yield a larger positive flow effect on the other dams in the series. On the other hand, although the reservoir to be created at Touba is located above Xiaowan, the Touba reservoir would have a smaller downstream flow-augmentation effect because of its relatively small size. This indicates that the Xiaowan Dam Project in the middle reaches of the river constitutes the most important dam-development project for the Lancang River system.

Determining the order of construction for the Lancang River dams and the outlook for development

The Electric Power Bureau of Yunnan Province undertook a study to determine which project in the Lancang River serial dams should be undertaken first (figs. 4.1 and 4.2). On the basis of such considerations as the relatively small size of the population to be relocated (3,042 people), the lowest construction cost per kW (estimated at 1,333 yuan/kW in January 1991) and the relatively low overall cost of construction (2.0×10^9 yuan), the Manwan Dam (initial power output of 1.25×10^6 kW) was chosen as the first project.

Following the Manwan Dam, the highest priority has been given to the construction of the Xiaowan Dam (installed capacity of 4.2×10^6 kW), which is projected to be located immediately above the Manwan Dam. This is because the completion of the Xiaowan Dam will significantly improve the status of the already created reservoir lake of the Manwan Dam and lead to an increase in the installed capacity of the Manwan Dam to 1.5×10^6 kW.[4]

However, the development of the Xiaowan Dam will not be an easy task because of its very large scale, the high cost of construction, and the large number of people who will have to be resettled (28,700 people). For this reason, the Electric Power Bureau of Yunnan Province has turned its attention to the construction of the Dachaoshan Dam (installed capacity of 1.35×10^6 kW and baseline output of 709,300 kW), which is to be located downstream from the Manwan Dam. A feasibility study has been completed and construction of the temporary diversion canal was started in December 1994. As in the case of the Manwan Dam, the Dachaoshan Dam will be domestically financed in full. Another similarity is that the population to be relocated is relatively small and estimated at around 5,200 people. For these reasons, the project can be expected to proceed without serious difficulty.

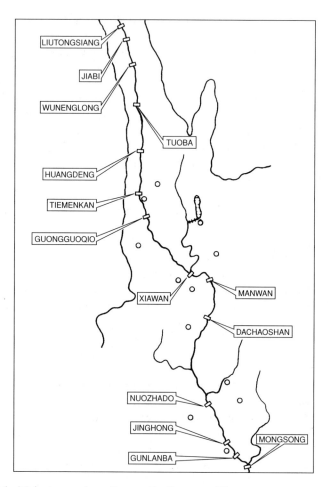

Figure 4.1. Mainstream dam sites on the Lancang River
Source: Yunnan Provincial Government 1995

The question remains whether the Dachaoshan Dam (installed capacity of 5×10^6 kW and baseline output of 2,267,100 kW) will be constructed before the Xiaowan Dam. This is because the position of the Japanese government in providing financial support for the construction of the Xiaowan Dam remains unclear. If work on the Xiaowan Dam is delayed, the construction of the Dachaoshan Dam may precede that of the Manwan Dam. It is possible that the Jinghong Dam (located upstream from the town of Jinghong with an installed capacity of 1.35×10^6 kW and baseline output of 847,400 kW) may be constructed before the Dachaoshan Dam (table 4.1).

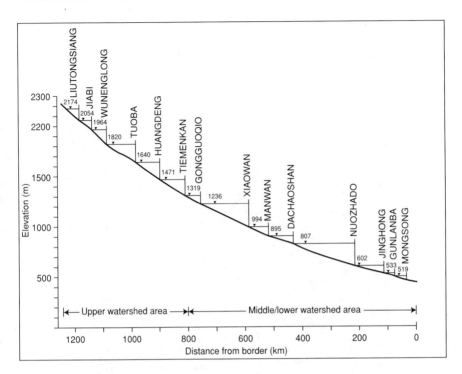

Figure 4.2. Profile of the Lancang River
Source: Yunnan Provincial Government 1995

Since 1993, the ADB has focused its attention on improving the transportation and communication systems in the area of Jinghong where the regions of eastern Myanmar and north-western Laos adjoin Xishuangbanna Province. In fact, regularly scheduled ferries already connect Jinghong with northern Thailand and the Laotian city of Luang Prabang. Following this lead, the MDX Corporation, a Thai firm with headquarters in Bangkok, cooperated with the Electric Power Bureau of Yunnan Province to undertake a feasibility study of the Jinghong Dam. (The prefeasibility study was to be completed in 1995 and the company is scheduled to embark on the feasibility study thereafter.) The Jinghong Dam will be built in a baserock area and a perfunctory survey of the site indicates that construction will be relatively easy. The population to be relocated, estimated at 1,700 people, is also small.

The MDX Corporation is currently aiming to develop the Nam Theun 2 Dam in central Laos in partnership with the French government, Italthai Corporation (Italy), Jasmine International, and the Transfield Co., Ltd (Australia). The company is working to improve the ferry link between Jinghong and Luang Prabang (through the introduction of high-

Proposed site of the Xiaowan Dam. Hiroshi Hori 1994.

speed vessels) and is also involved in hotel construction and the promotion of tourism in this region. There is no doubt that the company is very
interested in the electricity that will be generated by the Jinghong Dam.
These considerations indicate that the Jinghong Dam may be constructed
relatively soon.

To summarize this section, it can be said that work on the serial dams
of the Lancang River has now been initiated and, unless serious complaints are lodged by Viet Nam and other riparian countries, the construction of these dams will proceed smoothly.

Projected dams on the Lancang River

In the original Japanese version of this book, because there are very few people in Japan with any knowledge of the Lancang River dams, this section was devoted to a description of 3 of the 14 dams which have been planned for the river – the Manwan, Jinghong, and Xiaowan dams. However, in this English version, a somewhat more detailed description is given for the Manwan Dam, the first phase of which has already been completed. Drawings of the dam are not included, and the descriptions of the other two dam projects have been omitted in order to save space.

Manwan Dam Project

As stated above, work on the first phase of the Manwan Dam Project has been completed. This project is located at the border between the provinces of Yunshang and Jinghong in the middle section of the Lancang River and is situated 270 km west of the provincial capital of Kunming.

As already explained (pp. 201–204), the Manwan Dam was selected as the first dam to be constructed among the Lancang River serial dams because topographical conditions were favourable, the cost of construction was relatively low, and the size of the population to be relocated was relatively small.

The entire project was undertaken by China alone, including all phases of design, construction, and financing. The preliminary design work was completed by October 1984 and the project was included in the National Construction Plan in April 1985 with the approval of the National Planning Committee. In 1987, it was identified as a priority project. Various organizational changes were made at the time to support the construction project: specifically, the Bureau for the Management of the Manwan hydroelectricity project was created to function under the joint supervision of Yunnan Province and the former Ministry of Water Resources and Electric Power (now the Ministry of Electric Power). Although the headquarters of this Bureau are located at the dam site, the supervisory tasks associated with the construction of the dam have been delegated to the Yunnan Province Electric Power Bureau in Kunming.

The final installed capacity of the Manwan hydroelectric power station, when operating alone, is 1.25×10^6 kW (baseline output of 352,000 kW) and the annual generated energy is given as 6.3×10^9 kWh. However, with the completion of the Xiaowan Dam, to be situated above the Manwan Dam, installed capacity will rise to 1.5×10^6 kW (baseline output of 787,000 kW) and an annual generated energy of 7.66×10^9 kWh will be achieved. Given the scale of this project, the Manwan power station promises to be the keystone in the development of the hydroelectric potential of the Lancang River system. By the same token, this dam will have a significant effect on the economy of Yunnan Province.

The dam site is located at a narrow gorge where the river banks are asymmetric: at the right bank the slope is gentle and outcrops are seen everywhere. The right bank is steeply inclined; however, as faults and cracks were seen, it was thought that the site would be unsuitable for construction of a high dam. Nevertheless, as the geological survey revealed that the site was generally composed of homogeneous hard rhyolite, it was decided to construct a concrete gravity dam at this location.

The dam stands on a base rock (elevation 873 m). Its height and crest lengths are 132 and 418 m, respectively. Five radial-type spillway gates were installed almost at the centre of the concrete gravity dam. At the foot of the sides of the dam, wide openings were provided – the left to spill flood flow and the right to flush sediment. The length and breadth of the spillway tunnel are each 12 m.

The gross volume of the Manwan reservoir is 920×10^6 m^3 (elevation at the normal full water level, 994 m) and the effective volume is 252×10^6 m^3 with a 14 m drawdown. Compared with the volume of the annual inflow, the effective volume is fairly small and therefore the dam does not provide capacity to adjust to seasonal changes of inflow. Because long-term monitoring showed that the unit weight of sand and silt contained in the inflow was 1.21 kg/m^3, it was estimated that the volume of sand and silt flowing into the reservoir would be 47×10^6 t/year and the volume of sediment accumulating at the bottom of the reservoir would be 1.5×10^6 t/year.

The length and the area of the reservoir at its normal full water level are 70 km and 23.9 km^2, respectively. For the construction, 96 villages (3,042 local residents) had to be relocated, but it was considered that no important mineral resources would be submerged.

The annual precipitation, number of days of rain, and average temperature at the Manwan Dam site are 1,027 mm, 154 days, and 19.2°C, respectively. The annual average inflow volume at the dam site (the drainage area is 114,500 km^2) is 1,230 m^3/s. The previous maximum flood volume and the minimum low-inflow volume were 9,150 m^3/s and 273 m^3/s, respectively.

In May 1986, construction was started by installing two temporary diversion channels at the left bank; it was anticipated that flooding at a rate of 10,000 m^3/s (which corresponds to the greatest degree of flooding in 25 years) might affect the project site during construction.

The volumes of soil removed from the dam site and for the tunnelling works were 12.36×10^6 m^3 and 445,000 m^3, respectively. The total volume of concrete was 2.5×10^6 m^3 and total weight of steel used for the construction was 7,600 t. All equipment installed was manufactured in China, with the exception of the 500 kV switches, which were Swiss. Almost all the construction machinery is also of Chinese manufacture, although many machines were manufactured in cooperation with the Komatsu Manufacturing Co. of Japan and some were imported from Germany.

There were less than 20 fatal accidents during the construction work, and most of these occurred outside the construction site. The most serious incident during the course of construction was a landslide on the left bank on 7 January 1987, involving 150,000 m^3 of soil. This necessitated adjustment of the work schedule for construction of the powerhouse.

One of the problems for those engaged in the construction was the necessity to manufacture most of the concrete aggregates, as they were not available on site. This caused contamination of the river downstream of the dam during construction, giving rise to great concern in Laos. Another problem was the relocation of over 3,000 local inhabitants, including a considerable number of minority groups. The government regarded this as a prime concern, and not only built new villages and schools for them but also guided them in the exploitation of new, neighbouring farmlands.

Jinghong Dam projects (planned)

Jinghong is the capital city of Xishuangbanna Province, where members of various minority ethnic groups mingle in the streets. The total population of the city and its suburbs is some 60,000 (of the city itself, some 20,000). The city is only an hour or so by air from Kunming and there are also regular flights to and from Thailand. The Jinghong project site is located about 5 km upstream from the Jinghong Bridge over the Lancang. Hard rock outcrops are seen at the proposed dam site.

According to the information available, the dam will be of the concrete gravity type, with a height of 102 m and a crest length of 560 m. The gross volume and effective volume of the reservoir will be 1.02×10^9 m^3 and 230×10^6 m^3, respectively. The total volume of concrete used will be 2.6×10^6 m^3 and the volume of earth excavated will be 5.5×10^6 m^3. The installed capacity of the powerhouse will be $1.35–1.50 \times 10^6$ kW; the baseline capacity will be 847,000 kW; 7.6×10^9 kWh electricity will be generated annually if the project is operated jointly with other upstream dam projects.

Xiaowan Dam projects (planned)

The proposed Xiaowan Dam site is located on the Lancang just upstream of the tail end of the elongated reservoir of the completed Manwan

Dam. The dam site is composed of biotite schist and biotite granite. It is reported that there is neither fault nor fissure at the site. However, the weakest point of the planned project is that there are only small amounts of sand and gravel near the dam site. So far, what has been planned is a very high double-arch-type concrete dam with a height of 284.5 m. The type has been selected owing to the paucity of aggregate available and the existence of hard baserock at the site. However, severe earthquakes have occurred at places not far from the proposed site, so that great care is required in its design, construction, and maintenance. The powerhouse will be built on the right bank, where six turbines and generators of 700,000 kW each will be installed. According to information obtained from the Chinese government, $10\frac{1}{2}$ years may be needed to complete this project.

The gross volume and the effective volume of the reservoir will be probably be 15.34×10^9 m^3 and 10.28×10^9 m^3, respectively. The reservoir will provide, therefore, considerable capacity for adjustment of the annual run-off (about 38.3×10^9 m^3, 1,220 m^3/s).

Development of the Xiaowan Dam Project will certainly increase the output of the Manwan Project. When the two projects are operated jointly, not only Yunnan Province but also adjacent provinces, such as Chuan and Guizhou provinces, will be able to satisfy their energy requirements. Development of the Xiaowan Reservoir Project will enable all the hydroelectric projects located downstream to increase their power-generation capacities. Furthermore, the dry-season flow volume downstream of the dam, not only that of the Lancang River but also throughout the main Mekong down to the delta, will be increased. Water quality along the river to the estuary at the South China Sea will also be affected. The Xiaowan is, therefore, a key project for the development of both the Lancang and the Mekong, and the Chinese government has been enthusiastically explaining the advantages of the project to the governments of all the downstream countries (particularly Viet Nam), in an attempt to enlist their support.

With the aim of implementing this key project, the Chinese government has provided a document not only for the riparian government but also for governments of developed countries, including Japan, who might help to finance the construction. The gist of this document is as follows:

By the operation of the Xiaowan Reservoir, the total output of all the downstream mainstream projects which have been planned on the Mekong, such as the Luang Prabang, the Sayaboury, the Chieng Khan, and the Pa Mong, will be increased: the total increment will be some 810,000 kW, 2.3×10^9 kWh. Also, the completed Manwan Dam and other dam projects planned on the main stream of the Lancang are capable of adjusting river flow volume to a considerable degree; they will supplement the decreased dry-season flow during the construction of the Xiaowan Dam Project.

On the other hand, the drainage area upstream of the Xiaowan site is 113,000 km^2, which corresponds to only 15.2 per cent of the entire basin area, and the total inflow volume pouring into the Xiaowan Dam will be only 7.6 per cent of that of the whole river basin. As the Xiaowan Dam will be operated solely for hydroelectricity generation, water will never be taken for other purposes such as irrigation and domestic and industrial water supplies; the Xiaowan Project will not, therefore, reduce the flow.

After completion of the Xiaowan Dam Project, the dry-season flow downstream will be augmented from the present 275 m^3/s to 986 m^3/s. The depth of the river at the dry season will therefore increase; consequently, downstream navigation will be improved.

The paper provided by the Chinese government further sets out the benefits of the construction of the Xiaowan Dam Project to the downstream Mekong Basin as follows:

The water level of all the mainstream channels of the Mekong will be raised; moreover, sea-water influx into the Mekong Delta will be reduced. As the flow of mud and silt into the Xiaowan Reservoir will be blocked, the volume of sediment in the downstream water channels will be reduced and, as a result, navigation will be much improved.

Furthermore, by providing watershed management, the soil and water in the vicinity of the Xiaowan Reservoir will be better preserved and, therefore, the water quality downstream will be improved, By the operation of the reservoir, the amount of peak flood flow will be clearly decreased.

Meanwhile, the Mekong Committee studied the probable effects of the construction of the Xiaowan Project on the downstream area, on an unofficial basis in February 1989. The conclusion was as follows:

The effective capacity of the planned Xiaowan Reservoir will exceed that of the planned Pa Mong (that with a full-reservoir water elevation of 210 m). Almost 90 per cent of the inflow to the Pa Mong is from the Lancang River. Judging by the total amount of mainstream flow in 1981, by operating the Xiaowan Reservoir in the dry season the volume of flow into the Pa Mong will be increased by 800 m^3/s when the flow of the Mekong reaches its minimum in March. In other words, the minimum flow to reach the Pa Mong in the dry season will be considerably increased by operating the upstream Xiaowan.

If the Xiaowan Dam is completed before the Pa Mong, the following benefits may be expected:

• The firm output of the Pa Mong (with an assumed full water elevation of 210 m above m.s.l.) will be increased by some 250,000 kW;
• The amount of power generated will be increased;
• The required spillway capacity of the Pa Mong could be reduced;

- The volume of irrigation water could be increased;
- The low-flow volume in the dry season in the delta could be increased by some 800 m^3/s (i.e. by more than 30 per cent);
- The amount of sediment in the Lower Mekong Basin will be reduced;
- It will be possible to reduce the construction cost of the Pa Mong if the other countries (i.e. China, Thailand, and Laos) ratify an agreement jointly to operate both the Pa Mong and the Xiaowan;
- Because of the above-mentioned various benefits expected to accrue from construction of the Xiaowan, the social and environmental effects of the Pa Mong will greatly improve.

On the other hand, if the construction of the Xiaowan Project is delayed and is realized only after the completion of the Pa Mong, various problems may occur as a result of the impoundment of the Xiaowan Reservoir. However, as this impoundment will take place for only a few months during the rainy season, such problems may be negligible.

The two documents outlined above support the Shaowan Project. Is this optimism justified? In the author's opinion, those involved in development of the Mekong should provide opportunities for discussion with as many Chinese government officials as possible, in order to examine this important project with them in depth.

Future of the Lancang River development

For the future of China, the Lancang River development which has already been started is absolutely vital. The Chinese government regards the development of hydroelectric power, which will generate "clean" energy, as one of its greatest priorities in order to cope with the rapid expansion of energy needs in the Yunnan and its neighbouring provinces. Above all, the government is firmly determined to promote development of the Xiaowan Dam as its most important project. Even if negotiation with foreign financial sources to obtain necessary construction funds is ultimately unsuccessful, the government will find some means to accomplish this magnificent project in the long term; furthermore, it will promote the construction of other dam projects on the Lancang.

On the other hand, development of the dam on the Lancang will bring about a great improvement in navigation in the most upstream areas of the Mekong. Communications and transportation in the region will also improve rapidly and the current frequent flood problems in the lower basin will be mitigated, at least to some extent, and, in addition, the flow volume in the dry season at various locations on the main stream will be increased. Owing to the dam development on the Lancang River, the output of dams proposed on the main stream of the lower basin will be increased. Sooner or later, the transmission networks in China will be interconnected and, as a result, a grand power network connecting China to South-East Asia will be achieved.

All the above-mentioned concepts must be a vision held in the minds of all those involved in the development of this great river. The author wonders what effects there will be on the inhabitants' lifestyles as a result of the newly developing situation, in which there may be sweeping trading and cultural exchanges between the previously remote and almost completely closed societies in the Yunnan Province and the already partly or fully open downstream riparian countries in the Mekong Basin.

Notes

1. *International Herald Tribune*, 18 April 1995.
2. The Yangbi River is separated by a single mountain from the plains on the east (located south-west of the city of Dali) which serve as the source of the Yuan River which travels towards the south-east, eventually to flow into the Red River, the major river system of northern Viet Nam. From the watershed of the Yangbi River, the view of these vast plains which give birth to the Red River is truly magnificent.
3. According to the Indicative Basin Plan report published in 1970 by the Mekong Committee, the area of the basin is 186,000 km^2; according to the ADB Project Profile Compendium published in February 1995, the area of the basin is 189,000 km^2 (165,000 km^2 in China and 24,000 km^2 in Myanmar).
4. The left bank partially collapsed during the construction of the Manwan Dam. After this accident, it was feared that the installation of additfonal generators would be difficult.

BIBLIOGRAPHY

Information Research Institute MEP (1994). *Electric Power Industry in China*. Information Research Institute, Beijing.
Manwan Hydropower Project Management Bureau (1993). *Manwan Hydro-electric Power Station of Yunnan Province*. Manwan Hydropower Project Management Bureau, Manwan.
———— (1993). *Manwan Hydro-electric Power Project*. Manwan Hydropower Project Management Bureau, Manwan.
Yunnan Electric (1992). *Lubuge Pounding*. Yunnan Electric, Kunming.
———— (1994). *Power Industry in Yunnan*. Yunnan Electric, Kunming.

5

Environmental problems of dam construction in the tropics

As mentioned in the Preface, the author wrote in 1986 a thesis entitled "The Influence of Dam Construction on the Natural Environment and Society in River Basins Focusing on Dam Development on Rivers in Tropical Continents", which dealt mainly with problems concerning tropical continental rivers. In order to make this thesis easier for the general reader to understand, the author has revised part of his description of the situation in the tropics, focusing particularly on the development of the Mekong.

The present chapter comprises the main part of this thesis. The three main sections outline the author's ideas on dam development and the preservation of the environment, the impact on the natural environment, and the impact on the social environment.

Dam development and the preservation of the environment

The relationship between the development of dams and the preservation of the environment may be compared to that between the two sides of a coin: any development that does not give equal consideration to environmental impact is invalid. On the other hand, environmental degradation can occur even if no development is undertaken. Provided that one considers the preservation of the environment and plans development carefully, excellent results can be expected; however, skilful execution is required.

More than 30 years have passed since the importance of preserving the environment gained acceptance. Recent years have shown many examples of dam construction that was suspended and dam projects that were abandoned, shelved, or drastically altered, owing to belated awareness of negative environmental impact. The financial waste entailed by such an unenlightened approach is obvious and sad; clearly, such waste can be avoided if proper attention is paid at the outset.

Although the preservation of local ecosystems is of great importance, it is also imperative to promote orderly development to meet ever-growing social needs in a manner that is in harmony with the preservation of the environment, which sustains both life itself and local economies.

In regions such as most parts of the African continent and some less-developed areas of Europe, Asia, and Central and South America, where dams are not even an issue, forests are being randomly felled, rice fields thoughtlessly burned, and pastures rampantly overgrazed. The result in some places is erosion of the topsoil and the loss of many species of plant and animal life, including many insects, which, although considered by many to be comparatively lowly creatures, carry out life-promoting activities in ecosystems and are a source of food for higher forms of life. In other places, the damage comes in the form of desertification caused by overburdening the natural carrying capacity. Ill-advised agricultural methods, overconsumption of scarce plant and animal life, and sheer overpopulation all wreak havoc and hasten the degradation of our planet.

Dam construction is criticized by some as being too optimistic, when promises are made that a dam can be built without harming the environment. In truth, even with the most enlightened approach, hard and fast guarantees of a successful outcome cannot be given.

During the 1950s and 1960s, Japan built a number of large dams all over the country to meet the urgent need for food production and electricity generation. Such development was considered indispensable for the reconstruction of Japan's post-war economy and was, therefore, given top priority. Although, in promoting such projects, those involved did their utmost to avoid disrupting or degrading the environment, it cannot be denied that construction of some of the dams during this period had undesirable consequences.

Nowadays, criticism focused on environmental impact, particularly that involving large projects, is very severe. Dam builders have had to be much more sensitive and to undertake provisions and countermeasures to guard against damaging the environment, even accidentally. Planners must always be cognizant of "the law of unintended consequences," to borrow an American expression. The author believes that establishing a responsible balance between development and environmental preservation is highly desirable. In order to achieve success, one must examine all aspects of the relationship between development and the environment.

However, the impact of dam construction on nature and local communities varies. Each project is unique and must be dealt with on its individual merits. Every project must be tailored to the existing situation, taking all potential consequences into account. At times, even the scale of a project must be reduced in order to avoid creating a problem, or to avoid causing damage greater than the potential gain. All dams must be built with the utmost discretion.

Since the 1960s, as part of the effort to provide economic and technical assistance to developing countries, a number of large hydroelectric dams and irrigation and water-supply systems have been built in various parts of the world. This has contributed significantly to the economies of the developing nations. The improvement in food supply, energy access, and standard of living is beyond question. However, emphasis was often placed solely on raising economic output, while environmental impact was neglected. As a result, not a few dams seriously damaged their surrounding regions. The duty of those planning dams in developing countries should be to seek harmony between human needs, nature, and economic goals. This chapter summarizes the influence of dam development on nature and society in the lower Mekong Basin. To study the environmental effects of dam construction, the author has delineated the following four categories of concern:
1. Areas around the dam and the reservoir;
2. Interior of the reservoir;
3. Downstream areas;
4. Other concerns.

The results of this study can be seen in table 5.1. Those places where the impact was assessed are shown in figure 5.1. Among the important items of environmental impact shown in table 5.1, those in categories 1, 2, and 3 can be classified as effects on the environment. Category 4 includes "natural" items, such as fish and wildlife, and matters designated as "social," which includes the resettlement of groups of people, landscape preservation, preservation of adequate arable land, recreation, protection of archaeological relics, and waterborne/water-related diseases.

Impact on the natural environment

This section deals with the potential physical, scientific, and biological effects on the natural environment of the lower Mekong Basin and related river basins when a dam is built on the river. Of particular concern is the effect around the dam and the reservoir, the effect at the interior of the reservoir, and downstream (table 5.2); this is hard to predict; moreover, pinpointing cause-and-effect relationships is no easy task. It is,

Table 5.1 Major environmental effects of dam construction

Location, number, and abbreviation[a]	Full description
(1) Around reservoir	
1. WE	Lake-water evaporation
2. MC	Change of microclimate
3. WL	Water leakage
4. IE	Induced earthquake
5. LC	Land collapse
(2) Interior of reservoir	
6. TR	Turbidity of reservoir
7. ET	Eutrophication
8. ST	Sedimentation
9. WT	Change of water temperature
(3) Downstream	
10. RD	River-bed scouring
11. RE	River-bank erosion
12. AS	Accumulation of salts in soil
13. DF	Impairment of soil fertility in delta
14. SI	Sea-water influx
15. CE	Coastal erosion
(4) General	
16. RS	Resettlement
17. PA	Preservation of aesthetics
18. RC	Recreation
19. PR	Protection of archaeological relics
20. FP	Effects on fish productivity
21. PW	Preservation of wildlife
22. WD	Water-related diseases

Source: Original.
a. As in figure 5.1.

therefore, important to make a thorough and comprehensive scientific study of the potential problems caused by the dam. It is also necessary to investigate the possible consequences for each region of the entire basin, and such matters are highlighted in this section.

Reservoir evaporation

Wherever a dam is built, rainfall and wind activity increase and reservoir water evaporates. In some areas, mist becomes prevalent. In the frigid and temperate zones, the amount of reservoir evaporation is less than in the tropical zone: in Japan, for example, the amount of evaporation is said to be some 66–80 per cent of local precipitation; in the Tennessee Valley Development in the United States, the amount of water lost annually due to evaporation comes to roughly 57 per cent of annual pre-

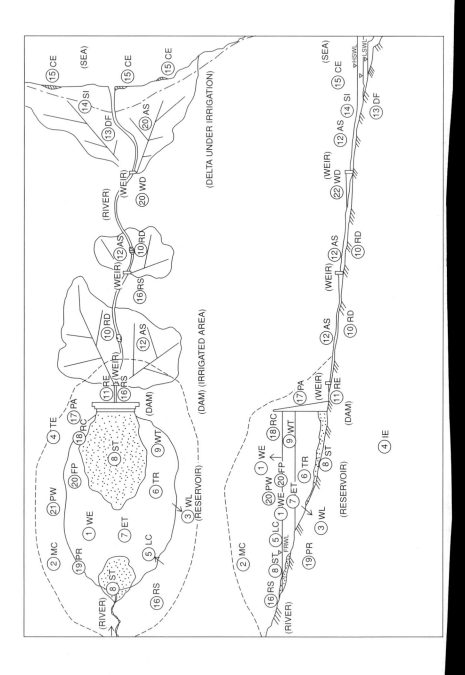

Table 5.2 Direct effects of dam construction upon the natural environment

Area surrounding reservoir and upstream	Interior of reservoir	Downstream from reservoir

Upstream:
(a) Accumulation of sediment in backwater of reservoir
 	
Surrounding area:
(a) Land collapse
(b) Induced earthquake
(c) Water leakage and change in level of groundwater
(d) Lake-water evaporation
	(a) Sedimentation
	(b) Eutrophication
	(c) Turbidity of reservoir water
	(d) Change of water temperature
		Immediately and further downstream:
		(a) River-bank erosion
		(b) River-bed scouring
		(c) Change in water level
		(d) Change in flow volume
		
		In delta and estuary:
		(a) Reduction in sediment
		  • Accumulation of salt in soil
		  • Impairment of soil fertility
		(b) Change in water level and flow volume
		  • Change in degree of sea-water influx
		  • Change in groundwater level
		  • Change in extent and depth of flooded area
		
		In coastal area of estuary:
		(a) Change in coastal drift
		(b) Coastal erosion

Effects on wild animals and plants
Water-related diseases
Effects on fish and shellfish
Change in landscape

Source: Original.

cipitation when the actual level of evaporation is added to the water that would have evaporated from the vegetation that stood within the areas inundated by the dams. That 57 per cent is equivalent to the total amount of evaporation from the Tennessee River prior to construction of the dam; in other words, the effect of the dam in this respect has been negligible.[1] In the tropical zone, on the other hand, mist sometimes occurs in dry, windless areas if water that has evaporated from a reservoir during the day cools at night. Furthermore, if the air around the reservoir becomes too humid, it can give rise (it is said) to lung disease. In the case of the

Aswan High Dam, heavy rain began to fall on and around the reservoir, something that had never occurred before construction; this is thought to be due to substantial evaporation from the newly created reservoir, a factor which must be considered when planning dam construction in the tropical zone.[2]

The amount of evaporation varies, influenced by the area of water; the surface temperature; the pressure, humidity, and temperature of the air over the surface; and other factors. Evaporation can be calculated by multiplying a coefficient by an actual value measured by an evaporator; alternatively, it can be assessed by calculation alone, without the aid of any measurements. When the Volta–Akosombo dam was built in Ghana, a study was undertaken that compared calculated (theoretical) data and measured data. The theoretical data were calculated using two formulae – one for calculating the amount of radiation energy needed to evaporate a specific volume of liquid and one related to air movement, of which air–vapour pressure and wind velocity are the parameters. The study showed that the theoretical data and the measured data were in agreement. Evaporation from the Volta–Akosombo reservoir is 1,650 mm/year ± 10 per cent (Harclow 1956). In the case of the Bhakra Dam, however, the theoretical figures were only 70–80 per cent of the measured figures.

The amount of reservoir evaporation and of precipitation both ultimately affect the amount of electricity generated by a dam. A survey by the Preparatory Committee for the Development of the Volta–Akosombo dam reported that the worst-case scenario would be a loss of 11.5 per cent of the output of the power station (617 MW) when annual evaporation was assumed to be 1,400 mm and the annual precipitation was estimated at 1,270 mm.

Expectations for the Pa Mong Dam

The proposed site of the Pa Mong Dam is located in a typical tropical zone. In the lower Mekong Basin, the south-westerly monsoon blows from May to October. Much of the precipitation falls on the western side of the Annam Cordillera, where temperature and humidity are both high. Cloud cover is dense, with the densest cover occurring in August and September. From November to March, when the north-easterly monsoon blows, the basin receives little rain, and both temperature and humidity are quite low; cloud cover is light. In the intervening period, between the end of the north-easterly monsoon and the start of the southwesterly monsoon – i.e. from the end of March to early May – the temperature is uncomfortably high.

The average monthly temperature in the lower Mekong Basin is 26–30 °C in the rainy season and 20–25 °C in the dry season. The highest temperature in the rainy season is about 45 °C, whereas the lowest temperature in the dry season is nearly 0 °C. The average monthly humidity is greatly affected by the wind and rain, but the relative humidity in the rainy season is 75–80 per cent. The level of evaporation peaks in March and April, when the temperature is at its highest and humidity is at its lowest. In a seeming paradox, evaporation is lowest at the peak of the rainy season.

Annual precipitation in the town of Khon Kean in north-east Thailand – located some 120 km due South of the proposed site of the Pa Mong Dam – is roughly 1,000 mm; however, the amount of rainfall in the area around the proposed site increases in reverse proportion to proximity to the mountain ranges of Laos: in Vientiane, annual rainfall exceeds 1,500 mm and the May–October rainy season accounts for roughly 88 per cent of that total.

According to the Interim Mekong Committee records published in July 1992, the following averages prevailed at the proposed Pa Mong Dam site:
- Temperature: 25.7–27.3 °C;
- Humidity: 66.9–74.2 per cent;
- Rainfall: 1,533–2,137 mm.

At this point, no change in climate is expected to result from the creation of the Pa Mong Reservoir or the construction of the related irrigation system, even if the High Pa Mong Dam Project is carried out. However, in the dry season, evaporation, evapotranspiration, temperature, humidity, and other factors may be affected to some extent in the immediate vicinity. Nevertheless, it is believed that the winds characteristic of the Pa Mong Dam district and reservoir might weaken those effects (Mekong Secretariat 1982).

If the height of the Pa Mong Dam is reduced, as is being discussed at the time of writing, the size of the reservoir would be much smaller than originally planned, which would mean less evaporation. However, if the dam is built to the height originally planned, it is estimated that less than 1 per cent of the reservoir's annual inflow would be lost to evaporation.

Water leakage

Water leakage from dams and reservoirs is one of the problems of dam maintenance. It can be classified into the following three categories:
1. Leakage through the body of the dam;
2. Leakage through the foot and/or the abutment of the dam;
3. Percolation and leakage from the reservoir.

Leakage through the body of the dam takes place at joints and cracks in the dam. The amount is somewhat affected by changes in the water level of the reservoir, but it is more significantly affected by seasonal changes in water temperature and air temperature. As time passes, a gradual decrease in this type of leakage usually occurs as the escape routes are slowly eliminated.

Leakage through the foot and/or the abutment of the dam, and percolation and leakage from the reservoir, change according to changes in the water level. Leakage can be a dam's mortal enemy: in the state of Idaho in the United States, the Tieton (fill-type) Dam collapsed in June 1976 because of water leakage due to piping phenomena that occurred where the dam core met the base rock; this was a rockfill dam.

Water leakage seriously undermines dam safety, but prevention of leakage is also important for the preservation of the volume and level of the water in the reservoir. At the planning stage, the importance of including measures to prevent leakage cannot be overemphasized: not only does leakage put at risk the original capital committed to build the dam, it also jeopardizes any additional capital that may be used to remedy subsequent defects. Leakage can also have a significant effect on the lives and livelihood of the people living around the dam and reservoir.

After the creation of a reservoir, changes in the groundwater level sometimes occur in the areas around the reservoir and downstream. This is particularly so in tropical regions, where people still utilize wells and fountains as important water sources for drinking and irrigation. A change in the groundwater level in the dry season can be very significant. Generally speaking, it can be said that water leakage favourably affects the people living around the reservoir, as it often raises the groundwater level. Apart from the groundwater level and volume, leakage also affects water quality. In tropical regions, reservoirs are sometimes biologically contaminated owing to the high temperatures of the water and the air.[3]

Finding, mitigating, and preventing water leakage is not an easy task, even in developed countries; it is much harder in developing countries.[4] To implement appropriate countermeasures, it is necessary first to find and trace the leaks. However, establishing observation networks around a large reservoir is expensive in terms of both time and labour, particularly in the tropics. The local population must be recruited and taught how to monitor accurately the indicators placed at private and public wells, but their cooperation (including the unsupervised discipline necessary to perform the task effectively) does not come easily, and their data are not completely reliable. Even under the best of circumstances, it is difficult to make continuous observations.

Water leakage from the Aswan High Dam

In the 1950s and 1960s, when the Soviets were helping to build the Aswan High Dam while the world watched, some of the Western engineers involved in the dam's construction warned that the reservoir would never fill completely, owing to loss through leakage. The reservoir started to fill in 1964; the water level continued to rise until it reached its planned peak of 175 m in 1975; as is universally known, the dam has continued to operate quite successfully and has attained its projected targets, including those regarding power generation, flood control, navigation, and irrigation. However, evaporation and some water leakage have occurred. For instance, leakage in 1975 and 1976 was reported to be 2.017×10^9 m^3, equivalent to 2.4 per cent of the annual inflow (roughly 85×10^9 m^3). Evaporation loss in the same period was reported to be some 14.257×10^9 m^3; this amounts to about 16.8 per cent of the annual inflow, although this figure is admittedly imprecise.

Water leakage in the lower Mekong Basin

In trying to estimate the probable amount of water loss from the dams and reservoirs that will be developed in the lower Mekong Basin, the author sought reference data on the existing dam projects in the Mekong River Basin: these include the Manwan Dam Project on the main stream in the upper basin, the Nam Ngum Dam Project, and other projects on the tributaries in the lower basin; unfortunately, current data were not available.

While engaged in the planning of dams in the lower Mekong Basin at the office of the Mekong Committee during the period 1964–1968, the author estimated the annual loss from each mainstream dam project as about 10 per cent of annual inflow. However, arriving at a reliable estimate can be complicated by many contributory factors, including the height of the dam and the extent of the drawdown.

Nam Theun 2 Dam Project

The area upstream of the proposed Nam Theun (tributary) 2 Reservoir (which is located on a plateau with an elevation above 500 m) is approximately 3,950 km^2. When the dam is built, the reservoir will have a total capacity of 6.7×10^9 m^3 and comprise an area of 530 m^2. The dam will be 40 m high, with a crest length of 330 m and a volume of 2.1×10^6 m^3. The plan is to allow the reservoir water to fall to the adjacent tributary, to generate some 600,000 kW cheap electricity, or 4,650 GWh year.

One of the major worries with this huge power project is water leakage from the dam and reservoir. The dam was to be built on a natural foundation of Mesozoic limestone through which severe water leakage was expected. Although a safer site for the dam has now been selected, the possibility of water leakage through the bottom of the reservoir and elsewhere still remains.

Earthquakes and the creation of reservoirs

As far as the author knows, no systematic research was made public before the 1940s concerning whether the creation of a reservoir can cause an earthquake. The study of induced earthquakes is relatively new.[5]

Dam safety is of paramount importance. It is particularly crucial when building a high dam with a correspondingly large reservoir, when building a dam in an area where earthquakes frequently occur, or when the geological conditions at the dam site are problematic.

Studies in Japan

Japan is situated in a geologically unstable zone where active tectonic movement causes frequent earthquakes. Consequently, the utmost care must be taken in every aspect of the design, planning, construction, maintenance, and operation of a dam. Particular attention must be paid to the geological conditions of the baserock in order to determine whether there are any faults. If faults are found, they must be carefully and precisely examined before construction can begin. The available records of past earthquakes and volcanic activity must be studied with the utmost care and caution.

A number of high-quality papers are available in Japan on safeguards against earthquake damage to dams. Nevertheless, it is not clear whether the creation of a man-made reservoir can induce an earthquake, or whether earthquakes are at all related to dam construction.

When he was chief of the Office for Research on Earthquake Activity at the National Disaster Prevention Science and Technology Centre, Professor Seiji Otake of Tohoku University studied records of past earthquakes occurring to a depth of 30 km. He utilized the catalogue of past earthquakes stronger than magnitude (M) 3 that had occurred in Japan between 1926 (when the first nationwide network of meteorological observatories was established) and 1984. Professor Otake concluded that there was a 90 per cent probability that tremor activity had increased in the areas around eight of the 42 dams that were more than 100 m high (Ohtake 1984). For example, although earthquakes had occurred in the area near the Makio Dam before the impoundment of water in the reservoir, they became unusually frequent after impoundment: 14 years after the reservoir was filled, a remarkable series of earthquakes, including one of M 5.3, occurred within a radius of 5 km. Despite this, no verifiable relationship could be established between the occurrence of the earthquakes and the filling of the reservoir (Ohtake 1984).

Professor Shunzo Okamoto of Tokyo University, while admitting that it cannot be proved conclusively which conditions might cause such earth-

quakes, has concluded that earthquakes are likely to be induced in the following circumstances:
1. When the depth of the reservoir exceeds 100 m;
2. At relatively shallow spots underground (but not of significant magnitude);
3. As a result of reservoir water impoundment (but not of a magnitude exceeding that of a natural, non-induced earthquake).

He also stated that an earthquake might be induced by reservoir-water impoundment where potential earthquake energy had already been building up. However, he claimed that the accumulation of more data would be necessary before any definite conclusions could be reached. According to Professor Okamoto, there are a number of fissures in the earth's crust to a depth of 10 km. Once impounded water enters such fissures, it can swell rock deposits, increasing the internal subsurface pressure. Furthermore, the swelling increases the fissure pressure, and the result can be widespread shattering of underground rocks due to the accumulated internal stress (Okamoto 1984).

Some examples in the tropics

Although it cannot be said with certainty whether there is a relationship between induced earthquakes and climatic conditions, the author's research has led him to conclude that there is probably no connection. There are several reports on so-called construction-induced earthquakes in both cooler (temperate) developed countries and in tropical countries. In the developed countries, there are several well-known examples, such as the Hoover Dam,[6] where successive earthquakes occurred after the reservoir was filled in 1935, and the Nurek Dam in Russia, where earthquakes occurred in 1972. More germane for our purposes, however, are accounts from tropical countries, as these are more relevant to the future development of the lower Mekong Basin.

Volta-Akosombo Dam (Ghana).[7]   Ghana is not an earthquake-prone country; however, on the coast near the capital Accra, five large earthquakes occurred in 1863, 1883, 1906, and 1939; one was very strong, of M8 or M9. Ten smaller earthquakes were also recorded after 1858 (Kumi 1973).

The impoundment of water in the reservoir of the Volta–Akosombo dam began in May 1964. An earthquake occurred in November of that year, when the water volume in the reservoir had reached 28×10^9 m³. At that time, no seismograph had been installed, but after experts reviewed seismographs in 11 neighbouring countries, they concluded that the epicentre would have been at Koforidoa, which is located some 40 km south of the Volta–Akosombo reservoir, and the intensity would have been about M5, a weak earthquake.

In December 1966, when the impounded water had reached 102×10^9 m^3, another earthquake of similar intensity occurred, and experts surmised that its epicentre was likely to have been in the sea off Accra. In February 1969, a slightly greater tremor occurred on the coast near Accra, coinciding with the attainment of a water impoundment of 160×10^9 m^3 (Kumi 1973).

Kariba Dam (Zambia–Zimbabwe).[8]   In this case, the impoundment of water began in 1958 and dam operations were officially started in May 1960. In 1959, the first earthquake occurred when the water volume reached 18.5×10^9 m^3 and the water depth reached 90 m. Five weeks after the reservoir reached its full capacity, a larger earthquake, of M6, struck; subsequently, nine tremors, of M5.1 to M6.1, occurred from August to November 1963. A number of weaker 'quakes, with magnitudes of less than M3, were also registered, their epicentres widely scattered. The biggest 'quake occurred 20 km upstream from the dam. Earthquakes registering M2.5–M3.8 occurred as many as 89 times from September to November 1971; All were centred within the reservoir area (Zein and Lane 1973). There are faults along the reservoir's perimeter: some are underwater, while others are under the dam itself. It is presumed that these faults might be the cause of the earthquake activity (Zein and Lane 1973).

Murchison Falls Dam (Uganda) (planned).   Murchison Falls is located on the Victoria Nile River downstream from the mouth of Lake Victoria, from which the Victoria Nile flows. The site has long been considered excellent for power generation because of the river's stable flow. However, owing to fears of harming the wildlife, which includes crocodiles, hippopotamuses, and many other species, and of despoiling the beautiful natural environment, the plan was shelved and is now considered defunct. The project had another fatal defect: a fault runs under the intended dam site, and the planned reservoir was to be situated at the northern edge of the Rift Valley, where earthquakes have occurred frequently.

Koina Dam (India).   The Koina River is a tributary of the Krishna River, which runs south-east of Bombay and flows to the Arabian Sea through the basalt Dekan Plateau. In April 1959, construction began on a concrete gravity dam with a height of 104 m, a crest length of 800 m, and a total water volume of 2.78×10^9 m^3. Water impound began in 1962; soon afterwards, successive earthquakes occurred; in 1964, three seismographs were installed.

The Koina Reservoir is 50 km long and lies on a north–south axis. The dam is situated at the south end. In September 1967, earthquakes of M5 to M5.5 occurred and continued for two days along a 15 km stretch of the western shore of the reservoir covering a breadth of 6–8 km at a depth of 6–8 km, causing local disasters. On 10 December, a severe earthquake occurred measuring M6.3, causing a major catastrophe: some 180 people died; horizontal cracks appeared in the dam wall on both the upstream and downstream faces; the cracks appeared where the sections of the dam interfaced and at the crest of the dam. The severity of the damage was exacerbated by the vulnerability of the mud and clay houses built near the reservoir. News of the tragedy shocked people everywhere and highlighted the need to study induced earthquakes.

Aswan High Dam (Egypt). A number of earthquakes have occurred since the reservoir of the Aswan High Dam reached its full water level in 1975; however, there is no record of an earthquake ever having occurred before 1980. On 21 March 1982, there was an earthquake of intensity M4.7 and, on 24 February 1983, a 'quake of intensity M4.3 occurred under the reservoir itself.[9] The Egyptian government then installed a seismological observatory at the dam site to study the relationship between changes in the reservoir water level and the occurrence of earthquakes. Some believed that the earthquakes may have been induced by the newly created large body of water, which might have disturbed existing faults; however, the government tried to reassure everyone that such disturbances would gradually end as the underlying earth recovered its equilibrium over time.

Pa Mong Dam (lower Mekong Basin) (planned). The Mekong Committee has long paid attention to the possibility of inducing an earthquake by the creation of a large reservoir on the Pa Mong. Its investigation discovered that faults, one of them the Ban Ang fault, ran under the proposed dam site. However, the committee dismissed the theory that an earthquake can be caused when an underground fault zone is subjected to pressure from reservoir water soaking into the rocks. Fortunately, the faults in question are ancient, so they were thought to be inactive. At the time of writing, there is no information available to show that earthquakes have, as yet, struck the Pa Mong area.

As there is no movement of the earth's crust in the lower Mekong Basin, there are few volcanoes or hot springs. However, the upper basin is active: it contains several volcanoes, and earthquakes are a frequent occurrence, especially in the area near the Arakan mountain range in Myanmar, where the Pu Fai Yai and the Pu Fai Noi volcanoes are situated. Both these volcanoes are 60–90 m higher than the Muang Hongsa-wadi Plateau.

An M6 earthquake occurred in the 1990s in the vicinity of the planned Xiaowan Dam on the Lancang River in the upper Mekong. The upper Mekong Basin is an earthquake zone; it borders Myanmar, where earth crust movement is considerable, and where, it is feared, earthquakes may occur if dams are built in the upper Mekong Basin.

Few such fears surround the proposed site of the Pa Mong Dam. In fact, during the designing of the Nam Ngum Dam near Vientiane, the effects of earthquakes were taken into account. Some minor earthquakes have occurred at the Mekong Estuary, and hot springs exist in the southern parts of Viet Nam, just outside the Mekong Basin. So far, however, it cannot be predicted with certainty whether earthquakes would be induced by dam construction.

Induced earthquakes: their classification and characteristics

From the above examples of dam building and coincidental earthquakes, so-called induced earthquakes can be classified as follows:

1. Active versus inactive zones of earthquake activity:
   (a) Reservoir water impoundment in an earthquake-prone zone increased the frequency of tremors. However, there is no proof that the increased frequency is related to changes in water level and/or storage volume in the reservoir;
      - Increased frequency that cannot be explained by changes in water level and/or storage volume in the reservoir
      - Increased frequency explicable in terms of changes in water level and/or storage volume in the reservoir
   (b) An earthquake occurs even though the reservoir is built in an area that has no history of earthquakes.
2. Presence or absence of earthquakes:
   (a) Earthquakes occur some time after the reservoir is filled, followed by more periodic earthquakes;
      - Intermittent small earthquakes
      - Earthquakes of increasing magnitude over time
   (b) A single earthquake takes place but is not followed by any others.
3. Location of seismic centres:
   (a) Seismic centres are located within the reservoir areas;
   (b) Seismic centres are located outside the reservoir areas;
   (c) Seismic centres are located at the perimeter of the reservoir areas.

It should be noted, however, that data obtained in areas or countries where observation systems are not well organized lack credibility and should be treated with caution: for example, in one case the actual seismic centre was found to be some 10 km away from where it was originally supposed to be located.

Although attempts have been made to classify induced earthquakes according to the categories described above, the relationship between the creation of reservoirs and the occurrence of earthquakes has not been entirely elucidated. It may be possible in future to predict earthquakes, but success and consistency will depend on taking into account factors such as topography, geology, existing faults, and internal tectonic tensions of the area occupied by the reservoir and its surroundings. All these must be related to the induced earthquake classifications described above.

Earthquake observation

As stated above, earthquakes may not be linked to differences between the various climatic zones; however, when an earthquake occurs in a developing country, greater damage often occurs because local dwellings are not always built to earthquake-resistant standards. In developing countries, although the relationship between earthquakes and the creation of reservoirs is not yet clear, it may be necessary to exercise greater caution when evaluating the possibility of inducing earthquakes than would be exercised in developed countries, where monitoring activities can be made more precise. Budgets for dams in developing countries should include funds for a network of properly equipped seismological stations. Moreover, monitoring should be continuous throughout the construction period or, at the very least, should be in place before water impoundment begins.

Recent studies on induced earthquakes

The above is a summary of information published before 1987 on induced earthquakes worldwide; however, it does not touch on the Mekong Basin. Since 1987, several more papers on the subject have been published: one of the latest came out in May 1993 and was written by Dr T. Vladut, who was part of a Canadian research team, the Hydro Environmental Research Group. According to Dr Vladut, more than 2,000 papers have been published on induced earthquakes. On the basis of those studies, he reached the following conclusions:

1. It was in 1932 that people first became aware of the possibility of the induced earthquake in association with the Qued Fodda Dam in Algeria. In 1945, people paid attention to the view that there might be some relation between the depth of the reservoir water of the Hoover Dam and the earthquakes which occurred.
2. It is believed that, at more than 120 dams built around the world since 1932, earthquakes occurred during water impoundment and/or during reservoir operation.

Table 5.3 Risk of induced earthquakes in relation to scale of reservoir

	Scale of reservoir								
	Water depth (m)			Reservoir volume (10^9 m^3)			Reservoir area (km^2)		
	50	100	150	0.2	1	10	5	50	500
Percentage risk of occurrence	14	52	83	20	55	80	35	72	95

Source: Vladut (1993).

Table 5.4 Examples of induced earthquakes

Name of dam (and country)	Dam height (m)	Reservoir volume (10^9 m^3)	Average water depth (m)	Magnitude of 'quake (M)	No. of years elapsed until 'quake
Koyna (India)	103	2,780	44.4	6.5	3
Kremasta (Greece)	165	4,750	53.3	6.3	3
Kariba (Zambia)	128	180,600	27.5	6.25	4
Marathon (Greece)	63	41	25.1	5.75	8
Aswan (Egypt)	111	168,900	42.2	5.5	17
Volta Grande (Brazil)	33	2,300	13.3	5.2	1
Porto Columbia (Brazil)	53	2,400	21.2	5.1	2
Hoover (USA)	221	34,852	73.8	5	3
Kurobe (Japan)	186	149	80.1	9.9	1
Akosombo (Ghana)	120	148,000	17.4	4.7	1
Nurek (Tajikistan)	317	11,000	126.8	4.5	3
Camarilas (Spain)	44	40	14.3	4.1/3.5	1

Source: Vladut (1993).

3. So far, it is believed that induced earthquakes occurred where the reservoir depth was more than 100 m, or when the storage volume reached more than 1×10^9 m^3.

The results of a study on 60 dams have shown that there is a greater possibility of inducing an earthquake if the depth of the reservoir exceeds 100 m or if the water storage volume is more than 1×10^9 m^3.

Table 5.3 summarizes such factors as maximum water depth, volume, and the surface area of water in reservoirs. It also illustrates the risk of inducing an earthquake by classifying and cross-referencing depth (50, 100, or 150 m), storage volume (0.2, 1, or 10×10^9 m^3), and surface water area (5, 50, or 500 km^2). Table 5.4 compares the magnitude of induced earthquakes and dam height, storage volume, and average depth.

Land collapse

Decision-making priorities when planning a dam project are dictated by its goals, based on development requirements, and include the optimum size of the dam, its design, and the operation programme necessary to attain those goals. Planners must collect data on topography, geology, hydrometeorology, and other matters; they must also study all possible alternatives. The geological study is particularly important, requiring exhaustive investigation to ensure the safety of the dam. Until the disastrous accident in 1963 at the Vajont Dam in Italy, very few planners worried about the possibility of land collapse in the area surrounding reservoirs.

Generally speaking, land collapse occurs when some or all of the following conditions are present:
1. Existing geological defect(s) in the area surrounding the reservoir;
2. Severe rainfall;
3. Extreme fluctuations in the reservoir water level.

Of these, the first is probably the most important. The land collapse at the Vajont Dam is an example from a temperate country, while that at the Furnas Dam in Brazil is an example of its occurrence in a tropical country. At the Vajont Dam, all three conditions were present. Unfortunately, those in control of operations compounded the disaster by completely misjudging the situation. In fact, the human error factor was as significant as it was during the Chernobyl nuclear disaster in the former Soviet Union in 1986, where mistakes quickly made a perilous situation completely disastrous. Although land collapse obviously does not threaten the global environment, as does escaped radiation, in both cases a lack of foresight and faulty contingency planning compounded the consequences. Regarding the Furnas Dam, it was concluded that conditions (1) and (2) caused the land collapse.

In tropical countries, the necessary basic data are not always available in sufficient quantity at either the planning stage or the execution stage; consequently, it is difficult to predict land collapse. Moreover, if the risk is not considered to be important enough to warrant attention during planning, those responsible for operating the dam will find it hard to justify expending the time, effort, and cost required to measure the conditions of the surrounding land once the dam has been built, and they seldom undertake to do so.

Examples of land collapse

Vajont Dam (Italy).   Vast sections of the left bank of the reservoir of the Vajont Dam on the Pieve River in northern Italy collapsed in October 1963, causing many deaths. The news shocked dam experts worldwide, prompting the Japanese government to order a thorough study of the area surrounding all existing dams in Japan. Pre- and post-construction examinations have been conducted at every project site since then. As a result of this thoroughgoing effort, Japan has suffered no property loss or personal injuries due to land collapse.

The causes of the land collapse at the Vajont Dam[10] were given as follows:

1. Geological conditions on the left bank of the reservoir were unstable;[11]
2. Heavy rain had fallen at the site the previous day;
3. The reservoir water level had been rapidly drawn down;[12]
4. Owing to overconfidence in the results of experiments involving models and the effects of grouting work, an unjustifiably optimistic judgement was made by those responsible for operating and monitoring the dam;[13]
5. There was delay in issuing an evacuation order.[14]

Causes (1) and (2) can be classified as natural, whereas causes (3)–(5) can be classified as man-made – specifically, judgement errors by management. However, on closer analysis, it is clear that the core of the problem was the failure to take a comprehensive view. Those who were engaged in planning and design focused on narrowly delineated categories, such as the strength of the base of the dam, but they failed to consider other related risks.

Furnas Dam (Brazil). The Furnas (rockfill) Dam was constructed solely for power generation in the early 1960s on the Rio Grande in the state of Minas Gerais, Brazil. The reservoir was full to capacity by May 1965: the output was 1.2×10^6 kW, the height 127 m, and the crest length 550 m.

The land collapse did not occur at the reservoir itself, but rather on the right bank of a tributary flowing into it. It occurred in January 1969, after a period of heavy rainfall. Although no serious accident resulted, as much as 7,500 m^3 of soil collapsed. Changes in the reservoir water level, the volume of outflow, and the timing of the land collapse were all investigated to determine if there was any relationship between them. Heavy rain had fallen from near the end of December until January 1968. The collapse occurred when the water level was some 10 m below the reservoir's maximum capacity. Unlike the situation in the Vajont Dam, the water level had not fluctuated greatly or rapidly, but there were geological weaknesses in the surrounding land. It was concluded that the prolonged heavy rain was the factor that triggered the collapse.

The report on the collapse gave the following four reasons to explain it:

1. The earth's surface layer, which had been eroded, became unstable owing to the extreme distortion of the existing faults in the underlying quartzite;

2. Rainwater flowed across and through weaker portions of the cliff rock on the right bank, gradually eroding the rock and opening destructive water paths;
3. The eroded cliff rock had collapsed previously in places;
4. Some water had permeated the rock, saturating it and causing an increase in the rock's internal water pressure.

Potential collapse in the area surrounding the Pa Mong Dam

If the plan that calls for the higher dam is approved and executed, there is a possibility of land collapse on the eastern and southern fringes of the reservoir because the rocks are badly weathered.

If land collapse were to occur and the soil fell into the reservoir, the consequences would be disastrous; however, this potential hazard can be avoided. Under the original plan, the reservoir's maximum water level would be 250 m. There has been a proposal to reduce this to 205–210 m, which would make the risk negligible. Under the original plan, the water level would be drawn down by some 12–18 m at a rate of 2.38 m/month at peak demand during the dry season, which runs from November through June. This reduction of the water level would be recouped at a maximum speed of approximately 5.43 m/month during the wet season, which runs from July to October (Mekong Secretariat 1976). There were fears that such a rapid fluctuation in the reservoir's surface might induce land collapse.

Whatever the final decision regarding the height of the Pa Mong Dam, trees at the edge of the reservoir should be cut down before water impoundment begins, since they may soon wither and die; this must be done around any reservoir where the water level fluctuates significantly. However, with the trees gone, the rise and fall of the water level can erode and expose the subsurface soil, making the shoreline unsightly, or worse, can contribute to land collapse.

People who take up residence around a new reservoir can be expected to engage in any number of activities, including fishing, farming, tourist-trade enterprises, and navigation services. Roads, piers, houses, shops, and other structures may be built beside the reservoir, and they are all susceptible to damage should any land collapse occur. Therefore, the utmost care should be taken; foolhardy activities such as the building of houses near the waterline should be strictly forbidden around reservoirs where the water level will be drawn down.

Water temperature in man-made lakes and reservoirs

When a dam is built in a country in the temperate zone, such as Japan, the temperature of the water downstream is often colder than it was before the dam was built. This can sometimes cause trouble for farmers who grow rice, since irrigating with cold water slows crop growth; however, such problems do not arise in the tropical zone, because the water temperature at the surface of man-made lakes and reservoirs in the tropics does not differ greatly from that at the bottom. Generally, the water temperature is roughly 30 °C at the surface and about 20 °C at the bottom in the deepest parts. In a tropical reservoir or lake, there is a progressive decrease in water temperature with increasing depth. In addition, the thermoclines of the water remain fairly stable throughout the year.

Dams in the lower Mekong Basin

The proposed site of the Pa Mong Dam is in the northern section of the lower Mekong Basin, where the air temperature exceeds 27 °C from March to September. During the floods, when inflow is significant, the water in a small reservoir would mix and blend, eliminating thermoclines for the most part; however, if the reservoir is large, even if it is shallow, thermoclines form during the very warm period and the resulting stratification stays constant all year round.

Nam Pong Dam (also called the Ubolrathana Dam) in north-east Thailand. The upstream catchment area of this dam is some 12,000 km^2 and the annual inflow volume is 2.6 × 10^9 m^3. The gross storage volume of the reservoir is 2 × 10^9 m^3 and its surface area is 410 km^2. The average water depth is 16 m (27 m in the vicinity of the dam and 20 m near the intake). The water in the reservoir remains stratified throughout the year (Mekong Secretariat 1976).

Table 5.5 illustrates water-temperature data obtained in the vicinity of the Nam Pong Dam and its intake. The difference between the temperature at the water surface and that at the bottom reaches a maximum of 3.2 °C in July. The air temperature reaches its peak during the period April–July. As the radiant heat of the sun intensifies, the water temperature of the Nam Pong Reservoir also increases, peaking in June and July. Thereafter, the difference between the water temperature at the surface and at the bottom gradually decreases.

Nam Ngum Dam, Laos. With a maximum depth of 40 m, the Nam Ngum Reservoir is much deeper than the Nam Pong Dam, and the water is naturally stratified. The water temperature varies from 18 to 32 °C. The air temperature, insolation, and wind around the reservoir influence the formation of thermoclines.

During the rainy season, the inflow increases, obliterating the thermoclines. The wind disturbs the water surface, causing some turbulence, but

Table 5.5 Water temperature (°C) of the Nam Pong Reservoir

Position in reservoir		Month					
Location	Depth	December	January	March	May	July	September
North-eastern end	Surface	29.3	25.5	29.3	30.5	28.4	29.4
	Bottom	28.0	24.5	28.1	30.0	27.0	28.8
Southern end	Surface	28.0	24.8	29.6	34.1	33.8	28.4
	Bottom	27.0	24.5	25.0	30.7	30.6	27.0
Eastern end	Surface	27.3	23.8	27.9	30.4	33.8	29.4
	Bottom	26.2	23.2	25.4	29.8	30.6	28.5

Source: Mekong Secretariat, Water Weeds and Studies on Fish, Nam Pong Environmental Management Research Project, Interim Mekong Committee, April 1980

the effect is moderated somewhat by the islands and inundated trees, which act as a barrier. This diminishes the amount of vertical mixing that takes place, resulting in quite significant temperature differences between the water at the surface and the deeper layers.

The water in the Nam Ngum Reservoir is divided into three distinct layers:

1. The surface layer is warmest, usually 23–32 °C, and runs to a depth of 6–10 m;
2. The middle layer, affected most by the inflow, is at a depth of 10–25 m and has a temperature of approximately 23 °C;
3. The deepest layer, at a depth of 30 m or more, has a temperature of 18 °C or less.

Pa Mong Dam (projected). According to the Interim Mekong Committee, the climatic conditions in the area around the projected Pa Mong Dam and Reservoir are as follows:

• Air temperature: 25.7–27.3 °C;
• Humidity: 66.9–74.2 per cent;
• Wind velocity: 2.1–3.9 knots;
• Evaporation (pan method): 1,533.7–2,137.3 mm;
• Evapotranspiration potential: 1,695.3–2,061.2 mm;
• Precipitation: 1,105.8–2,278.9 mm.

In the lower basin, the temperature of the main stream decreases in proportion to the distance from the Golden Triangle area. Measurements taken in 1972 and 1973 at various hydrological and meteorological stations established on the main stream of the Mekong gave the following data for average water temperature:

- Upstream: 27 °C;
- At Pa Mong: 24.6 °C;
- At Nahkon Phanom: 22.5 °C;
- At Mukdahan: 23 °C;
- At the point furthest downstream in the delta: 22 °C.

On the assumption that the Pa Mong Dam would be built to an elevation of 250 m above m.s.l., as originally projected in the 1960s, the following was expected:
- Reservoir surface area: 3,722 km^2;
- Gross storage volume: 98.3 × 10^9 m^3;
- Maximum water depth: 94.6 m;
- Drawdown: 14 m;
- Intake elevation: 183 m.

A thermocline was assumed to form at a depth of 6–15 m, and the difference in temperature between the water at the surface and the water below the thermocline was assumed to be 5–8 °C. It was thought that this difference would decrease during the cool season in January and February. In April, wind-initiated water circulation at a depth of 20–25 m would occur for a short while, breaking the thermocline, yet there was no change in the temperature of the water in the deeper layer. Below the thermocline, the water temperature gradually decreases, stabilizing at a depth of 35–40 m. The temperature of the Mekong stream flowing into the reservoir would be between 21 and 29 °C, but the temperature of the thermocline would be slightly lower, say, 20–25 °C. Since the dam was designed to accept inflow at an elevation of 183 m, the inflow would enter the stabilized part located below the thermocline; therefore, the released water would be a few degrees cooler than the inflow. The water temperature at the surface of the reservoir would vary as a result of air temperature and insolation, but it would still be 5–10 °C higher than the water temperature of the Mekong.

Such a detailed study regarding water temperature appears to be lacking from the preliminary studies on the alternative plan for the Low Pa Mong Dam Project. The following factors are relevant:
- Catchment area: 299,000 km^2;
- Inflow volume: 82.404 × 10^9 m^3 (2,613 m^3/s) in 1957 to 145.476 × 10^9 m^3 (4,613 m^3/s) in 1966;
- Gross storage volume: 8.5 × 10^9 m^3 (at an elevation of 210 m);
- Reservoir surface area: 560 km^2 (at an elevation of 210 m).

Compared with the High Pa Mong Dam, the gross storage volume ratio is 1:11.5, and the reservoir area ratio is 1:6.6. Compared with the annual inflow volume, the gross storage volume of the Lower High Dam would be less than 1:10, so that the effect on the temperature of the main stream of the Mekong would be inconsequential.

Eutrophication in dam reservoirs

Eutrophication

Artificial lakes in the torrid zone generally contain high levels of organic materials carried to them from the surrounding areas by the rivers which feed them. Tropical climates also encourage the decomposition of submerged trees. The combination of these factors contributes to the rapid growth of aquatic plants and algae.[15]

Of course, the growth of aquatic plants and algae (vegetable plankton) is not unique to tropical climates and the same phenomenon can be widely seen in the lakes and river-banks of countries in the temperate zone, such as Japan, as well in the countries of the frigid zone. Vegetation will flourish in wide areas of shallow water where the submarine plant life receives sufficient sunlight. Under these conditions, the growth of aquatic plants and algae will continue unchecked, unless the inflow of nutrients is cut off; failure to do so will eventually lead to the pollution of the lake and to a marked deterioration in water quality. Some forms of algae are particularly troublesome because they emit a noisome odour or affect the taste of the water.

The effective prevention of the eutrophication of man-made reservoirs in tropical climates requires advance planning and careful control. Steps to manage the regions above the dam and around the reservoir effectively must begin before water starts to be impounded.

Restraint of eutrophication

The restraint of eutrophication through effective management of upstream regions and the areas surrounding the reservoirs is currently recognized as a global challenge. For various reasons, effective management is more easily pursued in the underdeveloped countries of the torrid zone than in the advanced industrialized nations: this is because the former regions tend to be populated by primitive farming communities; secondly, political conditions in these regions may be more amenable to strong governmental action in managing the river basins.

Data.   Any attempt at effective control must be predicated on the collection of reliable data. For this purpose, observation posts must be established on upstream regions before work is started on the construction of a dam. Located at various points on the main river system and its tributaries, these posts are used to collect data on water quality and the amount and composition of pollutants contained in the current. Specific measurements to be taken include the following: phosphorus and nitrogen content; oxygen content and pH; heavy-metal content; and electro-conductivity. Needless to say, these observations must be continued during and after the construction of the dam. The observations can be facilitated by the use of Landsat, the earth-surveying satellite.

The effects of over-logging and slash-and-burn agriculture in the upstream areas and areas surrounding the reservoir can be readily observed in the data. The proper analysis of the eutrophication of reservoirs must be based on the careful examination of extensive data collected in this manner. Unfortunately, however, the collection of reliable data poses a difficult challenge in the case of most developing countries.

Removal of vegetation. The existence of vegetation in the surrounding areas plays a critical part in the supply of nutrients to reservoirs. As such, the water quality in reservoirs depends in great part on the prior clearing and proper removal of trees and other vegetable growths. It is obviously necessary to take some preventive measures before filling the reservoir. The problem, of course, is that the complete clearing and removal of vegetation is difficult from an economic point of view in the case of larger reservoirs.

Discharge of nutrients. The relation between the total volume of the reservoir and the total volume of water flowing into it constitutes an important factor to be considered in determining the extent to which the removal of vegetation makes economic sense. When the reservoir volume is relatively large, compared with the inflow volume, and several years are needed to fill the reservoir to capacity, there is a greater probability that eutrophication occurring during the filling period will give rise to algal growth. On the other hand, when the reservoir volume is relatively small and full capacity can be reached within a few months, it becomes possible to discharge the accumulated nutrients.

Reservoir control. Another important factor to be taken into account is the projected use of the reservoir. If it is primarily designed for supporting increased fishing and marine transport, trees left standing in the lake with their heads above the water pose no special problem. Because reservoirs serve a wide range of purposes and are constructed in areas of highly varied natural and social conditions, it is only to be expected that the necessary level of control of upstream areas and the area of the reservoir itself prior to the impoundment of water will also vary widely.

Prior to embarking on the construction of any of the proposed Mekong Basin dams, the above factors must be carefully examined and proper measures must be taken for the prevention of eutrophication to match the specific environmental conditions of the reservoir.

Water hyacinth

One of the particularly troublesome problems associated with the development of dams and waterways in tropical climates is the control of water hyacinth. These plants originate in the Amazon River Basin but have gradually spread to the tropical regions of Asia and Africa. The correct botanical name of the water hyacinth is *Eichhornia crassipes*. The water hyacinth has an internal duct structure and is found in two varieties – the rooted and the free-floating types.[16] The plant has flowers of varying size which are either white or blue. The seeds of the water hyacinth sink below the surface and float underwater when the water temperature is low. Ideal conditions for germination are attained when the water temperature stands at between 28° and 36 °C. Germination requires about two months, after which the plant grows to full size within a month and eventually bears flowers. While the plant is growing, older segments gradually die and fall away. The water hyacinth reaches its full growth potential during the summer and autumn (rainy season) months, when there is plenty of rainfall and sunshine. During the colder winter and spring (dry season) months, the plant is reduced in size. It is thought that water hyacinth seeds are carried to new locations by fishing vessels and water birds. Once in place, the water hyacinth grows at a tremendous rate and covers large areas of water and the surrounding banks in a very short time.

It is reported that *E. crassipes* first arrived in the Congo River Basin in 1952; within 3 years, it had spread to almost all parts of the basin. Similar reports abound: in the case of Mexico, it is reported that a lake of area 4,000 m^2 became completely covered within a span of only 10 days. In the case of a certain African variety, two water hyacinth plants had multiplied to 1,200 plants within a period of 130 days (Guadiana 1976). Water hyacinth was first seen in the Aswan High Dam some 30 years ago; today, the plant is widely observed in the irrigation and drainage canals throughout the Nile River Delta.

A newly completed reservoir in a tropical environment undergoes the following type of cycle. During the early stages after completion, various aquatic plants and algae begin to grow in the lake. During the first few months, these plants and algae quickly absorb the available nutrients while releasing large amounts of oxygen into the lake. In the case of the normal varieties of aquatic plants, this is essentially a benign cycle; however, if water hyacinth of the free-floating type is introduced into the lake, an accelerated outbreak of the plants will quickly cover the entire surface of the lake with a thick mat of vegetation. This development upsets the oxygen cycle in the lake in the following manner: because the rate of evaporation of the water hyacinth is anywhere from 1.5 to 5 times higher than the normal rate of evaporation of an uncovered lake surface, the lake becomes cut off from the regular supply of oxygen generated by photosynthesis; furthermore, the dense surface cover prevents the winds from promoting water circulation in the surface stratum of the lake; finally, the surface cover deprives the vegetable plankton of sufficient sunlight to maintain photosynthesis; consequently, the surface stratum of water also becomes rapidly depleted of oxygen.

Yet another problem relating to the water hyacinth is that older segments are constantly dying and falling away and sinking to the bottom of the lake, where they consume large amounts of oxygen as they gradually rot and disintegrate. In tropical climates, lake-bottom temperatures are high enough to accelerate the oxygen intake of this process substantially; in this way, the various properties of the water hyacinth lead to the complete exhaustion of oxygen in the lake. What makes matters worse is that water circulation in tropical lakes comes to a halt during 10–11 months of the year. During this period, thermoclines effectively deprive the lower strata of any opportunity to mingle with the upper strata: consequently, zero-oxygen conditions persist at the bottom of the lake for extended periods. Although the agitation effect of winds can promote water circulation and prevent zero-oxygen conditions at the bottom of relatively shallow reservoirs, under normal conditions the bottom becomes completely depleted of oxygen once a mat of floating-type water hyacinth has fully covered a lake. At the same time, for the reasons already cited, oxygen levels in the surface stratum also fall too low to support fish, which then rapidly die out.

A surface layer of water hyacinth plants generates a very large volume of dead and rotting debris. Floating-type hyacinth can generate as much debris as 600 t/ha of surface area (60 kg/m^2). Approximately 10 per cent of this sinks to the bottom to create debris equivalent to 60 t/ha (6 kg/m^2) (Guadiana 1976). Because it takes only roughly 60 days to repeat this cycle, the presence of water hyacinth plays a critical role in shortening the effective life span of a reservoir.

Negative and positive effects. The negative impact of water hyacinth can extend even to the fishing communities located in the vicinities of the reservoirs. Floating-type water hyacinth can also cause the following types of damage to reservoirs:
1. Obstruction of the operation of dragnets, drift nets, and trawl nets;
2. Obstruction of the passage of vessels;
3. Obstruction of the flow of water through the dam by covering intake ducts;
4. Water loss from evaporation in water hyacinth (1.5–5 times higher than the normal rate of evaporation at the reservoir surface);

5. Snails carrying schistosomes inhabit water hyacinth.

On the other hand, rooted water hyacinth does not cause any of the above problems and instead yields the following benefits:

1. Reduction of surface evaporation by creating a buffer against winds;
2. Increase of the food supply for fish by providing a platform for insects, larvae, and algae;
3. Prevention of soil erosion on the banks by taking root.

However, if the water level of the reservoir changes frequently and is allowed to fall substantially, the roots will become exposed and the plant will die. The water hyacinth does not easily take root on steep banks and grows best in flat and shallow areas where the water level tends to be constant.

By the same token, the floating-type water hyacinth is not completely evil and does have the following positive effects to offer:

1. Generates methane through anaerobic dissolution;
2. Provides material for pulp and paper fibres;
3. Provides feed for livestock;
4. Is effective as an additive for maintaining soil humidity levels.

Of these advantages, the most generally exploited uses of the floating-type water hyacinth are as feed for livestock and as an additive in reforestation and general soil improvement. Furthermore, both types of water hyacinth have some very unexpected benefits to offer. Specifically, the water hyacinth plant is an efficient absorber of mercury, chromium, cadmium, silver, strontium, lead, phenols, detergents, and other troublesome pollutants. According to research conducted at the University of São Paulo, Brazil, water hyacinth is capable of reducing *Escherichia coli* contamination by 99.1 per cent and can lower biological oxygen demand by as much as 10 per cent. In the future, perhaps these outstanding properties will be utilized in combating water pollution.

Controlling water hyacinth. In general, the spread of water hyacinth is not a desirable phenomenon. Methods available for control of its growth are outlined below. The most important factor in effective control is the launching of a thorough programme for the restraint of eutrophication well before the first signs of colonization become apparent. If, however, the plant becomes established in an environment, it is necessary to cut and remove the plants as soon as possible. Methods for the removal of water hyacinth are outlined below (Guadiana 1976).

Manual removal. Plants are cut and removed by hand, then dried and used as fertilizer. Because of the presence of surplus labour in developing countries, the manual control of water hyacinth cannot be said to be uneconomical or inefficient. Manual methods are sufficiently effective in the early stages of an outbreak.

Mechanical control. Control using machinery can be justified if the income from the sale of the hyacinth exceeds the cost of purchasing and maintaining the machinery. In other words, the choice between manual and mechanical control must be made on a case-by-case basis. In any case, both methods are said to be effective only if applied in the early stages of an outbreak.

Measures to restrain further spread. To restrain further spread, for example, netted areas can be built around the water intake to prevent the water hyacinth from reaching the water turbines. This, of course, is only a partial measure for protection.

Chemical control. Herbicides offer a highly effective method of control; however, there will be damage to fishery resources in the reservoir and downstream areas. There will also be a negative impact on downstream irrigated crops. In addition to short-term damage, it is believed that the use of herbicides presents a long-term risk to both people and livestock, but the extent of this harm is not clearly known.

Biological control. Biological control methods include the introduction of viral agents and parasites, the use of sea cows to eat the hyacinth plants, and the release of tilapia and a species of carp. Since tilapia and carp readily multiply in tropical lakes, this is thought to be the most effective approach. On the other hand, although sea cows willingly feed on the water hyacinth, they have a low reproductive rate and do not multiply sufficiently rapidly to keep up with the spread of the hyacinth. The introduction of snails which eat the hyacinth plant is thought to be risky, because snails can carry schistosomes.

Eutrophication of reservoirs in the lower Mekong Basin

Nam Ngum Dam. One of the few examples of eutrophication of reservoirs in the lower Mekong Basin is the Nam Ngum Dam in Laos, where impoundment started in 1971 and was completed in 1972. The reservoir is about 370 km^2 in area, holding some 7×10^9 m^3 of water.

The pH of the reservoir is 5.2–8.5 and the water temperature is a constant 20–32 °C. The geological make-up of the area surrounding the reservoir consists of limestone and sandstone. The vegetation mainly consists of evergreen trees and bamboo. The drawdown topography (between the maximum water level and the lowest water level) is steeply inclined, with sparse vegetation and few trees.

For a long while after its completion, trees were left standing inside and outside the reservoir, but most of them were felled recently. Trees which had stood at the deeper sections of the reservoir were entirely submerged and have completely rotted, but there are still some that stand high enough in the reservoir to impede water circulation. As a result, the reservoir water lacks oxygen below a depth of 10 m.

Although the surface water is for the most part uncontaminated, sediment which has piled up at the lowest part contains sodium hydroxide, nitrogen, phosphorus, and other elements. The surfaces of the concrete dam and attached power station have turned dark brown, perhaps owing to the effect of hydrogen sulphide. Eutrophication is progressing within the reservoir. Aquatic plants such as weeds and algae are proliferating, resulting in evapotranspiration from the leaves of these plants.

Pa Mong Dam (planned). What can be expected if the High Pa Mong Dam is built? The results are, of course, difficult to predict. However, there are several cities upstream of the dam site, such as Luang Prabang and Sayaboury (Laos) and Chieng Rai and Chieng Sen (Thailand) in the lower Mekong Basin, and Jinghong in the Lancang River (upper Mekong) Basin. The development of all these cities and their adjacent areas is progressing rapidly. In addition, the planned development of several Chinese dams is under way, as mentioned in chapter 4. Therefore, once the Pa Mong Dam is completed, eutrophication is certain to occur in the reservoir unless preventive steps are taken.

In addition to the above-mentioned upstream influences, one must also consider the development of areas near the Pa Mong, particularly the urbanization of Loei in north-east Thailand, and the surrounding agricultural development. These factors comprise a eutrophication threat that worries residents.[17]

Another concern is the impoundment of water. The oxygen balance is almost certain to be disrupted by oxygen emission from inundated trees and plants and from organic matter trapped in the inundated soil. If a lack of funds prevents the felling of trees prior to impoundment, and if trees and bushes are burned instead, a problem will result because the residual ash contains much organic matter. One thing seems certain: the presence of trees and plants in the reservoir proper will cause some measure of eutrophication.

According to a report entitled "Preparatory Environmental Study for the Low Pa Mong Project" (Mekong Secretariat 1992) presented by the Interim Mekong Committee in July 1992, the alga which grows around the proposed dam site is a species commonly found throughout north-east Thailand, but no water-weeds are found in the upstream area. However, it is possible that water hyacinth might flourish as part of the eutrophication of the reservoir after the Pa Mong Dam is completed.

Sedimentation

Rivers in tropical regions usually flow through relatively old, weathered land subjected to a hot, humid climate. In the flood season, erosion occurs in the river basins, and the river bed and its banks are scoured as well. Consequently, the flood water contains a great deal of earth and sand.

The topography of the middle and lower river basins in the tropics is generally flat, and the inclination of rivers is gentle, both vertically and laterally. Therefore when a dam is built, the reservoir generally assumes a shape that is long, wide, and shallow. The earth, sand, clay, and silt in the river flow are deposited in the reservoir and usually remain there without being discharged downstream; this is termed reservoir sediment.

Efflux of earth and sand

In the tropical zone, owing to the hot, humid climate, the surface soil is usually weathered. In the rainy season it is eroded and, as a result, tropical rivers convey vast quantities of earth and sand downstream. Under such conditions, certain human activities can greatly increase the efflux: slash-and-burn agriculture, timber harvesting, livestock overgrazing, and large-scale mining are the principal offenders.

In developing countries, poor control of the river basin in the upstream area and at surrounding reservoirs allows considerable earth, sand, and silt to flow into the reservoir (fig. 5.2; table 5.6). The sediment is composed of various inorganic substances such as phosphorus, iron, and silver, and of undissolved organic materials. The larger particles of earth and sand usually accumulate furthest upstream, the so-called backwater area of a reservoir, where some of the finer-grained earth and sand also piles up. The rest flows downstream along the river bed, mainly towards the dam site, and forms a muddy lake. When such a lake forms at the back of a dam, the drain holes at the bottom of the dam become covered by mud. Once this mud solidifies, it hinders the operation of the drain. If the layer of mud rises around the dam and enters the intake, it disrupts the normal functioning of the water wheel. Once this happens, it often causes other problems, such as immobilizing the spillway auxiliary apparatus.

Sedimentation can have several adverse effects, such as decreasing the retaining capacity of a reservoir. For example, within 25 years of its completion, the *15 de Noviembre* Dam in El Salvador had lost as much as 60 per cent of its water-holding capacity. However, not all the effects of sedimentation are negative: living downstream is considered desirable because the discharged water is much cleaner than the water was before the dam was built; when the reservoir has been built to supply water locally, the fact that suspended loads sink, producing cleaner water, provides a welcome benefit. The turbidity of water in a reservoir, which is sometimes of long duration, is due to the suspension of fine particles (generally less than 0.05 mm in diameter) (table 5.7). In the case of long, wide, shallow reservoirs created in the middle and lower reaches of tropical rivers, such suspended loads gradually sink and pile up on the

Upstream region:

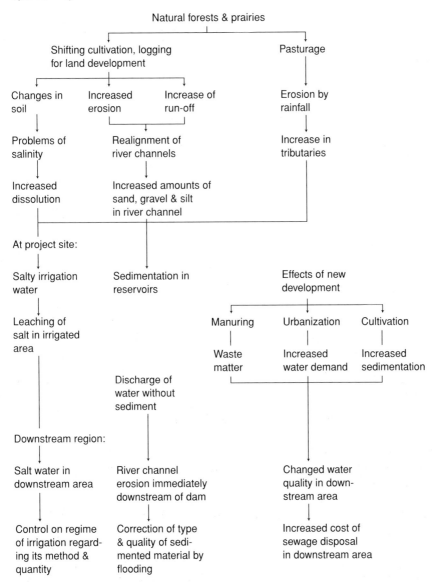

Figure 5.2. Problems resulting from development of tropical river basins
Source: Original

Table 5.6 Annual discharge of earth, sand, and silt into major world rivers

River	Drainage area (1,000 km^2)	Weight of sediment[a] discharged per year (10^6 t)	Weight discharged per unit area (t/km^2)
Yellow	672	2,080	3,090
Yangtze	1,941	550	283
Indus	969	480	495
Irrawaddy	430	330	767
Red (Viet Nam)	119	140	1,176
Mekong	795	190	238
Colorado	637	150	235
Mississippi	3,221	340	105
Missouri	1,370	240	175
Amazon	578	400	69
Nile	298	120	40

Source: K. Ashida. "Landslide Disaster and Its Countermeasures." Morikita Publishing Co., Japan, July 1983
a. Earth, sand, and silt.

Table 5.7 Weight of suspended material in major world rivers

River	Weight of suspended material (t/km^2/year)	River	Weight of suspended material (t/km^2/year)
Yellow	2,643	Mekong	151
Yangtze	492	Colorado	559
Indus	503	Mississippi	94
Irrawaddy	817	Amazon	59
Brahmaputra	1,302	Nile	18

Source: Y. Takahashi, Y. Sakaguchi. "Rivers in Japan." *Science* Vol. 46–8, Aug. 1976

bottom over a long period, partly because a thermocline exists in such reservoirs nearly all year long.

As mentioned on page 232, water-temperature variances in tropical reservoirs are relatively small compared with those in the temperate zone. However, below the mid-point of their depth, the water temperature in such reservoirs changes markedly. A thermocline is formed at the mid-point, which hinders the mixing of the water above with the water below. Since the reservoir water is stratified, and the temperature is stabilized in general, the water is not agitated. Thus, the earth and sand are likely to settle more than they do in reservoirs in temperate or frigid zones.

Outside the tropical zone, when the water temperature of a tributary differs from that of the main stream, and it flows into a reservoir, the water temperature of the reservoir becomes unstable. If the volume of the inflow from the tributary is large, the water in the reservoir is pushed both upward and downward, and mixing results. In the tropics, since the water temperature of the inflow equals or approximates that of the water in the reservoir, the result is a stable body of water, with little or no mixing. Thus, the inflowing earth and sand gradually accumulate over the bottom of long, flat reservoirs.

Human causes of upstream erosion

As previously described, sedimentation in reservoirs is caused by both natural and man-made events. Some of the degradation of tropical reservoirs is due to certain types of action or inaction – for example, failure to treat the sedimentation, or mishandling it. The following cases are among the more dramatic examples.

Malaysia.   The annual precipitation in West Malaysia[18] is some 2,400 mm. The subsurface soil tends to erode easily; nevertheless, if the natural, tropical forest is dense and the soil is covered by the usual thick grasses, soil erosion is rare. However, 80 per cent of the earth and sand efflux from the main river basins is caused by human activities (Fatt 1985).

In West Malaysia, the four types of human activity that cause efflux of earth and sand from the river basin are tin mining, logging, expansion of farmland, and urbanization.

Tin mining. The development of tin mines in river basins in West Malaysia has caused a major efflux of earth and sand. There are four rivers, each of which produces earth and sand at a rate of some 600 t/km^2/ year.

Logging. Research shows that logging, which has become quite intense, has caused a sharp rise in the efflux of earth and sand in West Malaysia. The resulting environmental damage has been exacerbated by the ruts worn in the earth by the trucks which haul the logs.

Expansion of farmland. In the woodlands, the land is eroded severely when it is cleared and tilled. In the Cameron Highlands, for example, as the tilling has progressed in what was formerly forest, the efflux has increased rapidly (Catakli 1973). Lately, in order to restrain this erosion, the removal of weeds has been abandoned and rubber trees have been densely planted so that their roots secure the soil and the fallen leaves provide protective ground cover. This has considerably decreased the efflux of earth and sand, particularly that of silt.

Urbanization. The increased presence of earth, sand, and silt caused by urbanization is noteworthy. In cities such as Kuala Lumpur, the construction of houses, office buildings, factories, and roads has progressed rapidly, yet there has been no detailed study of how much earth and sand will be dislodged and end up in the waterways. Furthermore, no study has yet been made of how much sediment has accumulated in each reservoir, and what impact that has had on the quality of life of city residents.

The Malaysian government maintains that the amount of sediment accumulating in reservoirs is negligible, and that it has not caused any problems at dams built for power generation, water supply, or flood control (Fatt 1985). The government has, in fact, made efforts to control various aspects of the river basins, but, owing to economic and financial constraints, has not entirely succeeded in its soil-conservation efforts. Until recently, the lack of funds prevented the government from building sabo dams (sand-obstruction works) and anti-landslide structures along the river-banks. Moreover, as is common in tropical countries, substantial bureaucratic confusion prevails, with several overlapping ministries engaged in these activities. Their respective areas of jurisdiction and responsibility are not clear; lack of cooperation and of combined effectiveness, therefore, has prevented them from meeting the challenge.

India. The Nizamsagar Dam in India was completed in 1930. In the 37 years until 1967, 52 per cent of the gross volume of the reservoir was lost as a result of the build-up of sediment (Yiicel 1976; Chaduick 1976). By 1972, 60 per cent had been lost. This was caused by the fact that the river basin has a broad area of land that lacks natural foliage, and no soil-conservation measures were taken to compensate for this. In time, this combination of factors resulted in an unacceptable level of sedimentation (Chaduick 1976).

Sedimentation in the lower Mekong Basin

General geology of the lower Mekong Basin. The mountainous region in the northern parts of the lower Mekong Basin is mainly composed of crystalline and other rock of the Archaeozoic, Palaeozoic, and Mesozoic eras, such as limestone, shale, slate, and quartzite.

In the central part of the lower basin, which extends from the Vientiane Plain to the Korat Plateau, there are numerous gently warped Mesozoic formations consisting of thick beds of sandstone and siltstone. The surface has turned into laterite, a great deal of which is washed away during the rainy season.

In the Annam Cordillera mountains, which separate the river basin from the coast of the South China Sea, volcanic rock of basalt and andesite extrudes from the Mesozoic baserock. The Boloveng Plateau, which is part of the Annam Cordillera, is composed of a fertile red soil (*terre rouge*).

The southern part of the basin is formed extensively by diluvium and alluvium, which embrace the Great Lake and spread into the vast delta.

Table 5.8 Amount of suspended material and run-off in some Asian rivers[a]

	Mekong	Chao Phraya	Red river (Viet Nam)	Irrawaddy	Naktong	Indus
Suspended material (10^6 t)	80	12	130	350	10	400
Run-off (10^9 m^3/ year)	378	31	76	428	11	175
Silt component (g/m^3)	210	390	1,700	820	910	2,300

Source: US Bureau of Reclamation (1956).
a. Measured in 1911.

Density of matter suspended in the Mekong. A considerable portion of the mountainous area in the Mekong Basin is still forested. This plays an important role in fostering water resources, as well as in protecting the soil against erosion. On the other hand, erosion progresses where trees have been cut down and subsurface soils exposed.

The flooding of the Mekong begins in the vicinity of Vientiane. The laterite content in the water gives it a reddish colour during the flood season and a red-brown colour in the dry season. Nevertheless, the amount of silt and clay contained in the flow of the Mekong is relatively small compared with that in other large rivers around the world. Among Asian rivers, the Mekong can be included in those containing the least silt and clay (table 5.8).

Generally speaking, the flow velocity of the main Mekong diminishes in a downstream direction, and the unit amount of sand and silt conveyed by the flow is also reduced. Nevertheless, the total amount of sand and silt transported increases, because the flow volume increases as the Mekong moves downstream (Mekong Secretariat 1976).

The amount of sand, silt, and clay contained in the Mekong in the rainy season (June to October) is 250–300 g/m^3; in the dry season, which begins in December, it is 50–100 g/m^3. These figures vary according to where the measurements are collected (fig. 5.3) and from one year to the next. As shown in table 5.9, the measurement at Stung Treng, for example, indicated 172 parts per million (ppm) on average from May to September; however, the measurement at the same place for September alone was 650 ppm (US Bureau of Reclamation 1956).

Of the sand, silt, and clay in the water of the main stream of the Mekong, 20 per cent is composed of organic matter and the remaining 80 per cent is mineral substances. The pH of more than half of the silt is 6.8–7.3, which is almost neutral; however, the pH of the sand, silt, and clay piled

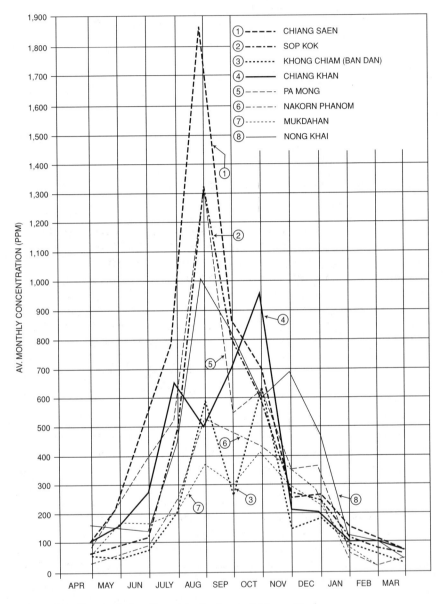

Figure 5.3. Suspended sediment in the lower Mekong Basin
Source: Mekong Secretariat (1976)

Table 5.9 Average turbidity of the main Mekong and its tributaries in Cambodia, May–September 1911

Location	Average turbidity (ppm)
Stung Treng	172
Kompong Chhnang	280
Pursat	140
Kompong Thom	97
Stung Seran Bicu-Kam	77
Stung Thli Niramo	70
Stung de Krang Pavly	85
Stung de Chnerty-Bok	65
Stung de Phsor	86

Source: US Bureau of Reclamation (1956)

Table 5.10 Annual flow volume and weight of suspended material in the lower Mekong, 1974

Location	Distance from estuary (km)	Drainage area (km^2)	Annual peak river flow (m^3)	Annual flow (10^6 m^3)	Wt of suspended material per year (10^6 t)	Unit wt of suspended material (t/km^2)
Chiang Saen	2,330	186,000	10,700	89,870	67.21	361
Luang Prabang	2,010	264,000	16,600	132,510	–	
Vientiane	1,584	301,000	16,300	144,500	109.14	363
Thakhek	1,216	366,000	26,100	241,530	–	
Mukdahan	1,125	187,000	29,200	263,760	–	
Pakse	869	540,000	39,000	301,850	132.20	244
Stung Treng	668	624,000	39,000	450,310	–	
Kratie	545	636,000	52,700	452,890	–	
Phnom Penh	232	653,000	43,700[a]	428,530[a]	97.55	149
Phnom Penh	332	653,000	50,000[b]	512,000[b]	–	
Delta	0	774,000	44,000	550,000	–	

Source: Mekong Secretariat. "The Water Situation in the Lower Mekong Basin." Interim Mekong Committee, April 1976
a. Flow volume measured within river channel.
b. Flow volume including overflow water.

up along levees of the main stream is said to be 8.0–8.5, which is on the alkaline side.

Table 5.10 shows the volume of suspended matter in the water of the

Table 5.11 Main features of the Nam Pong Dam

Height of dam (m)	32
Length of dam (m)	800
Reservoir elevation (m above m.s.l.)	182
Reservoir volume (10^3 m^3)	1,920
Installed capacity (MW)	25
Energy generated (GWh/year)	65
Irrigation area (ha)	53,000

Source: Mekong Secretariat, Annual Report, Interim Mekong Committee, 1983

main Mekong.[19] The density of suspended silt is lowest between Chiang Saen and Nong Khai, at less than 50 ppm of the maximum (see fig. 5.3). The density at Chiang Saen reaches its peak in August, but downstream of Chiang Saen the peak occurs 2–3 months later. The peak decreases as the flow distance increases: before the dam development was started on the Lancang River, at the border between Myanmar, Laos, and Thailand, it was at least 1,500 ppm, but only 1,000 ppm at the proposed site of the Pa Mong Dam. Although figure 5.3 does not indicate this, there is a further decrease in density towards the delta, dropping to about 500 ppm. The following are measurements (g/m^3) of the density of suspended matter in the Mekong: 748 at Chiang Saen; 755 at Vientiane; 483 at Pakse; and 227 at Phnom Penh.

Sedimentation in the Nam Pong Reservoir. It would be meaningful to examine the sedimentation in the Nam Pong Reservoir and the upstream area, as this would be a useful reference regarding the life expectancy of future mainstream dams.

The Nam Pong Project, completed in 1965, is the furthest upstream from the Nam Chi, one of the largest tributaries on the Korat Plateau. This was the first water-resource project built in the lower basin (table 5.11). Since its completion, the upstream area of the Nam Chi has been rapidly developed: forests have been randomly cleared and the subsoil has consequently eroded, causing an increase in sediment in the Nam Pong Reservoir; eutrophication has also taken place.

The topography is relatively steep in the upstream area. As is typical of the tropics, the subsoil is scanty and poor, making the effectiveness of irrigation marginal. However, a number of illegal irrigation projects have been developed there, contributing to further soil erosion and the rapid increase of sedimentation in the Nam Pong Reservoir.

Before 1964, prior to the completion of the dam, more than 950,000 t of earth, sand, and silt were found in the Nam Pong at the dam site each year. However, a post-impoundment sediment survey carried out in 1978 by EGAT (Electricity Generating Authority of Thailand) showed that

Table 5.12 Land-use changes in the Nam Pong Basin, 1965–1982

Land-use categories	Area (km^2)			Area changes (km^2)
	1965	1975	1982	1965–1982
Urban and built-up land	17.50	18.75	26.25	+8.75
Agricultural land	2,499.50	6,954.00	7,801.50	+5,302.00
Forest land	9,998.50	5,300.00	4,533.00	−5,465.50
Water resources	95.25	288.00	200.00^a	
Total watershed area	12,560.75	12,560.75	12,560.75	

Source: S. H. Johnson III and Shashidhara Kolavalli. "Physical and Economic Impacts of Sedimentation of Fisheries Activities, Nam Pong Basin." *Water International* Vol. 9-4, IWRA Dec. 1984
a. LANDSAT images obtained during dry season (January)

some 2×10^6 t of sediment, more than double that found in the pre-impoundment inflow, were being deposited in the reservoir. The authors of this survey expressed grave concern, stating that if the Royal Forestry Department of Thailand did not carry out reforestation in the devastated areas and did not prevent random and unauthorized development (table 5.12), the annual deposition of sediment would increase. The best that could be expected, the authors claimed, was for sediment inflow to stabilize at an annual average of about 3×10^6 t; with no action, the rate of sedimentation would increase indefinitely.

Since the completion of the Nam Pong Dam, a number of tributary dams, such as the Nam Pung and the Lam Don Noi, have been built in the lower basin in north-east Thailand. In all the tributary basins, the increased efflux of earth and sand due to human activity has become more severe every year (table 5.13), sending a definite warning of shortened life expectancies of completed projects.

The Mekong Commission is scheduled to carry out several studies, including one on increasing sedimentation in the Nam Ngum Reservoir in relation to logging activities and agricultural development. The Commission is also scheduled to study soil erosion, sedimentation, changes in run-off, and other developments regarding the Lam Praplerung Reservoir in north-east Thailand.

The Pa Mong Dam (planned).   Before development of the upper Mekong Basin was started, it was expected that, if the Pa Mong Dam were built, the accumulation of sediment would gradually raise the river bed at the point where it meets the reservoir, since most of the suspended load would pile up at that location (Mekong Secretariat 1976).

Table 5.10, drawn up before the start of development of the upper

Table 5.13   Average[a] sediment inflow (tonnes) into the Nam Pong Reservoir from the watershed

Sub-watershed	Year of conditions			
	1969	1978	1982	1990
I	66,299	87,747	117,762	145,059
II	246,615	357,349	480,411	562,518
III	494,586	649,071	790,585	1,085,807
IV	90,878	129,458	165,155	191,557
V	336,417	500,353	676,590	923,918
Total	1,234,795	1,723,978	2,230,503	2,908,859

Source: S. H. Johnson III and Shashidhara Kalavalli. "Physical and Economic Impacts of Sedimentation of Fisheries Activities, Nam Pong Basin." *Water International* Vol. 9-4, IWRA Dec. 1984
a. Average based on 10 years of climatological and streamflow records.

Mekong Basin, shows that the estimated amount of suspended matter in the main stream was 109.14×10^6 t/year; if all the suspended matter were to be deposited in the reservoir, the amount of sedimentation would reach some 64×10^6 m³/year; this is based on the assumption that the unit weight of the sand, silt, and clay is 1.7 t/m³.

Another paper asserts that if the Pa Mong Reservoir trapped 100 per cent of the inflowing suspended load, the amount of sedimentation in the reservoir would reach some 16×10^6 t/year if the full-reservoir water elevation is assumed to be 250 m (Simons and Shen 1973). In that case, the average annual amount of sedimentation would be 94×10^6 m³, or 219×10^6 t/year.

As we have seen, various predictions were made regarding the probable amount of suspended matter and sedimentation, before the development of the upper basin was started. The Mekong Committee predicted that, even in the worst-case scenario, the amount of sedimentation accumulated in the Pa Mong Reservoir 100 years after its completion would not exceed 15 per cent of gross storage capacity; sedimentation was not, therefore, considered to be a serious problem. However, it was also anticipated that the sediment might solidify after lying for a long period inside the reservoir, causing a change in the behaviour of the groundwater. These concerns may well be merely academic, as it now seems very unlikely that the High Pa Mong Dam will ever be built.

In addition to the aforementioned considerations concerning the lower basin, another factor has intervened: the development of large dams has recently been started in the upper basin, and sedimentation has already begun in the Manwan Reservoir. It may, therefore, be possible to study the sedimentation accumulated there; if so, future estimates of the amount of sedimentation in mainstream dams in the lower basin are likely to be more accurate.

Changes in downstream flow

In order to clarify the various aspects of changes caused downstream when a dam is built on a tropical river, the author has divided the downstream reaches into two parts – the area immediately downstream, and the furthermost reaches downstream (i.e. the delta).

In the tropics, if a long, flat reservoir is created on a river such as the Mekong or the Nile, the inflowing sand, silt, and clay will settle in the reservoir and the river water immediately downstream may become clearer. The water discharged gradually increases in velocity and erodes river-banks. As a result, the downstream river bed changes, and the amount of material suspended in the flow increases as the river flows onwards.

When a dam is located far upstream from the delta, its influence on the delta will be minor; however, if the dam is located in the relatively lower reaches, it will definitely influence the formation and development of the delta. If the soil from upstream is fertile, the delta's fertility will be increased by the dam. As a dam project may reduce the peak flow of flood water, the delta may be partly or fully protected from flood damage, and its depth of inundation may be shallower than before. As a result, the cultivated land in the delta may become more stable; agricultural development will probably be enhanced, with a concomitant surge in industrial and commercial development. On the other hand, during the dry season, owing to the water available in the reservoir, the flow reaching the delta may be increased, assuring adequate water for agriculture. In addition, the increased volume of river water can partly or completely prevent the influx of salt water from the sea into the delta, which would assure greater stability and success for agriculture and industry.

The topography of tropical regions is generally flat, so the middle and downstream reaches of tropical rivers contain fewer rapids than rivers in Japan, for example, where the middle and downstream reaches are steep, and the flow is more rapid. In general, the amount of upstream earth, sand, and silt flowing into the reservoirs may also be greater in Japan. However, in tropical rivers, for economic and social reasons, scant effort is made to protect against erosion, or to repair and restore erosion damage. Neglect of such problems is common and erosion is often progressive. Very few data are available on river-bed changes and river-bank erosion, so precise understanding of such changes is often lacking.

Predicted changes immediately downstream from the proposed Pa Mong Dam

Current situation.   Above the proposed site of the Pa Mong Dam, the Mekong flows through rocky terrain; downstream from the site, it meanders gently through the alluvial plain. Travelling downstream by boat on the Pa Mong, one sees a number of rocky outcrops in the river channel for a fairly long distance. Both banks are covered by soft clay, muddy soil, and gravel. Here and there, the soft river-banks have been eroded and scoured by the swift flow of the river during the rainy season. Serious erosion and scouring are seen along the banks between Vientiane in Laos and Nong Khai in Thailand. The stretch most at risk is that near the new Friendship Bridge, especially a section of the Laotian bank extending for a distance of about 32 km and a stretch of the Thai bank running for a distance of 28 km.

Along the river-bank that fronts the city of Vientiane, a large amount of sand and gravel has piled up, forming a giant sand bar which extends to and joins a small island located a short distance downstream, near the city's outskirts. This has had quite a beneficial effect, protecting the river-bank along the city by causing the Mekong to flow through the resulting narrow channel between the sand bar and the Thai bank on the opposite side. As a result of the barrier effect of this sand bar, Vientiane has been safely protected against river-bank erosion for the past 100–150 years. The Laotian government has taken the utmost care to preserve this crucial sand bar.

Meanwhile, owing to the efforts of the Interim Mekong Committee, assisted by the Australian government, some extensive protective works have been built along the Laotian river-banks over the past 10–15 years. The Thai government has also built strong concrete structures along the opposite river-bank. However, the results of a survey conducted by the Economic and Social Commission for Asia and the Pacific (ESCAP) several years ago (in which the author participated), revealed that the protective structures completed on the Laotian side were fairly flimsy and unlikely to survive a serious threat. Moreover, the crest level of the river-bank, which was raised by topping it off with solid fill (earth, stone, and rubble) intended to meet the most serious flood recorded so far, was nevertheless lower than that on the Thai side. This means that the city of Vientiane and environs is vulnerable to flooding in the event of a major flood.

Predicted risk of river-bank erosion and river-bed change downstream from the Pa Mong Dam.   Prior to the start of the development of the Lancang River (upper Mekong), several studies were conducted to predict the probable influence of the proposed Pa Mong Dam on Vientiane.

Tentatively, it was confirmed that the above-mentioned small island protecting Vientiane would become eroded in the immediate downstream reaches, as would the river-bank in this area on the Thai side. As for the threat of possible landslides on the Vientiane side, the river-bank would probably be safe, although the river bed itself might be degraded. The study also predicted the following: (1) the load of earth, sand, and silt would decrease immediately downstream of the Pa Mong Dam as they are halted by the newly created reservoir and, as a result, the river bed would be degraded; (2) however, the suspended load would gradually increase as the river flowed onwards, owing to the degradation of the river bed and the erosion of the Thai river-bank.

In 1968 and 1973, the US Bureau of Reclamation studied river-bank erosion and river-bank changes in the area downstream of the proposed site of the Pa Mong Dam as far as 160 km from Vientiane. The results revealed that the breadth of the river immediately downstream would roughly double, and the depth of the river bed would be increased by some 8 m soon after the dam was built. In the 1974 report, a group of experts concluded that, 60 years after the dam's completion, the river would widen to about 100 m and the river bed would be further lowered by 3 m on average. Furthermore, although the area downstream would be stabilized, the river might gradually meander.

However, in 1975, the above prediction was criticized by the Hydraulic Laboratory of Delft in the Netherlands, which contended that the previous estimates were too high, and that the river bed downstream would be lowered by an average of about 2 m over 133 years. In July 1992, the Interim Mekong Committee published a report entitled "Preparatory Environmental Study for the Lower Pa Mong Project." The report stated that the Pa Mong Dam would scour the river bed by some 3–4 m immediately downstream. However, this study was made without considering the effect of the dams to be constructed upstream, including those on the Lancang River.

In any case, it is very difficult to make such predictions at the time of writing, as the grade and extent of both river-bed changes and river-bank erosion depend upon such factors as the quality of the soil along the river-bank, the magnitude of the river flow, the flora growing along the river-bank, and the structures to be constructed.

Flow changes.   The flow volume of the Mekong at the proposed site of the Pa Mong Dam is at its lowest in April, but the volume seldom falls below 1,000 m^3/s. The flow reaches its peak in August or September, when the volume ranges from 10,000 to 26,000 m^3/s. It is reported that the average flood volume between 1913 and 1968 was 12,900 m^3/s.

Assuming that the High Pa Mong Dam will be built and that the power generated will be a minimum of 800 MW and a maximum of 4,800 MW, the volume of discharge for power generation would vary between 1,400 and 8,500 m^3/s. This amount of discharge would cause enough river-bed scouring and river-bank erosion to produce 30.7×10^6 t of gravel, earth, sand, and silt per year (i.e. 84,100 t/day would be carried by the river flow.

Flow conditions at the downstream reaches furthest from the Pa Mong Dam. The changes in flow caused by the discharge of the power station, as well as the scouring of the river-bed resulting from the construction of the High Pa Mong Dam, would have a definite influence on navigation between Vientiane and Savannakhet. Navigation upstream from the confluence of the Nam Ngum and the main stream would be particularly affected, but the impact of river-flow fluctuations would be minor as long as the Pa Mong Project is operated mainly to supply the base load of energy. If the power plant attached to the High Pa Mong Dam is operated at the peak load, the change in flow conditions between the dam site and the confluence with the Nam Ngum tributary would not be negligible; however, flow conditions would not be unduly affected by building of the Low Pa Mong Dam.

Effects on downstream areas: delta, estuary, and coast

Under natural, unregulated conditions, rivers have long brought both benefit and disaster to inhabitants in the delta. Rivers were formerly beyond control: their behaviour, whether beneficial or adverse, had to be accepted. However, large dam development has made it possible to control the natural flow conditions to a certain extent. When a large reservoir is constructed, the earth and sand contained in the river flow are fully or partly halted. As a result, the formation and development of the downstream delta is hindered or retarded, and there is a change in both the quantity and quality of the downstream flow. People living in the delta, who previously suffered from floods and droughts, are now spared such disasters as the flooding is drastically decreased, which in turn reduces the breadth and depth of inundation.

Dams also provide increased river flow during the dry season, enhancing agriculture. Although this might simultaneously degrade soil fertility, the problem can be solved by the correct use of chemical fertilizers. Any possible influx of sea water can be controlled by discharging stored water from upstream dams. Therefore, in the delta, as a result of water control and electric-power generation, the dams have enhanced agriculture and encouraged industrial development. Thus was born the myth of the dam as a panacea – a boon to food production, a preventer of flood disasters, and a provider of power for industrialization.

Until recently, this dam success story continued to grow to mythological proportions and prevailed as a New Age gospel. Nations all over the world built dams one after another, but then the sobering reality became apparent: it was learned that deltas were suffering serious adverse effects from dam development. On the Volta River, a large dam upstream has damaged fisheries at the delta's water channels and creeks. In Egypt, the Aswan High Dam has caused broad and extensive degradation of soil fertility, making it necessary to apply substantial amounts of fertilizer in order to maintain successful agricultural yields. Moreover, the amount of salt in the soil has increased, owing to the poor drainage system, and the coastal shorelines have begun to show considerable signs of retreating. Still worse, Mediterranean sardine fishing has suffered extensive damage due to an interruption of the normal supply of fertile silt and clay previously deposited by the Nile on its way to the sea. Another adverse effect in the delta is the spread of schistosomiasis.

For better or worse, most parts of the lower Mekong Basin have so far been left in their original, primitive state. It has long been optimistically believed that dams would not greatly degrade the soil fertility of the delta as the relatively thin soil from upstream had been carried downstream to accumulate in the delta. This was assumed to occur even if dams were built in the main stream. However, if the frequent delta flooding could be stopped by building a number of upstream dams, it might be necessary radically to change existing traditional farming practices developed in harmony with the natural flow conditions. It might also have a detrimental effect on the fisheries in both the Great Lake and the delta. Finally, it would erode the coastline along the South China Sea, owing to the retention of the earth and sand deposits normally brought from upstream.

Below, the author describes the problems that will be caused in downstream areas – including the delta, the Cambodian plain, the Great Lake, the estuary, and the coast – by the construction of large mainstream dams such as the Pa Mong Dam.

Probable effects of planned mainstream dams

Delta and Cambodian plain.   *Topography.* The main stream of the Mekong flows over the Khone Falls in Laos and meets its two major tributaries, the Se Kong and the Se San, at Stung Treng in Cambodia. The Mekong then passes the Sambor Rapids and reaches the town of Kompong Cham, at which point it begins to display the characteristics of an alluvial river. It then flows on gently to Phnom Penh; a number of small islands and sandbanks are scattered along this stretch of the river.

For about 200 km between Phnom Penh and the estuary, the river flows through an area that is quite flat. The water elevation at Phnom Penh is only 1 m above sea level during the dry season, so even the areas upstream from Phnom Penh are affected by the tides.

The Tonle Sap river serves as a channel connecting the main Mekong with the Great Lake. Immediately downstream, the Mekong divides into two – the Mekong, which runs down the east side, and the Bassac River, which flows down the west side. The two rivers are interconnected more or less midway downstream. Still further downstream, the Mekong divides into three major branches, all of which pour into the South China Sea.

The extensive broad area downstream from Kompong Cham is divided into the Mekong Delta and the Cambodian plain, which surrounds the Great Lake. Between the two regions there is a small, low hill composed of granite and andesite. The hill is located near Kompong Chhnang, a small fishing port situated along the Tonle Sap river. At the apex of the topographic configuration of the Mekong Delta is the city of Kompong Cham. The western edge of the configuration runs from Phnom Penh and Ha Tien along the Gulf of Thailand; the eastern edge runs from Prey Veng to Gokong.[20] The Mekong is divided into three major branches downstream from the town of Vinh Long.

The region beyond the town of Sadec, which is situated at the dividing point, is called the flood plain, while the region in the vicinity downstream – the towns of Vinh Long, My Tho, and Can Tho – is known as the New Delta. The material that pours into the South China Sea from the edge of the New Delta is conveyed westward by the coastal sea flow, which is caused by the south-east wind; it forms complex shorelines as a result of the equilibrium between the river flow and the coastal sea flow. At the coast, the natural levees extend along the Mekong and the Bassac rivers. A broad area of swampy land stretches behind the shoreline, together with the flood plain and the New Delta.

In the alluvial era, the areas of the Great Lake and the surrounding lowlands are said to have been under water. Most of the Mekong Delta was also covered by the sea, which retreated in later eras. As the earth and sand carried by the Mekong accumulated, the sea became shallower and sand banks were formed around the hill near Kompong Chhnang. The Great Lake area was thus originally under water and the present adjacent Cambodian plain was formed by the earth and sand conveyed by the Mekong and the small rivers which now flow into the lake.

Precipitation. With regard to precipitation, the Mekong Delta and the Cambodian plain can be divided into the following two sectors: (1) the delta's flood plain and its immediate downstream areas such as the New Delta, the coast, and the Cambodian plain, which receives some 1,500 mm/year; (2) the western coast, which receives 1,500–2,000 mm/year.

The monthly rainfall during the rainy season, from May to October, shows different patterns at each measuring station, but the total sum of each registers 150–300 mm. Compared with the rainfall in north-east Thailand, which has a short dry period during the wet season, most parts of the delta (sector 1) have a climate favourable for paddy agriculture. In the south-western region of the delta (sector 2), both monthly rainfall and annual rainfall are ample, so that even the rain-fed paddies can produce sufficient rice.

However, as, for rice farming, the amount of monthly rainfall should be at least 300 mm during the wet season, it is necessary to irrigate using pumps in all areas of the delta except the south-western region during the wet season. In years when precipitation is lower than average, this pumping is even more essential. Above all, at Siem Reap in the Cambodian plain, where the range of fluctuation in annual rainfall is greater than average and reaches as much as 67.5 per cent of normal, the irrigation requirement is quite high. Consequently, a number of small and medium-sized reservoirs have been created, and water-wheel irrigation has become very popular. On the other hand, little rain falls during the dry season, so that rain-fed production is essentially not feasible anywhere in either the delta or the Cambodian plain.

River flow and depth of river water. The flow volume of the Mekong in the downstream areas increases from May or June and ordinarily peaks in September. It rapidly decreases in November or early December, then subsequently decreases gradually until it reaches its lowest level. At Kratie, the annual average flow volume is 14,800 m^3/s, while the monthly average volume in April is 2,000 m^3/s; in September, it rises to 41,500 m^3/s. The volume of flow of the main Mekong in the wet season is reduced by roughly 10 per cent just before entering the delta, owing to the influence of the Great Lake.

With regard to the present flow conditions of the Mekong in the downstream areas, the flood water overflows the river-banks, spreading widely every year between September and November; as a result, some $3–4 \times 10^6$ ha of Cambodian and Vietnamese lands in the delta are inundated. The relatively coarse particles of silt carried by the flood water are deposited along the river-banks owing to the sudden reduction in velocity caused by friction between the flow and the river-bank, where they add to the height of the ridges along the natural levees that border the river.

Behind the natural levees, the land slopes away and downwards quite gently for a distance of 1–2 km to depressions termed "bengs," the soil of which contains very fine clay particles and is therefore quite fertile (fig. 5.4). People living on the outskirts of Phnom Penh and downstream in Kandal Province have cut into the natural levees to draw water from the Mekong in June or early July through excavated creeks leading into

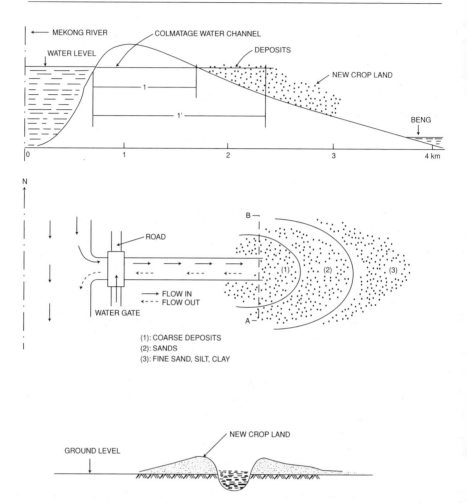

Figure 5.4. Sedimentation model of colmatage system in Cambodia
Source: Mekong Secretariat 1975

the beng depressions; this has expanded the bengs sufficiently for profitable tilling. The local people call this traditional farming custom the
colmatage system. The entire system of bengs is inundated when the
Mekong rises to a sufficient height.

Joncs Plain. Figure 5.5 illustrates the extent and depth of inundation
and the limitation of salt-water influx. When the river approaches the
maximum water level, the flood water overflows the left bank and moves
southward into the Joncs Plain[21] in southern Viet Nam, where part of the
flow finds its way into the East Vaico and West Vaico rivers. On the right
bank of the Bassac, the flood water flows over the land.

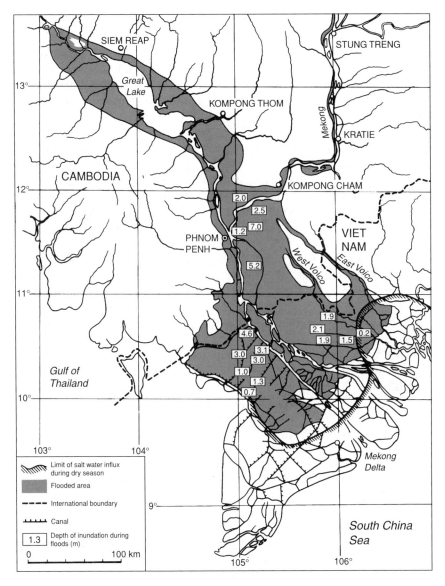

Figure 5.5. Inundated area of the lower Mekong Basin
Source: UN 1957

In the delta, both on the Mekong and its branches, the higher areas
along the banks are usually too high to be submerged. Typically, they are
sandy, and therefore devoted to orchards and vegetable farming. The
low-lying lands, however, are not suitable for farming, as the soil tends to

be highly acidic with a high sulphate content due to poor drainage. There are some 1.9×10^6 ha of such barren areas, which include the Joncs Plain, most of the coastal areas along the South China Sea, and the south-western coastal wetlands. The South China Sea coastal areas suffer from sea-water influx and from problems of salty groundwater.

As of 1988, excluding the above-mentioned barren areas, some 2.4×10^6 ha of a total of 3.9×10^6 ha are being tilled.

The Great Lake. The Great Lake and the Mekong are connected by the Tonle Sap river. Thus, the Great Lake influences both the flow volume and the depth of water of the main Mekong in the areas furthest down-stream. In June every year, when the water level of the Mekong rises, part of the flow of the river begins to enter the Great Lake. This lasts until September or October, some 100–120 days, raising the water level of the Great Lake from 2 to about 11 m and increasing the area of the lake from 3,000 to 10,000 km^2. After October, the lake-water flow reverses through the Tonle Sap to the Mekong. The total volume of the inflow and the outflow of the Great Lake for 1 year between April 1962 and March 1963 was 45×10^9 and 72.9×10^9 m^3, respectively. This difference is due to the outflow from surrounding tributaries towards the Great Lake. The outflow volume from the Great Lake towards the main Mekong, which amounts to some 6,000 m^3/s, reaches its peak in December.

In the low-lying areas which surround the Great Lake, and in the flood plain to the north of the delta, which extends from Phnom Penh to Sadac through Long Xuyen, the flooding is so deep that the rice variety known as "floating" rice is grown.[22] In the other areas around the Great Lake where the depth of flooding is relatively shallow, and in the wetlands located south of Long Xuyen, late-sown rice is grown in May or June after the primary rainfall and is harvested before the flood season.

The Great Lake expands to about 1×10^6 ha at the flood peak, but the total area of inundation, including the vast adjacent areas, reaches some 3×10^6 ha altogether. The lake thus plays the role of a gigantic flood-control reservoir. Because of this, remarkably, the increase in the water level downstream of the delta slows down and, as described earlier, the flow volume itself decreases by some 10 per cent compared with the main Mekong upstream of the confluence with the Tonle Sap river. This makes possible various types of paddy farming, including the cultivation of "floating" rice.

Rice cultivation in the delta. The Mekong Delta is well known for its rice cultivation. Nevertheless, the extent of flooding is quite irregular: on average, an unusually severe flood occurs once every 9 years. What is worse, the water level rises before the fields are prepared for planting, and before the harvesting of the rice sown earlier, which normally happens in September. Sometimes the rise even outstrips the growth of the "floating" rice, destroying crops over wide areas. Before the outbreak of the Viet Nam War, the delta customarily exported its excess rice production; however, the war brought about labour shortages and, from 1965, Viet Nam began to import rice. This period of negative production continued for a long time, the worst year being 1967, when rice production fell to 4.7×10^6 t and rice imports peaked at 770,000 t.

Soil in the delta. Generally speaking, most of the soil in the delta is composed of heavy clay and is therefore better suited for rice farming than for other crops. However, as mentioned above, there are some broad areas, such as the Joncs Plain and the Ha Tien Plain located along the Gulf of Thailand, which are not tillable because of the very acidic soil. The soil is also acidic in the territory between the Bassac and the Mekong.

According to a report submitted by the University of Hawaii to the Mekong Committee in January 1974, the fertility of soil in the Vietnamese Mekong Delta is on the whole rather poor, of medium grade. The report also mentioned that the annual deposition of sediment does not measurably increase the fertility. This means that a combination of fertilizer and water control would be required to increase agricultural productivity. Furthermore, the study reported that soil near the river contained the same levels of mineral nutrients as soil further away from the river (University of Hawaii 1985).

The same study included a sampling survey carried out at Long Xuyen, Can Tho, and My Tho. It is not clear whether the conclusions are applicable to the entire Mekong Delta and they should be carefully re-examined by some other qualified body. If the assertions can be proved, it could safely be concluded that the development of the upstream dams and the probable resulting suspension of sediment in the reservoirs would not greatly reduce the fertility of the delta soil, but the use of fertilizer would definitely be necessary.

High-yielding rice varieties and water control. The unit productivity of traditional varieties of rice in the Vietnamese Mekong Delta was only 1–1.8 t/ha. In the Cambodian delta, the traditional variety yielded only 1 t/ha. In 1968, a revolutionary variety of rice was imported to the Mekong Delta from the Philippines, which yielded 3.6–4.7 t/ha. In the Cambodian delta, the new high-yielding variety resulted in 2.2–2.5 t/ha, on average.

The high-yielding rice varieties are usually sown after mid-November, when the floods begin to recede, and harvested at the beginning of April. The second planting takes place immediately after the first harvest, and the second crop is harvested at the end of August or September. These varieties are, of course, very attractive to farmers, but success requires a supply of water sufficient for irrigation, intensive fertilization, and incessant weeding and application of insecticides; these needs are labour-intensive and expensive.[23]

More importantly, the success of growing high-yielding varieties depends greatly on the whims of nature, since the second harvest is in September, when the delta often suffers from flooding. In order to secure an appropriate supply of water for irrigation, the irrigation canals (which were extensively developed in the Mekong Delta in the past) have been expanded and well equipped by a number of simple low-head water pumps. Small boats which provide pumps with low-horsepower engines have been widely used.

When irrigation canals, locks, and levees are more widely installed and flooding is properly controlled, it will certainly be possible to have as many as three harvests a year in much broader areas of the Mekong Delta. There is no doubt that, in the future, the cultivation of high-yielding rice varieties in flatlands where the irrigated water is well drained would make the Mekong Delta the leading rice-producing region in Viet Nam.[24]

Salt-water influx. One of the problems in the downstream area of the delta is the influx of salt water into the coastal plain and the inundation area near the estuary. The difference in the sea-water level between high tide and low tide in the South China Sea is 3–4 m at its maximum. (The water level of the Saigon River at Ho Chi Minh City, located just outside the Mekong Basin, rises to 2.2 m at high tide.) Therefore, the inflow reaches beyond Phnom Penh and as far as Kompong Cham in Cambodia, located 410 km upstream from the estuary. Consequently, the sea water flows in through rivers and water channels of the delta as far as 20–65 km from the coast. As a result, some 500,000 ha of the coastal areas of the delta are affected by salinity.

The maximum salinity that still allows for rice cultivation is 1–2 g/litre in the planting season, 5–6 g/litre during the growing period, and 4 g/litre at the end of the growing period. However, without human intervention, the salinity of the water invading the delta exceeds 4 g/litre at high tide. (The extent of the encroachment varies according to the depth of the river bed at the estuary, the tide level, and the rate of flow of the river.) On the other hand, the maximum salinity acceptable for drinking water is 0.4 g/litre; however, in the delta, at high tide, the salinity sometimes exceeds this. If the current level of salt-water influx is not rectified, and if a large volume of river water is pumped upstream for irrigation, the degree of salinity will increase. As shown in figure 5.6, the affected areas in the delta where salinity is within acceptable limits vary during the dry season. The flow of the Mekong immediately downstream from Phnom Penh varies from 1,900 to 7,000 m^3/s.

The Vietnamese government and the Interim Mekong Committee have paid attention to the problems caused by pumped irrigation in various parts of the delta, and since 1978 they have made efforts to improve water control. For example, weirs and gates have been built at the north sides

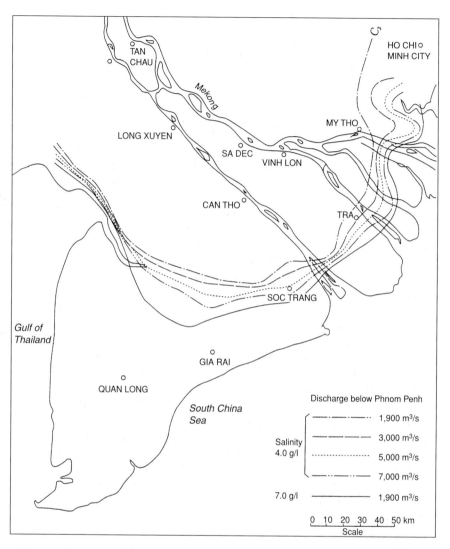

Figure 5.6. Degree of salinity in the Mekong Delta and dry-season flow volume of rivers at Phnom Penh
Source: Mekong Secretariat 1974

of the main water channels and in the vicinity of creek inlets in order to block or adjust salt-water influx (Mekong Secretariat 1985). If water is taken from rivers in various upstream areas for large-scale irrigation, or if large dams are developed in the future and the discharge from the dams in the dry season is restricted in order not to exceed the inflow volume, the influx of salt water will be more serious. On the other hand, if the dams are operated in such a way that the flow in the dry season is increased, such problems will be reduced.

About once every 5 years, the flow of the Mekong at Phnom Penh falls to less than 1,250 m^3/s. Experience shows that salinity increases markedly when the flow is less than 1,500 m^3/s. If a large dam, such as the High Pa Mong Dam, is built upstream, and if the reservoir is operated to increase the flow during the dry season, the problem of salt-water influx could be dramatically alleviated. If the flow of the Mekong is increased during the dry season, it will not only hinder salt-water influx in the coastal area but also certainly promote irrigation during the dry season. Assuming that the required flow volume in the dry season is 1 m^3/s/1,000 ha (1 litre/s/ha), 2,600 m^3/s will be needed to irrigate 2,600,000 ha. If the High Pa Mong and Stung Treng dams are built and dry-season irrigation is carried out in the delta, the remaining river flow will be sufficient to reduce the influx of salt water.

Flooding and upstream developments. As described in the section on rice cultivation in the delta, the fertility of the soil in the delta seems not to be measurably related to the deposition of sediment, and therefore the suspension of sediment by the dam will not have major adverse effects on the fertility of the delta.

A decisive factor in promotion of agricultural development in the delta is reduction of the extent and depth of the flooding that often occurs during the rainy season. One effective way of reducing flooding and salt-water influx in the delta is, of course, to build dams upstream. However, there are several other means, including the provision of dykes along the Mekong and the Bassac, the use of gates to adjust the flow (into water channels), and the prevention of salt-water influx by the construction of coastal dykes. A combination of such countermeasures should successfully reduce flooding without the use of upstream dams.

As all dam sites proposed for the Mekong Basin are outside the territory of Viet Nam, the government has been concentrating its efforts on dyke construction and on improving and expanding water channels in the Vietnamese Mekong Delta, without taking into account the construction of any upstream dams. However, the Mekong Delta would not be truly and stably developed unless several large dams are built to control floods and increase low flows. The results of surveys carried out from 1965 to 1969 by SOGREAH, a French consulting firm, has confirmed the above conclusion. Since the total flow volume of the upstream Lancang River is roughly no more than 20 per cent of that of the entire Mekong, the present flow conditions in the delta may not be much improved even if all the proposed dams in the upper Mekong Basin are completed.

If both the High Pa Mong and High Stung Treng dams are built, even without river dykes, the water level will be lowered by as much as 1.42 m immediately downstream of Phnom Penh, by 42 cm at the town of Can Tho on the Bassac River, and by 59 cm at the town of Sa Dec, located along the Mekong. If only the High Pa Mong Dam is built, the water level at Phnom Penh would be reduced by some 45–75 cm and the flooding area of the delta would be reduced by some 441,000 ha, i.e. about 10.6 per cent of 4,170,000 ha at the natural flow conditions occurring during the 1961 flood.

Coastal erosion. As previously described, in the coastal area of the delta, the earth and sand discharged into the sea from the apex of the New Delta have been transported westwards by the coastal ocean current caused by the south-easterly wind over the South China Sea. This volume of earth and sand has piled up along the western coast for a number of years and has gradually expanded the shoreline.

If dams are developed in the future, and the earth and sand are held, the westward ocean current and the effect of the waves of the South China Sea will cause the shoreline to retreat. The strong ebb and flow of the sea will further erode the shoreline. If dams are not developed in the upstream areas of the delta – and if the population continues to grow, cities expand, roads and highways are extended, agriculture is developed, and forests are rapidly logged – all of this will cause the subsoil to be eroded and washed away and the seashore will be extended. However, the estuaries will gradually silt up, making navigation difficult in such cases.

Effect on the fishing industry

When constructing dams in the lower Mekong Basin, in addition to providing fish ladders in the dams and spawning grounds in reservoirs, it will be necessary to formulate policies for the protection of fish in important places on the flood plains. It is also vital to ensure that the discharge of water from reservoirs does not adversely affect the spawning, hatching, and proliferation of the fish downstream from the dam.

Fishing industry in the lower Mekong Basin

Fish are essential for the lives of the inhabitants of the lower Mekong Basin: the basin populations rely on fish for 50–80 per cent of their total protein intake. In countries such as Cambodia and Laos, the diet consists of rice and fish, supplemented by a small amount of vegetables. At least one-quarter of the population make their living by fishing, or by fishing combined with farming. In the riparian countries in Indo-China, the fishing industry accounts for 2–8 per cent of the gross national product. If dams are not developed and the Mekong and its tributaries remain as they are, with no improvements made in the fishing industry, it is predicted that there will be problems in maintaining the supply of protein to these populations. As the development of the fishing industry is such an important issue for the inhabitants of the lower Mekong Basin, it is essential in developing the basin to conduct a thorough study and to get a clear grasp of the ecology of the fish inhabiting it.

Fortunately, over 85 per cent of the 400 species of fish inhabiting the lower Mekong are the same as those in the rivers of neighbouring regions, and the environmental conditions in the lower Mekong Basin are similar to those of the fish in the rivers in these areas.[25] It is, therefore, possible to make use of data gathered in regions outside the Mekong Basin in Thailand, Myanmar, Viet Nam, India, and Pakistan.

For the inhabitants of the lower Mekong Basin, fish constitutes the most important foodstuff after rice. Although the data are slightly out of date, according to a report published by the Mekong Committee in 1976, the average annual intake of fish per person was 22 kg in Thailand and Cambodia, 12 kg in Viet Nam and 13 kg in Laos (Mekong Secretariat 1976). According to the 1976 report, as annual sales of fish from the lower Mekong and its tributaries totalled 200,000 t and another 100,000 t were consumed by inhabitants for subsistence, the total fish production was estimated at 300,000 t. As 250,000 t of fish were caught each year in the South China Sea estuary and coast, the total annual production of fish in the lower Mekong Basin was 550,000 t (Mekong Secretariat 1976).

Assuming that there are 60×10^6 inhabitants in the lower Mekong and each consumes, on average, 10 kg of fish per year, the total annual fish consumption would be about 600,000 t. Therefore, if the current catch of fish in the lower Mekong Basin were the same as the 1976 level, it would not suffice.

Fish in the lower Mekong

Of the estimated 400 or so species of fish in the lower Mekong, 54 per cent are species of carp (Cyprinidae), 19 per cent are catfish (Claridae, Pangasiidae, and Bagridae), and 8 per cent are murrel (Ophicephalidae and Channidae). The remaining 19 per cent include feather backs (Notopteridae), climbing perch (Anabantidae), and gourami (Belontiidae).

Examination of the ecology of the fish in the lower Mekong Basin shows that spawning, growth, and reproduction all take place in accordance with the natural conditions of the Mekong. During the rainy season, as the water level rises, flooding begins in the paddy fields, waterways, marshlands, and other lowland regions. These floods spread from the Mekong and its tributaries to quite distant places, and the fish move with the flow. Some fish remain at these destinations for a long time, whereas others soon return to their original streams.

In the tropics, some fish make their way upstream (those which swim from the sea to rivers to spawn, e.g. salmon) and some move downstream (those which swim down to the sea from rivers, e.g. eels). There are also many species of migratory river fish, which move over short or long distances within the rivers.

The Mekong estuaries provide a habitat for many types of fish that can live in a mixture of salt and fresh water, and many of these migrate between the sea and freshwater streams. Among those that live in the freshwater regions upstream from the Mekong estuaries, many migrate seasonally between different freshwater regions. These fish migrate, spawn, and hatch their eggs at certain current velocities and temperature conditions. It is not easy to reproduce such conditions artificially and, when migratory fish exist in such large numbers and are low in commercial value, such aids as fish ladders are not profitable.

It has been pointed out that, because there are relatively few migratory fish of economic value in the tropics, there will fortunately be few problems in tropical rivers such as the Mekong (Mekong Secretariat 1976). In fact, however, as already mentioned, among the 400 or so species of fish in the Mekong, there are many migratory fish that repeatedly move upstream and back over various distances between the upper and middle reaches of the river. According to one report, of the various species which migrate in the Mekong, eight species – three species of carp, four of catfish and one species of herring – are important as fisheries products, but it is clear that there are many other species that migrate seasonally to various extents. Among these (although very few are caught every year) is the well-known migratory fish, the Pla Buk (*Pangasianodon gigas*), a large fish which grows to a length of 10 feet (3 m) and weight of 1,000 pounds (450 kg). The Pla Buk, which is known as the "master of the Mekong" or "Holy Spirit fish," is a type of catfish. The female was said to move freely between Phnom Penh in Cambodia and Lake Erhai in Dali, Yunnan Province. Since it exists only in small numbers, it has been pointed out that, even if it becomes extinct as a result of dam construction, this would not be an economic problem (although it is surely wrong to make such a decision on economic grounds alone).[26] Recently, however, as a result of the construction of several dams, including the Manwan Dam on the Lancang River in Yunnan Province, the passage to Lake Erhai has been cut off. There is still concern about the ecology of the Pla Buk, but it seems certain that it still swims downstream from Chiang Saen as far as the Khone Falls.

Incidentally, the Irawaddy dolphin, the subject of recent articles in newspapers both in Thailand and Japan, lives directly upstream from the Khone Falls. These mammals also live in the Se Kong River, which flows into the Mekong at Stung Treng. In the author's opinion, this means that the Irawaddy dolphin inhabits not only the Se Kong River but also the

Se San and Srepok rivers, because all three rivers flow into the main Mekong at Stung Treng, which is located downstream of the Khone Falls; however, this is only conjecture.

Several species of fish escape to the flooded areas on the shores of the river at the peak of the flow and return to the original streams when the waters withdraw. Dam planners optimistically predict that downstream conditions for these short-distance laterally moving species will actually be made more stable by dam construction and that they will be able to live more safely in the river.

Of the fish inhabiting the lowest reaches of the Mekong Delta from where the river enters the South China Sea and its branches, where salt water and fresh water mix, 27 species are able to continue living there. These species include members of the Carcharinidae, Pristidae, Clupeidae, Syngnathidas, Scombrosocidae, Polynemidae, Centropomidae, Sciaenidae, Scombridae, Solidae, and Tetradonitidae. Among these, *Macrobrachium rosenbergui*, a species of giant prawn, spawns in the lowest reaches of the freshwater river. At its larval stage, it moves to the estuary where the water contains salt. It subsequently returns to the fresh water after maturing. This prawn has commercial value and is being farmed.

In the vicinity of the Mekong estuaries, there are also crustaceans such as shrimps, prawns, and crabs; shellfish such as clams, mussels, and oysters; as well as amphibians such as frogs and reptiles such as alligators (Mekong Secretariat 1976).

Fish zones

One study divides the main stream of the lower Mekong into three fish zones. These zones, and the fish living in each or migrating between them, are described below.

Fish zones. *Lower part of the area upstream from Kratie.* In this zone, the slope of the river is relatively steep, particularly in the tributaries. As a result of bank erosion and scouring of the river bed, the banks have taken the form of a gorge. In the region with this topography, the total length of the main stream is about 2,200 km, and it is met by 25 tributaries. The Vientiane Plain is flooded at the peak of the rainy season, at downstream areas of the tributaries where they join the main stream; however, this flooding is not a problem compared with that of the delta.

Submerged area of the delta. In the delta in the lower reaches of the Mekong, an area of 53,600 km^2 may be submerged in exceptionally large floods that occur once every hundred years or so, but in the dry season the waters withdraw and usually only about 4,000 km^2 of land are covered with water. The slope of the river here is very gentle, with no gorges or shallows.

The area of the Great Lake in Cambodia, which occupies a special position in relation to the delta, is 3,000 km² in the dry season, swelling to about 10,000 km² during the rainy season. The Great Lake is a veritable treasure-store of fish.

South China Sea coast. In the area from about 100 km upstream from the South China Sea to the estuaries, marshlands where mangroves flourish are found in various places. The floods from the main stream of the Mekong influence the surrounding areas from the river estuary to about 30–40 km upstream. This mangrove area provides an important place for fish to reproduce. In recent years, however, the mangroves have been cut down in increasing numbers to make room for shrimp farming, mainly for export to Japan. This has been giving rise to considerable concern about the future of the area.

Fish in each zone. The following is a summary of the findings of studies of the influence of dam development on fish living in the above three zones and those that migrate between them.

Fish in the zone upstream from Kratie. The total length of the Mekong and its tributaries upstream from Kratie is at least 10,000 km in the lower basin alone. According to the author's observation, it seems that fishing in the lower Mekong is mainly "one-throw" (round-up) fishing such as seining, drift-netting, or trapping. Along the main stream, even at night, fishing boats with lights can be seen casting their nets everywhere on the river. Despite this, with the exception of Thailand, there are no clear statistics: we know only the number of fish caught at the few existing dams on the tributaries.

Among the dams on the tributaries, drag-nets and seine nets are mainly used at the Nam Pong Dam in north-east Thailand. However, the local fishermen face several difficulties: submerged trees in the reservoir impede the fishing boats and hinder fishing; moreover, the operation of the dam causes the lake to expand and contract, resulting in fluctuations in the distance to the lake from the surrounding areas and thereby creating difficulties for the fishermen.

Fish in the flooded delta zone, particularly the Great Lake and Tonle Sap River. The flooding of the Mekong Delta varies considerably from year to year and from place to place. In the lowland zone downstream from Kratie, 20 per cent of the whole area of Cambodia (including the Great Lake) is flooded. The area and level of water in the inundated zone also vary from year to year. The upstream area of the Vietnamese delta is completely submerged every year.

If a large dam were constructed on the main stream or on a tributary of the Mekong, the submerged area in the delta would be reduced. This reduction would be particularly marked in the Great Lake: the lake's depth currently ranges from 1 to 10 m; however, if a large dam were built (and assuming that its effective capacity was 60×10^6 m³), the maximum water level of the Great Lake would be 80 cm lower than the current level once every 30 years and 40 cm lower once every 4 years. The shallow basin-shaped belt around the lake would be affected by this drop in the water level. The slope around the lake is used by fish as a spawning ground and, since this is also where the fertile mud carried from the Mekong to the lake via the Tonle Sap river is deposited, it can easily be appreciated that the effect on the reproduction of fish in the reservoir would not be negligible.

For Cambodia, the fishing industry of the Great Lake is as important as paddy-field farming, if not more so.[27] While the lake becomes shallower every year, owing to the deposition of mud carried in from the Mekong, it is also adversely influenced by accumulation of mud from the tributaries which flow into the lake through soil erosion caused by the destruction of forests and by paddy farming around the lake. However, the extent of this influence is not known.

As previously discussed, the development of a dam upstream would considerably alter the volume of water, sand, and silt entering the Great Lake and the surrounding area. As this would affect the reproduction of fish, due care must be taken in this development.[28] This is the personal opinion of the author, but it should also be noted that, from the outset, the Cambodian government has made repeated statements opposing the development of an upstream dam.

Fish in the vicinity of the South China Sea coast. For the inhabitants of the Mekong Delta, the question of the fate of fishing along the South China Sea coast, including the estuaries of the main Mekong and its branch, the Bassac River, is second in importance only to the future of freshwater fishing in the delta embracing the Great Lake.

Once large dams have been constructed upstream on the Mekong and/ or its tributaries, the flow volume containing sand and silt discharged from the estuaries of the Mekong and Bassac rivers flowing to the South China Sea coast would undoubtedly change markedly from the original natural conditions.

The water in and around the estuaries of the Mekong and the Bassac is, of course, shallow and contains abundant nutrients. The mangroves around the estuary contribute significantly to the rearing of fish and crustaceans, as does the concentration of salt from the water entering the estuaries from the sea. Because many of the carnivorous animals and parasitic plants and animals that constitute a danger to the growth and survival of young fish have a low degree of tolerance to salt, it is considered desirable for certain species of fish and shellfish that the fresh water is mixed with a small amount of salt (about 1/1000–15/1000). Indeed, some such species can grow more quickly in water that contains a certain amount of salt. These are the conditions under which the fish and shellfish inhabiting the estuaries are living and proliferating. The dam planners must therefore take great care regarding the extent to which the construction of a dam upstream would disrupt these natural conditions.

According to current statistics, about 250,000 t/year of fish and shellfish are caught around the estuaries. However, these statistics were compiled during the Viet Nam War, when fishermen relied on traditional small-scale fishing techniques. As more modern, large-scale, drag-net and seine-net fishing methods come to be employed, the volume of the catch may increase dramatically: indeed, there is concern that such thorough fishing methods might even ruin the fishing grounds, resulting in devastation after temporary large profits. Appreciating the virtues of the traditional, comparatively moderate, fishing methods, experts should therefore reassess what would be the most appropriate and rational scale and methods of fishing. At any event, since many of the fish that can be caught at the estuaries (which include mackerel, shrimp, and crabs) are commercially very attractive, it is very important to conduct research on the influence of dam construction and the appropriate fishing methods.

Migratory fish. So far, we have considered the fishing industry in each region of the lower Mekong Basin. We also have to consider the influence of dam construction on the migratory fish mentioned above (pp. 268–270), which migrate over various distances, spawning and rearing their young throughout the riverine channels in the lower Mekong Basin. Unfortunately, however, we do not have any clear idea about the ecology of these migratory fish and no measures have been taken to investigate the influence of dam construction on them.

As described, many fish migrate around the Khone Falls and Tonle Sap River in the lower Mekong Basin. In the vicinity of the Khone Falls, during two periods – November to February and April to July – there are altogether 40 species of migratory fish. (In the period from November to February, 45 species of migratory fish are said to have been caught in weirs in the Tonle Sap river.) In addition, as the aforementioned cases of the Pla Buk and Irawaddy dolphin show, there seem to be more species migrating up and down the lower Mekong than was originally thought. It is of considerable concern whether dam construction, particularly at the Khone Falls or the entrance to the Great Lake, might result in the extinction of these species. The only clear information so far concerns the fish at the Nam Pong Dam: the carp species *Cirrhinus jullieni*, one of eight confirmed migratory fish in the Nam Pong Lake, lives in the reservoir, accounting for 20 per cent of the catch there. It has been inferred from this that many other migratory fish live in the reservoir.

Effect of dam construction on the fishing industry

It has always been stated that the ultimate aim of developing the lower Mekong is to bring the greatest benefit to the inhabitants of the lower Mekong Basin by constructing dams in appropriate places on the main stream and its tributaries and controlling these dams centrally. It appears that this dam construction might have both positive and negative effects on the fishing industry.

Let us examine this question in the light of the effect on fish of the development in the tributaries, which has been conducted in advance of dam development on the main stream of the lower Mekong.

Dam development on tributaries. Several dams have already been built on tributaries of the lower Mekong, such as the Nam Pong and the Nam Ngum dams, and it is generally held that dam construction provides habitats (lakes) in which fish can thrive. The reason given for this is that the fluctuations in the level of the lake due to reservoir operation differ from those that occur under natural conditions. Although some ecological alterations occur and the numbers of fish change, the fluctuations at small or medium-sized dams are usually less than those that occur naturally and, because care is taken in this operation, the adverse influence on fish is said to be small. Let us examine an actual example to see whether this rather simple logic holds true in practice.

With regard to the effect on fish of the construction of the Nam Pong Dam in north-east Thailand, before the dam was built there were 85 species of fish living in the river; 1 year after construction, this number had decreased to 54. As many as 31 species of fish, therefore, were eliminated from the region as a result of this environmental change.

The main species eliminated with the construction of the Nam Pong dam included the carp species *Labeo bicolor* and the catfish species *Wallagonia attu, Begarius bagarius, Kryptopterus cryptopterus, K. apogon*, and *K. bleekeri*. Regarding the change in the number of fish caught, only a few people were fishing before the dam was constructed, so the catch was almost negligible; after the dam was built in 1965, the annual catch (t) increased dramatically, as can be seen from the following figures: 1967, 1034.80; 1968, 1354.35; 1969, 1917.57; 1970, 1785.50; 1971, 2443.15. If the figure for 1971, for example, is divided by the effective water area of 400 km^2, the catch per unit area was 61 kg/ha – over five times the average yield of 12 kg/ha in tropical rivers under natural conditions.

Changes also occurred in the numbers of different species: most of the fish that survived the construction of the dam were species of carp, but the catch was considerably lower than under natural conditions; on the other hand, the number of mullet increased after the reservoir was built.

Subsequently, the catch at the Nam Pong Dam lake was well above 1,700 t/year until 1978. In monetary terms, this was equivalent to over US$650,000 year, while the electric-power production of the dam was said to bring profits of almost US$750,000/year (according to figures released in August 1976). As a result, there was a kind of fishing "gold rush" for a while in the area surrounding the dam.

In 1967, soon after the dam was completed, the number of fishermen registered with the government was only 265; however, in every subsequent year, farmers turned to fishing and, by 1971, the number of registered fishermen was over a thousand. More nets were cast and the catch per fisherman, of course, decreased.

This increase in the number of fishermen was not the only consequence of dam construction. In the area around and upstream from the Nam Pong Dam, development forged ahead after the dam's completion in 1965. As a result of the destruction of forests and unplanned, random use of land, soil erosion occurred and the sand build-up in the reservoir became acute. The total area of the region upstream from the dam is almost 16,000 km^2, but a huge area of 47,000 km^2 of forest, almost three times the size, was destroyed in just 11 years from 1965 to 1975.

According to a survey, the amount of sediment entering the reservoir was 950,000 t in 1964, the year before the dam's completion; by 1978, however, this amount had increased to 2×10^6 t and has now reached over 3×10^6 t/year. This build-up of sediment exacerbates the imbalance between plankton, herbivorous fish, and carnivorous fish, which already has an adverse effect. Consequently, fish productivity has fallen and the total annual catch has also been falling since it reached a peak of 2,060 t in 1978: it was 1,560 t in 1982, and by 1990 it was as low as 1,130 t. Finally, in 1984, a study concluded that, unless efforts were made to regulate the number of fishermen and to conserve the forest in the region, the local inhabitants would no longer be able to subsist.

According to Thailand's Fisheries Bureau, if the number of fishermen at the Nam Pong Dam does not exceed about 2,500, each can make an annual income of US$200 (assuming a total annual catch equivalent to US$500,000), which is just about sufficient income to live on. Consequently, it has recently become very difficult to get a fishing licence in this region.

Large-dam construction on the main stream. In assessing the impact of the construction of a large dam on the Mekong, we can refer to the experience of dams already constructed on tributaries and to the example of dams in India, where conditions are similar.

Fishing in reservoirs. If a large dam were constructed on the main stream of the lower Mekong, it was once estimated that the average level of fish production in the dam lake would be 20 kg/ha, or 8,000 t/year. However, in order to ensure this volume of catch, the submerged trees in the reservoir would have to be completely removed. In addition, fish suited to the environment in the reservoir would have to be released into it. Furthermore, it would be necessary to establish a suitable fishing village for the region around the reservoir, and to prepare the required fishing facilities, such as landing platforms, fish-processing plants, fishing boats, and nets.

Judging from past experience, the total catch tends to increase for a few years after reservoir construction, then to reach a peak and subsequently to decrease. Before a reservoir is stocked with fish, the adaptability of the fish already present should be studied. If it is then concluded that fish from another location should be transferred to a reservoir, fish should be selected according to the conditions of the reservoir. However, the influence of species introduced from outside upon those already present naturally gives cause for concern. According to one document, if a dam with a deep reservoir were built on a main stream, the species transferred there would include carp from outside such as grass carp (*Ctenopharyngodon idellus*), silver carp (*Hypothalmichthys molitrix*), bighead carp (*Aristichthys nobilis*), and common carp (*Cyprinus carpio*), two species of local carp and one species of catfish (*Pangasius* sp.). However, there is concern about the effect of fluctuations in the water level of the dam lake on fish reproduction in the reservoir.[29]

Regarding the effect of fluctuations in the lake surface, the reservoir water level is highest at the peak of the floods, and the 1–2 months when the water is highest is the peak time for carp spawning. If operations that substantially reduce the water level are avoided during this period, the damage can be kept to a minimum. Adjustment of this water level should be as gradual as possible, to minimize the adverse effect on fishing. This also has the advantage of hindering the proliferation of undesirable organisms (including fish) and particularly of algae.

In large and deep dam lakes, it is thought that a thermocline (a zone of rapid temperature change with depth) is formed when the water level is high. However, when the level is lowered through water discharge, the resulting elimination of anaerobic decomposition products (such as hydrogen sulphide and methane) and the reduction of highly nutritious substances is said to be good for the fish.

In large dams, the deeper the reservoir, the lower the catch per unit area.

Directly downstream from the dam. Water discharged from a large dam on the main stream of the Mekong (in the lower basin) would have various effects on the river directly downstream over a considerable distance. The fish directly downstream, in particular, would be affected by the water discharged for electricity generation and the resulting scouring of the river bed.

It has been argued that the hydroelectric outflow from the dam should not lower the original water depth directly downstream and should always be kept above a certain depth, but there is currently no guarantee that such efforts will be made after a dam is developed. It would be possible to maintain the downstream depth of the outflow by building a secondary dam further downstream, but this may not be economically viable. It is thought that the greater the distance directly downstream, the more the water level will decrease through the influence of natural storage in the river channel, but fluctuations over a considerable distance are, in any case, unavoidable.

In these circumstances, if fish that swim upstream to spawn were to lay their eggs immediately downstream from the dam, there is a danger that repeated water outflows from the dam would deprive newly laid eggs of sustenance and kill them. Also, if the amount of water flowing into the reservoir were less than expected and the reservoir water level dropped so far that it fell below the thermocline zone, oxygen-deficient water containing anaerobic decomposition products would be discharged, which might well harm the fish immediately downstream from the dam. (However, this oxygen deficiency would be corrected by the aeration effect a little further downstream, so the problem would be very localized.)

In view of this situation and the danger of disturbing spawning and hatching places, it would probably be better not to fish directly downstream from a large dam.

Immediately upstream from the dam. If a large dam were developed on the main stream in the lower Mekong, the effect on fishing directly upstream from the reservoir would depend on factors such as the size of the dam, its location on the river, and the flow conditions. As soon as impoundment of the reservoir was started, profits from fishing in this part of the river would fall.

A considerable distance downstream. In the region of the river a considerable distance downstream from a large dam, the extent of flooding would be greatly reduced by the operation of the dam, and the amount of the stream would be higher during the dry season. It might be expected, therefore, that better conditions than usual would be created for fish. However, it should be noted that, for migratory fish, dam construction would hinder their progress upstream.

Inundated areas in the delta. The effects on fishing industry on the Great Lake and Tonle Sap river if a large dam were constructed upstream on the Mekong and its tributaries have already been described (pp. 271–272). In other areas, the extent to which the dam would lower the flood peak; would reduce the inundation of the lowlands, paddy fields and waterways; and would deprive fish of important areas of water, would depend on the effective capacity of the reservoir and discharge volume. Generally, if the reservoir had a large effective capacity, the floods inundating the delta would be greatly reduced compared with natural conditions, and the submerged area in the delta would be considerably smaller. However, regardless of how much the inundated area decreases, it is thought that the resulting loss in fish production would not be greater than the increase in the catch through the creation of the dam lake. Moreover, as a result of the reduction of inundated land, the production of rice and vegetables would increase on the resulting stable arable land, and it is said that this would more than compensate for the loss of income caused by the decrease in the catch.

Regarding the influence of dam construction on fish production in the Great Lake and the Tonle Sap river, since the total combined catch in the lake and river is 50,000–100,000 t/year, the effects on this catch of the operation of a dam lake on the main stream will have a significant impact on the whole fishing industry in the lower Mekong Basin.

The Mekong and Bassac estuaries and South China Sea coast. The effect on fishing in the Mekong and Bassac estuaries and the South China Sea coast has already been outlined (pp. 272–273) and is not discussed further here. However, one negative effect, not mentioned above, is worthy of comment. Because the flood peak would be considerably lessened if a large dam were constructed on the lower Mekong, the flow of the river would become comparatively stable. In line with this development, modern agriculture using chemical fertilizers and insecticides would be developed in the Mekong Delta. Once a large dam was successfully constructed in the lower Mekong, additional large dams would be successively built at other sites. As a consequence, this trend towards modern agricultural methods would increase, inevitably affecting the river around its estuary. The use of agricultural chemicals and insecticides, together with the changes in the river flow and salt concentration caused by the operation of the dam, would be likely to have an adverse effect on fish and shellfish. However, scientific study of this effect and the formulation of rational policies is difficult at the present stage of preparatory planning; it appears, therefore, that it will be necessary to wait until a large dam is actually constructed and then to take measures to solve such problems or to devise means of making improvements.

There has also been concern that, if a large dam were built on the main stream or a tributary in the lower Mekong Basin, the fertile soils from upstream might silt up in the reservoir, adversely affecting agriculture and fisheries in the delta. However, in view of the fact that most of the soils in the Mekong Delta have a low or moderate fertility level and that, even now, the use of fertilizer is necessary for farming in the delta to maintain higher production levels, there seems to be no great cause for concern.

Mekong River development and the fishing industry

In tropical rivers, fish are of great economic importance. In the lower Mekong Basin, there is opposition in Cambodia to dam construction on the main stream or tributaries due to fears concerning its possible adverse effects on the fishing industry around the Great Lake and in the delta. Although the author believes that those opposing dam construction have good reasons, it should be noted that the construction of artificial lakes on tributaries of the Mekong, such as the Nam Pong Dam, has brought considerable success through the new fishing industries which have grown up on these lakes. The income from the fishing industries on the lakes of the Nam Pong Dam in north-east Thailand and the Nam Ngum Dam in Laos cannot be ignored. Considering that the income from the fishing industry at Nam Pong is almost as high as the income from electricity generation (originally the main purpose of the dam), it is clearly very important seriously to study the upstream, surrounding, and downstream areas of dams in relation to artificial lake fishing, as well as the questions of reservoir operation and lake control. Raising the potential of the fishing industry enhances the economic merits of dam development as a whole. However, in the case of artificial lakes, it should be noted that the total catch eventually stabilizes at some time after the dam is completed and, since the number of fishermen increases unless regulated, the catch per person will decrease.

To sum up, the question of whether to place importance on the downstream fishing industry when planning a dam varies greatly according to the social, economic, and topographical conditions, so no broad generalization can be made. Furthermore, there are very few documents from which one can scientifically and quantitatively ascertain the relationship between the operation of reservoirs and the ecology of fish. For these reasons, as far as the development of dams in the lower Mekong Basin is concerned, the necessity for, and effects of, fishing-industry management have not been studied when formulating plans for dams up to the present, but the opinion of the author is that they should be very seriously considered in the future.

Effect on wild animals

Deforestation and loss of tropical forests

It is well known that wetland zones and neighbouring areas are rich in tropical forests. They also contain a wide variety of vegetation serving as the habitat for various animals. Most of the world's bird species and a great many mammals, reptiles, and amphibians live in such humid zones and, since ancient times, almost half of the world's land animals and plants have inhabited the humid tropical forests of the Mekong Basin and other parts of South-East Asia, as well as those of West and East Africa and Central and South America.

It is said that there used to be 1.6×10^9 ha of tropical rain forest in the world, but that nearly 40 per cent had disappeared by 1978 with the recent rapid increase in the human population and the advance of modernization. Even the remaining 950×10^6 ha is still dwindling by 1.15 per cent (i.e. 11×10^6 ha) every year. Whereas the world's total forest resources stood at about 2.5×10^9 ha in 1978, this is decreasing by between 0.7 and 0.8 per cent (i.e. $18-20 \times 10^6$ ha) annually. This indicates that tropical rain forests are dwindling much more rapidly than other types of forest. In other words, an expanse of tropical rain forest equivalent to nearly one-third of the total area of Japan (almost 36×10^6 ha) is vanishing every year, turning into deserts or semi-deserts. If tropical rain forests continue to decrease at this pace, some time after the year 2000 there will be 250×10^6 ha less than in 1978. The number of animal species inhabiting tropical forests is therefore also expected to fall sharply.

This depletion of tropical rain forests in developing countries of the continental tropics is taking place for reasons that include the following:
• Rapid population growth and uncontrolled deforestation;
• Export of timber by governments to improve their financial situations;
• Slash-and-burn farming (for example, such farming methods are still being used over an area of $3-4 \times 10^6$ ha in northern Thailand);
• Forest fires;
• Consumption of wood and charcoal for fuel;
• Uncontrolled hunting;
• Development of roads due to improved means of transportation by road vehicles, which have replaced navigation by boat;
• Destruction by armies and battles.

Such loss of forests for the above (and other) reasons aggravates soil erosion in river basins, causing loss of valuable topsoil and devastating whole basins. This leads, of course, to a loss of diversity in the tropical rain forests and thus, as mentioned above, to the disappearance of the animal species whose habitat they provide.

Forests and wild animals in the lower Mekong Basin

This section is concerned with the forests in the lower Mekong Basin. Since 1960, the population of the lower Mekong Basin has increased sharply and, in many places, the natural environment has suffered as a result of the successive wars fought in the area. Consequently, the total forested area is now thought to be less than 50 per cent of what it was during the 1950s.

In the Mekong Basin's northern highlands, the natural evergreen forests have been diminished by forest fires and the slash-and-burn farming practised by four to five million non-sedentary inhabitants (mostly so-called hill tribes). Although the region is vast and there are still expanses of dense forest, more and more of what was once forest is being turned into grassland.

In the Korat Plateau in north-east Thailand, however, few forests remain.[30] Where forests once flourished, there remain now only scattered copses or grasslands where virtually no trees are left; this exposes the plateau to floods and drought. In the highlands in eastern Thailand, which used to be covered with thick tropical rain forests, slash-and-burn farming has destroyed the forest cover and caused progressive soil erosion. The very rainy, hilly area in the south is the only area still covered with tropical forests. This, then, is the present state of the lower Mekong Basin. Owing to the depletion of their forest habitats, wild animals are now in danger of extinction.

Surveys of wild animal populations.   The first surveys of wild animal populations living in the lower Mekong Basin were made and several reports were published from the early 1970s. These surveys were conducted to examine the potential problems caused by the construction of such dams as the High Pa Mong Dam. In November 1975, the Mekong Committee issued a paper entitled, "Wild Animals and National Parks in the Lower Mekong River Basin." Its objective was to designate areas in the zone for national parks. In the comprehensive surveys on wild animals conducted thus far, the zone was divided into five habitats – the northern highlands, the Korat Plateau, the Annam Mountains, the delta, and the southern hilly area. The surveys indicated that each differs from the others in terms of its vegetation, geological features, climate, and land-use patterns.

Animal Species.   According to the document *Wildlife and the Mekong Project* (Mekong Secretariat 1976) published in April 1976 by the Mekong Committee, there are a total of 1,121 animal species living in the lower reaches of the Mekong River – 212 species of mammals, 696 species of birds, and 213 species of reptiles. Of these species, 34 live exclusively in the lower reaches of the river or its surrounding zone.

The most numerous animal species are the apes and monkeys, accounting for 13 species in total – three species of gibbons (*Hylobates*, long-armed ape) and 10 species of monkeys. Five of these monkey species are macaques; the other five are langurs. The douc langur, which inhabits the evergreen forests of the Annam mountain range and northern Laos, is said to be the most beautiful ape in the world, but is now thought to be on the verge of extinction.

Table 5.14 Wild animals and birds inhabiting the lower Mekong Basin

Animals	Birds
Northern smooth-tailed treeshrew	Chestnut-headed partridge
Marshall's horseshoe bat	Sooty babbler
Bourret's horseshoe bat	Imperial pheasant
Douc langur	Short-tailed scimitar babbler
Fanois	German's peacock-pheasant
Langur	Grey-faced tit-babbler
Lesser slow loris	Red-vented barbet
Pileated gibbon	Bar-bellied pitta
Lao marmoset rat	Yellow-billed nuthatch
Owston's civet	Collared laughing thrush
Kouprey	Grey-crowned croicias
	White-cheeked laughing thrush

Gibbons do not normally form large groups but live in family units, each with a territory of no more than 100 ha. It is feared that the disappearance of forests could mean their extinction.

Of all the animals inhabiting the lower Mekong Basin, the greatest importance is placed on elephants, which are used for many purposes, including ceremonies, labour, and transportation. Quite a large number of elephants are said to inhabit the mountainous regions of Cambodia, but the veracity of these reports is doubtful. Elephants have disappeared from Viet Nam owing to the war, and few remain in Laos and Thailand. In Thailand, however, elephant festivals are still held quite often in the Surin District in the north-east.

A species of wild ox, the kouprey (*Bos sauveli*), inhabits northern Cambodia and southern Laos. Indigenous to the lower Mekong Basin, this large mammal is also on the verge of extinction and in need of protection.

The wild animals and birds inhabiting the lower Mekong Basin are listed in table 5.14. In addition, deer, pheasant, wild ducks, and many other wild animals and birds inhabit the region. Those species the author has seen himself, and his vivid memories include crocodiles in the central highlands of Viet Nam and in the Mekong tributary, the Srepok River; a giant lizard in the dense forests of the Se Kong River in north-western Cambodia; a splendid peacock in the jungle of central Laos; and a giant snake swimming up the Mekong in northern Laos. The lower Mekong Basin appears to have a truly dazzling variety of fauna, including unique wild animal species whose existence remains unrecognized officially.

Hunting of wild animals.   Until the seventeenth century in the lower Mekong Basin, elephants were killed in large numbers for their ivory, deer for their antlers, and other animals for their furs, which were exported. As late as the eighteenth century, 1000 rhinoceros were killed annually in Thailand and their horns exported. Even today, one can see large quantities of crocodile, frog, and lizard skins being sold in Bangkok, but these animals are bred domestically, not hunted. Until the late 1960s, live animal exports flourished, peaking around 1970 at an estimated 500,000 animals per year (worth US$4.5 $\times$ 10^6 in foreign exchange). From 1967 to 1971, 245 elephants were exported and in one year (1968), 240,000 birds were sold at Sunday markets in front of the Royal Palace Square in Bangkok (Mekong Secretariat 1976).

However, overhunting and forest depletion have brought about reduced catches of wild animals and, because of the concern that species may become extinct, active campaigns are now under way to oppose dam construction and other forms of development. More and more restrictions have been imposed on exports, and today pet-animal exports are banned, although live experimental- and zoo-animal exports are allowed.

Influence of dam construction. River development changes the ecosystems of river basins, especially along river-banks, with various effects on wild animals. Although dam construction constitutes a menace to these creatures, dams also contribute significantly to their survival once they have been completed. The following cases may help to clarify the dangers for animals if dams are developed in the lower Mekong Basin.

Threat from man-made lakes – the African example. One extremely well known case is that of the Kariba Dam on the Zambezi River separating Zambia and Zimbabwe in Africa. When this dam was built and the reservoir's water level began to rise in December 1958, a vast variety of mammals and reptiles began to flee the rising water from the far reaches of the basin, converging on the dam site and presenting a major threat to villagers. The animals' plight sparked a rescue campaign by animal protectionists around the world. The author visited the dam site twice, once in 1970 and again in 1980; tourists were delighted at the huge elephants that wallowed in the shallows surrounding small islands, the high spots that the immense dam's water had not flooded. Yet in retrospect, life in such an environment seems hardly salutary for these stranded elephants on small islets.

The Koka Dam in Ethiopia is a similar case where the appearance of the reservoir stranded animals on islands, as it had in the Kariba Dam, forcing these animals to cohabit in close quarters with species of completely different habits. For this reason, some animals attempted to escape from the islands and drowned, though it is reported that a certain number were rescued.

The edge effect of reservoirs. However, the appearance of the man-made Kariba Reservoir did generate a desirable edge effect in terms of the conditions for survival of animals in the area surrounding the dam. For aquatic and migratory birds in particular, the resulting artificial lake created an excellent habitat. It profoundly affected the lives of wild animals in the area, protecting them and providing breeding grounds. After 1963, when the reservoir water reached its maximum level, repeated fluctuations began, ranging between two and four metres over a year. When the water level rose, the plants and the animals' droppings left all around the shallows at the edge of the water were submerged, causing eutrophication of the water at the reservoir's surface; when the water level fell, water plants began to thrive on the banks where these nutrients had been left, attracting herbivores to the area. As a result, animal droppings accumulated again on the banks, causing eutrophication when the water level rose again. In general, the speed of change in water levels in an artificial lake and the extent of the rise and fall affect the growth and species of aquatic plants found there. In the case of the Kariba Dam, a close causal relationship can be seen between the changes in the water level of the reservoir and the ecosystem of the plants and animals around its banks.

Downstream from the dam. Cases have been reported where a dam has had a considerable adverse effect downstream as well. In the case of the Kariba Dam, 130 km downstream from the dam site the flood plain known as Mana, 75 km in length, spreads from the former river bed of the Zambezi River. Before the Kariba Dam was built, this flood plain was submerged in about 50 cm of water each time flooding occurred; this left alluvial deposits and led to a thick growth of aquatic plants throughout the flood-plain region, providing a habitat for elephants, wild oxen, large serow, impalas, zebras, and other herbivores. These animals used to migrate each time that there was a flood, allowing the aquatic plants to regenerate; eventually, the animals would return to feed again on the vegetation. After the dam was built, however, the river no longer flooded the plain and no longer deposited alluvial silt and clay. The water released from the dam not only stabilized the river's flow but also regularized its timing, leading to the disappearance of the wildlife that lived there.

Dam construction in the lower Mekong Basin. It is said that dam construction and a reservoir on the main stream of the lower Mekong would have little effect, because (a) few wild animals inhabit the area upstream from the Pa Mong Dam scheduled to be flooded by the reservoir, and (b) most of these animals live in the hilly areas overlooking the reservoir and thus would not be displaced (Mekong Secretariat 1976). It is thought that the effect on wildlife would be much less at other points in the river's main course where dams are scheduled to be constructed than around the above-mentioned Kariba Dam.

In the tributaries of the lower Mekong, fish as well as crocodiles and frogs are bred in some of the reservoirs. However, none of the man-made lakes constructed so far has had a serious effect on wildlife there, because all these reservoirs (except Nam Ngum) are small and originally few wild animals and birds were living there. It seems that wild boars, waterfowl, and other species still inhabit the banks, adjusting themselves to the operations of the dam. Downstream, however, dam construction has destroyed many water pools, forcing most of the wild animals and birds to withdraw to the reservoirs. The problem is hunters: quite a large number of foreign hunters gathered around the Nam Ngum Dam and overhunted the reptiles and waterfowl that had gathered there. The construction of the Nam Ngum Dam also reduced the extent of forest around the dam, with further adverse effects on wildlife.[31] On the other hand, the dam's construction has also had positive indirect effects on wildlife: it is said that the far upper reaches of the Nam Ngum River became easier for wild animals to inhabit, because the people who had originally lived in the area moved closer to the dam, whereas the population scattered throughout the basin has not markedly increased (Mekong Secretariat 1976).

Designation of national parks and wildlife reserves. As stated in the previous sections, dam construction influences the growth or diminution of forests and affects wildlife both directly and indirectly. We must therefore create better environments for wild animals, alleviating such negative effects and, if possible, giving positive assistance.

In the final analysis, dam construction helps people living around the reservoir in the lower basin to make good use of land; the highly productive land downstream from the dam is likely to be brought under intensive irrigation, resulting in the regulation of highland slash-and-burn farming. An additional merit of dam construction is that hydroelectric power stations decrease the need for charcoal and firewood fuel and, hence, the felling of trees.

The objectives of developing the region and preserving wildlife will both be achieved if dam construction leads to positive management of the Mekong Basin, if forests are protected and soil erosion is prevented on the upper basin, if poaching and overhunting are controlled, if wildlife preservation and ecological research facilities are established, and if the entire area surrounding the dam is designated as a national park or a wildlife reserve with the aim of further developing it for tourism and recreation. In developing countries located in the tropical lower Mekong Basin, the national governments' strong leadership makes the attainment of such objectives much easier than it would be in developed countries. Whether dam construction becomes a deadly weapon endangering the lives of wild animals on the verge of extinction due to neglect, or a turning point in the campaign for their protection and preservation, will depend on the perceptions and determination of dam planners and entrepreneurs in these tropical developing countries.

The objective of setting aside land for national parks and wildlife reserves is a matter of preserving nature in its touristic, biological, and historical dimensions through monitoring the effects of human activities on wildlife ecosystems, through research, and through protection. The designation of parks and reserves may impose constraints on dam construction, but appropriate implementation of plans from a long-term standpoint can make it possible to allow dam construction to function as a facilitator and harmonizer of interaction between human beings and wildlife.

Dam-construction planning should draw on prudence and ingenuity with the attitude that to protect forests and wild animals is to protect the entire vast natural ecosystem, including mankind. The Mekong Committee has therefore proposed the establishment of parks and wildlife reserves in 29 locations in the lower Mekong Basin (fig. 5.7).

Environmental effects on society

Whether a dam is developed for a single purpose or for multiple purposes, the development itself has broad implications and is related to various social and economic activities in complicated ways, both directly and indirectly. These relations are sustained for a long period and involve further development.

The influence of dam construction on human societies can be classified into the following categories:
1. Local only, influencing inhabitants in the neighbourhood of the dam as well as its immediate upstream and downstream areas;
2. Regional, influencing people not only in the neighbourhood but also in distant regions.

(There are various ways of utilizing a dam beyond its locale: for example, irrigation, navigation, and hydropower generation. The most extensive of these is hydroelectric power. The transmission of power from a dam site to other areas clearly influences the areas serviced, facilitating commerce and industry. Societies receiving the electric power are affected by the dam, regardless of their remoteness. Likewise, the transmission of water for irrigation and drinking water has an impact beyond the watershed boundary.)

Thus the development of a dam causes various socio-economic changes – directly and indirectly – including electric power, irrigation, fishing,

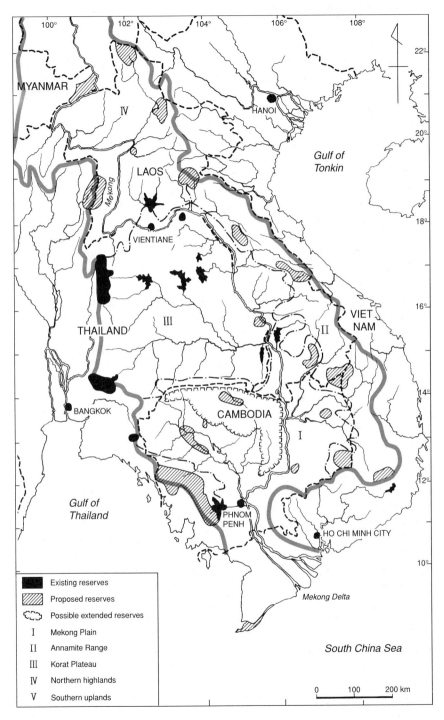

Figure 5.7. Wildlife habitats and existing and proposed national reserves, 1970
Source: Mekong Secretariat 1976

287

forestry, and industrial and domestic water supply. These have beneficial effects on transportation and communication and are related to each other in complex ways. Therefore, it is very difficult to grasp and analyse the socio-economic changes caused by dam construction from a wider, interregional perspective.

Even if we confine socio-economic effects to those which occur within a river basin, they are still complex. Viewed regionally, these effects can be classified as those which occur (1) at the dam site, (2) at the reservoir and its surrounding areas, (3) immediately downstream, (4) further downstream, and (5) upstream. The effects which occur at the dam site, at the reservoir and the surrounding areas, and immediately downstream are directly related to whether the dam could be developed.

Figure 5.8 illustrates these social effects. In this section, the author first discusses the most problematic issues related to dam construction – problems related to the submersion of dwellings of local inhabitants and their relocation, for example, and then turns to other important issues, such as the conservation of relics, development of tourism, and recreation. Finally, the difficult problem of water-related diseases is addressed.

The Mekong Committee has dealt with the issues related to submersion separately from environmental problems. However, the author takes the view that submersion is an important environmental issue which ought to be a primary concern in both tropical and non-tropical countries. Those involved in dam construction have recently begun to pay a great deal of attention to the conservation of relics. Water-related diseases have been treated with the utmost care in tropical areas.

Submersion and relocation

The development of a large dam sometimes encounters the serious socio-economic problem of large-scale submersion of valuable land, including a number of local dwellings. This problem is, of course, not limited to tropical countries: owing to the extreme difficulties involved in solving such problems, a number of dam-development schemes have been shelved or abandoned in both developed and developing countries. However, in some developing countries, in spite of the problem of submersion, the construction of large dams is still undertaken at the insistence of local government, in direct opposition to their own people.

Generally speaking, in developing tropical countries it is difficult to collect reliable data for planning and executing submersion. Moreover, neither the governments nor the inhabitants are always well prepared to take the necessary measures to counter effectively the various problems caused by submersion, which often results in tragedies striking the local people.

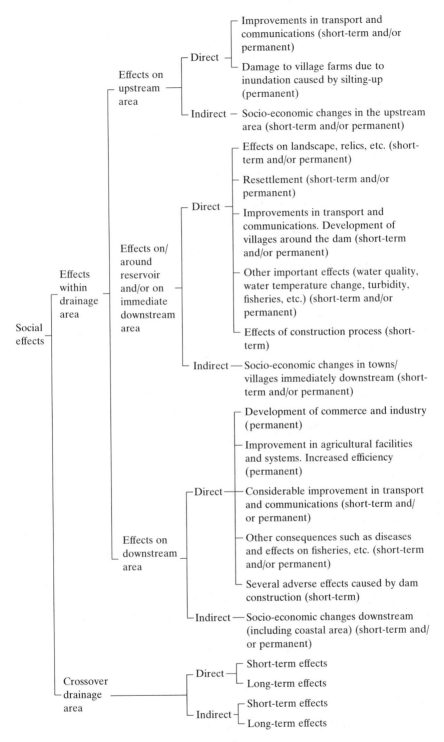

Figure 5.8. Social effects of dam construction
Source: Original

Submersion problems derive from many causes and are quite difficult to solve. These causes can be roughly classified as follows:
• Scarcity of reliable data;
• Insufficient investigation;
• Shortcomings in understanding and collaboration among local inhabitants;
• Faults attributed to local governments.
The problem can be identified as follows:
• Defects in the matter of investigation;
• Mishandled compensation;
• Relocation problems.
The size of numerous populations, some of them quite large, relocated from various project sites, is described below, followed by an outline of the various problems caused by submersion and relocation.

Relocated population

There are a number of examples of large dam construction in developed democratic countries. Most of these did not require the relocation of numerous local inhabitants, but there are exceptions: for example, in the development of the Tennessee Valley Authority dams, about 12,000 inhabitants had to be relocated away from the 20 dam project sites. In the development of the Bigge Dam Project on the Ruhr in Germany, some 2,600 local inhabitants had to be moved. In Russia, approximately 23,000 people were relocated owing to the construction of the Sarafor Dam on the Volga River. In China, more than a million people were forcibly evacuated from the Three Gorges Dam Project site on the Yangtze River. The relocation of such a great number of people is possible only where political power is irresistibly strong and determined.

However, an examination of past examples of large-scale dam construction in tropical regions shows that there are numerous surprising instances where vast numbers of inhabitants were forced to relocate from the project sites. The construction of dams was possible because of the existence of political and quasi-political foundations. Most were realized through financial aid from foreign countries or international lending agencies. Prior to dispensing the funds, of course, financial and economic feasibility studies were conducted, and the results were deemed economically justifiable even when financially problematic. This was partly because the estimated costs involved in compensation, particularly the compensation costs for population relocation, were erroneously or intentionally underestimated. The government estimates of compensation costs for private property and investment were quite low. On the other hand, there were very few public and private installations to be compensated in developing countries. The inadequacy of compensation made it relatively

Table 5.15 Number of people relocated for the construction of major world dams

Dam	Year completed	Number relocated[a]	Country
Aswan High	1968	120,000	Egypt and Sudan
Bhakra	1963	36,000	India
Brokopondo	1971	5,000	Surinam
Damodar (four projects)	1959	93,000	India
Kainji	1969	42,000–50,000	Nigeria
Kariba	1963	52,000–57,000	Zambia and Zimbabwe
Keban	–	30,000	Turkey
Kossou	1971	75,000	Ivory Coast
Lam Pao	–	30,000	Thailand
Thailand's 11 projects[b]	1963–1971	140,000	Thailand
Nam Ngum	1971	3,000	Laos
Nam Pong	1965	25,000–30,000	Thailand
Nanela	1967	90,000	Pakistan
Netzahualcoyoti	1964	3,000	Mexico
Tarbela	1974	86,000	Pakistan
Volta (Akosombo)	1966	80,000–84,000	Ghana

Source: Biswas (1982).
a. The number of people was calculated by multiplying the number of families relocated by 6.
b. See table 5.16.

easy to carry out the development of large dams and reservoirs in Asia and Africa from the late 1950s until the early 1970s (table 5.15). In Africa, inhabitants were moved from the following sites: the Kariba Dam Project site on the Zambezi river (50,000 people); the Akosombo Dam Project site on the Volta River (80,000); the Kossou Dam Project site on the Bandama Blanc River (75,000); the Aswan High Dam Project site (100,000); the Kainji Dam Project site (42,000). In Asia, some 30,000 inhabitants were relocated from the Keban Dam Project site on the Firat, a tributary of the Euphrates River in Turkey; 93,000 were relocated from the Damodar Valley, which included both hydroelectric and thermal power plants; and 30,000 were relocated from the Nam Pong Dam Project site in north-east Thailand.

It is, perhaps, instructive to compare these numbers with the 5,329 people relocated from Japan's 49 dam-project sites. On average, only 109 households per dam were relocated. Among these sites, that of the Yanba Dam Project was notorious for the number of the people relocated; nevertheless, even there, only 1,170 inhabitants (340 households) were moved.

Table 5.16 Numbers of people and families relocated for dam construction in Thailand

Project	Year completed	Major objectives of construction	Number of families relocated	Number of people relocated[a]
Bhumipon	1964	Power	4,035	24,210
Nam Pong	1965	Power	5,012	30,072
Lam Pao	1965	Irrigation	5,459	32,754
Lam Takong	1966	Irrigation	444	2,664
Kiew Lom	1969	Irrigation	496	2,976
Lam Nam Oum	1969	Irrigation	1,639	9,834
Lam Dom Noi	1969	Power	1,317	7,902
Nam (Sirikit)	1971	Irrigation	2,797	16,782
Huai Luang	1975	Irrigation	612	3,672
Krasiaw	1975	Irrigation	313	1,878
Kwae Yai	1977	Power	1,200	7,200
Total			23,324	139,944

Source: Lightfoot (1981).

a. The number of people was calculated by multiplying the number of families relocated by 6.

In both developing and developed countries, relocation has a serious social and natural environmental impact. Yet governments have underestimated – or ignored – the hardship of the relocated inhabitants, and the large dam developments have been forcibly carried out.

Problems of submergence and relocation of local people caused by dams in north-east Thailand

As shown in table 5.16, the booming construction in Thailand initiated by the completion of the Bhumipon Dam in northern Thailand and the Nam Pong and Lam Pao dams in north-east Thailand in 1965 continued until the late 1970s. This resulted in the relocation of some 140,000 local people from their homes, most of whom were forced to change their livelihoods as well. These people were mostly poor inhabitants who had eked out a living by traditional farming and whose losses were irrecoverable. Moreover, the value of the land given to them as compensation was on the whole much less than that of their original land, and their income at the relocated sites was much lower. As a result, not a few of these people were made even poorer than before.

The inhabitants moved from submerged areas in Thailand suffered greatly (fig. 5.9; see also Lightfoot 1981). The basic causes of their suffering, as identified earlier (p. 290) were as follows:

• Defects in feasibility studies;
• Problems deriving from inappropriate compensation;

Figure 5.9. Dam projects in north and north-east Thailand
Source: R. P. Lightfoot (1981)

- Selection of inappropriate relocation sites.

These causes are reviewed in more detail below.

Defects in feasibility studies. In the past, the development of large dams in Thailand involved the employment of foreign firms of consultants. Ostensibly in consultation with the Thai government, foreign governments or international lending agencies selected foreign engineering firms to conduct studies of the projects. Meanwhile, the socio-economic aspects of the projects were appraised by the foreign governments or international lending agencies providing assistance, before final decisions were made on whether to release the funds. Subsequently, selected foreign consultants, mostly engineering experts, carried out technical studies and the necessary economic and financial analyses. This procedure was routine for the implementation of such projects. Although the most important issue was to deal with the problems intrinsic to relocation, these were internationally omitted from the project studies!

Although careful studies were made regarding major aspects of public infrastructure, such as the highways and railways to be submerged by dam construction, studies regarding relocation of the inhabitants were left to the Thai government. The costs of replacing homes lost to submergence were, almost without exception, to be paid in local currency rather than hard currency; as a result, the assisting foreign governments, international lending agencies, and the foreign consultancy firms selected simply did not deal with the problem and behaved as if there were no submergence and relocation problems. This differed radically from procedures in some other developing countries; for example, during construction of the Kpong Dam in Ghana, relocation problems were regarded as part of the whole package. The foreign consultants selected by the assisting governments or lending agencies involved in the construction of dams in Thailand never did carry out studies on this important issue.

In the case of the construction of the Nam Pong Dam in north-east Thailand, the amount of compensation was kept to a low level; as a result, the relocated inhabitants were greatly impoverished. Even so, compensation costs still reached as much as 29 per cent of total construction costs. In spite of the fact that the project required such a large amount of compensation, no relocation study was carried out by the consultants, yet very precise calculations were made to work out the necessary costs of civil engineering. The Nam Pong Dam Project was studied by a French consultancy firm, which conducted the feasibility study on the basis of an overall reconnaissance survey of the entire lower Mekong Basin by a Japanese government team. Subsequently, the German government provided financial assistance for the dam's construction, which was carried out under the supervision of German consultants.

Similarly, in the feasibility study on the Lam Dom Noi and the Kwae Yai dams, both of which were assisted by the Japanese government, problems likely to result from submergence and relocation were not considered. Compensation for submergence and relocation involved such complications that the Japanese government decided not to make any appraisal or appropriate disposition, deferring to the Thai authorities. It was unfortunate that the Thai government did not include sufficient budget allowances for submergence and relocation costs, and that the Japanese government did not audit the budget or lodge a protest.

Among the 11 dam projects included in table 5.16, the aid-giving government and lending agency asked the Thai government to submit information on the relocation of local inhabitants with regard to only two projects. When the Lam Nam Oum Dam, one of these two Thai projects, was constructed, the American government urged the Thai government to present a relocation programme prior to rendering aid. The other project was that of the Kwae Yai Dam, for which the World Bank asked the Thai government to submit a relocation programme as part of the protocol to render financial aid. Incredibly, however, the contents of the submitted relocation programme were not examined by the World Bank because submergence and relocation were regarded as problems to be solved by the Thai government itself.

Problems deriving from inappropriate compensation. In Thailand, the compensation costs for dam construction are decided by the Compensation Committee, which is composed of representatives of seven ministries and government agencies. Actual payments are made by the government agencies responsible, such as the Royal Irrigation Department, the Department of Energy Development and Promotion, and the Electricity Generating Authority of Thailand (EGAT).

However, in the above-mentioned dam projects, there were defects in the appraisal of land values: since most of the farmlands had been sold within kinship groups, land prices customarily were kept low as a courtesy, without considering the monetary value of the crops produced, or the intrinsic value where upgrades in land usage might become important. This led to an artificially low price structure on which land values were based, and on which compensation was calculated. To make matters worse, the actual payment was postponed for years. Ordinarily, inflationary trends are taken into account, but not in this case. In addition, profiteers exploited the delay as middlemen of sorts, lending money to tide the farmers over during the waiting period, at high rates of interest.

Selection of inappropriate relocation sites. The Land Settlement Division of the Thai government was responsible for the selection and arrangement of relocation sites. The sites designated were to be as close as possible to the dam sites, and to be developed by clearing forested areas owned by the government. Unfortunately, such government land was usually quite barren and unsuitable for agricultural production; moreover, the selection was made hurriedly in order to meet the construction schedule. Even if somewhat better sites were available in forested areas, they often had already been illegally occupied by squatters, who were virtually impossible to evict.

Governmental efforts to rectify the defects. According to the author's recent interview with the Mekong Committee, the Thai government, the aid-giving governments, and the lending agencies have already reflected on the situation and rectified the resettlement system. The compensation system has also been improved: nowadays, compensation is made not only to local professional fishermen but also to local inhabitants who have pursued fishing as an occupational sideline.

Compensation payouts are expected to be made before submergence begins. Furthermore, the Thai government has decided to start the selection of relocation sites as soon as possible, for the benefit of local inhabitants who are obliged to move from their homes. In addition, it has decided to install all the basic infrastructure, such as municipal water supplies, needed to sustain life and livelihood in the relocated sites. The government's attitude has changed to include careful examination of the whole spectrum of resettlement questions so that those displaced are made as comfortable as possible.

More than half of the people who were obliged to move from their original locations near the dam construction in north-east Thailand refused to move to the designated settlement sites and instead moved elsewhere after receiving their money settlements. However, most of these have ended up in somewhat better circumstances than those who did move to the designated relocation sites.

Problems of submergence and relocation caused by the construction of the Nam Ngum Dam

In 1971, 7 years after the completion of the Nam Pong Dam in north-east Thailand, the government of Laos completed the Nam Ngum Dam in the Vientiane Plain. Some 3,000–3,500 local inhabitants were displaced by the construction. The success of the government's treatment of the relocated inhabitants cannot be evaluated easily: some say that the relocation was well conducted; others disagree. However, the author (who visited the Nam Ngum Project site a number of times before and after construction) is of the opinion that the relocation seems to have been fairly successful.

The following is a summary of the report which was provided by the government of Laos on the settlement of the local inhabitants relocated from the Nam Ngum site.

In 1967, 4 years before the completion of the Nam Ngum Dam in 1971, the government evacuated 579 households from the dam and reservoir site scheduled to be submerged. However, some time later, an appeal was made to the government, complaining of hardship suffered by evacuees who had been unable to sustain a stable livelihood in their new circumstances. Therefore, in 1974, the government selected other sites to resettle the previously relocated people and conducted social and economic studies. In January 1975, the studies were completed and the government adopted a revised plan to clear woods near the reservoir and to provide paddy fields and farms with an irrigation system. Minimum public facilities were also provided to sustain the lives of the people relocated to the site. By March 1975, an investigation was concluded into the existing facilities of former occupants living in the relocation sites of at least 1,300 ha. The clearing and parcelling-out of land, construction of irrigation and drainage systems, and other related works was nearly completed. By this time, there were some 700 households to be relocated to the new site.

By the end of 1980, Laos had completed roads, bridges, schools, clinics, public offices, silos, and other structures, including an agricultural centre. The government also provided a pilot farm at the site. Through these efforts, irrigation was successfully undertaken. A cooperative association was also set up, which fostered brickmaking, pottery production, weaving, and rice polishing, among other things. An irrigation canal 3 km long was built and, although the irrigated paddy fields could produce only 1 t/ha, two-crop paddy irrigation became established. The government carried out experiments on various varieties of rice, corn, soybeans, and other crops, and announced the results of experiments to educate relocated farmers. Together with other local inhabitants, some of the relocated inhabitants began fishing in the Nam Ngum Reservoir.

According to the government study in 1974, 3 years after the completion of the dam, only 2 per cent of the local inhabitants were making a living as fishermen; however, according to the 1981 study, the percentage was as high as 33 per cent, although most of these were not necessarily dependent on fishing for all of their livelihood. They set up a fishermen's association but have not yet matured enough as a group to profit sufficiently from fishing as a primary occupation.

In parallel with these studies, many administrative actions were taken, such as the improvement of sanitary conditions. A population census was conducted, showing how, and to what extent, evacuees had benefited from the various socio-economic measures taken. On the whole, the development of the relocation site has been carried out by the joint efforts of the government and the inhabitants. Developments for and by the people have been made quite steadily and successfully.

The above information, provided by the government, sounds ideal, but it is impossible to judge whether it can be accepted at face value. However, the author has been impressed by his frequent visits since 1984: the standard of living of those in and around the project site seems to be much better than it was before 1971. They appear to be better dressed and well fed, and most – particularly the young – seem in good spirits.

Problems of submergence and relocation expected to be caused by the construction of the Pa Mong Dam

More than 30 years have passed since the study of the High Pa Mong Dam Project was undertaken. Needless to say, the project is one of the most important in the lower Mekong Basin, not only in scale but also in magnitude of development potential. It was expected that, once the high dam was completed, neither citizens in the city of Vientiane nor inhabitants of the opposite bank immediately downstream would be affected by floods, and all the inhabitants along the main Mekong down to the Vietnamese delta could enjoy the beneficial effects of the dam. These would include the improvement of navigation, as well as the increased productivity of irrigated agriculture, resulting from the stabilized and increased discharge from the dam during the dry season. Moreover, the power generated by the dam project, which was estimated as 4.8×10^6 kW (when the maximum water level was set at 250 m), was expected to bring tremendous benefits not only to Laos and Thailand but also to those living in the adjacent riparian countries.

Regarding irrigation benefits, it was planned that the Vientiane Plain of 10,000 ha, and the Korat Plateau of 40,000 ha would soon be irrigated and double-cropping achieved. In the distant future, a much broader area, possibly as much as 1.0×10^6 ha in Laos and Thailand, could be irrigated with the completion of an extensive irrigation canal system. However, there was a very difficult problem hindering the immediate realization of such benefits: this was the submergence and relocation of local inhabitants caused by dam construction. Even in 1976, it was estimated that some 460,000 people would have to be relocated on creation of a reservoir with a maximum water level of 250 m.

Nevertheless, the High Pa Mong Dam Project was regarded as one of the world's most promising and noteworthy projects until the early 1970s, when the feasibility study of the huge project was completed. Owing to the Viet Nam War, the Americans left the region and the chance was lost. Later, the Mekong Committee was reorganized as the Interim Mekong Committee, and in 1983 Thailand, Laos, and Viet Nam held a meeting in Chiang Mai in Thailand. At this meeting, despite the earnest wishes of both the Interim Mekong Committee and the Thai government, the Laotian government strongly opposed the construction of the high dam. The major reason for the opposition was that some 110,000 people living along the main stream of the Mekong in the Vientiane Plain would have to be relocated.

Meanwhile, a movement against the construction of large dams gained momentum among people all over Thailand. As a result, a proposal was presented by the Interim Mekong Committee to reduce the height of the dam by nearly 50 m, and to scale down construction. However, the original development plan and any alternative development plans have all been shelved. To date, nothing has been achieved, mainly because neither the assisting government nor the international leading agency has shown interest in funding dam construction, given the attendant circumstances in the lower Mekong Basin. This was partly because of disputes over maximum water level: no common ground could be reached on the reservoir's water level, which determined the extent of the relocation.

Estimated number of people to be relocated. Several estimates were made of the probable number of local inhabitants to be relocated by the proposed dam and reservoir. According to the report of the Mekong Committee in August 1976, and the results of research carried out by the University of Michigan in 1978, the number of people to be relocated through the construction of the High Pa Mong Dam was estimated as more than 400,000 if water impoundment started in 1991.

Lands and houses to be submerged by construction. The Mekong Committee estimated in August 1976 that, if impoundment started in 1995, some 550,000 local inhabitants and 1.165×10^6 rai (186,400 ha) of land would be submerged.

Relocation costs. In its 1978 report, the University of Michigan announced that the total costs involved in submergence and relocation in the first stage of development of the High Pa Mong Dam would be about US\$626 $\times 10^6$. This estimate was made on the assumption that the full-reservoir water depth would be 250 m and that the submerged area would extend to some 172,400 ha.

Examining the breakdown of the total costs involved in the development of the High Pa Mong Dam Project, 80 per cent of the relocation costs are in the local currency; those costs make up roughly half of the total local-currency portion of the construction costs. On the other hand, these relocation costs comprise more than 30 per cent of the total construction costs. Nevertheless, division of US\$626 $\times 10^6$ by 398,646 – the number of local inhabitants who will have to move from the project site – gives US\$1,570 per person. Barring subsequent cost increases due to inflation, this is the level of funds that would be provided as compensation. Even given the fact that the above figure is 20 years old, the amount seems to be unrealistically low.

The breakdown of the compensation (in US dollars, assuming that the relocation was carried out in 1990) would be as follows:
• Compensation for the submergence of lands, etc., 400,582,000;
• Cost of persuading inhabitants to evacuate the site, 13,985,000;
• Compensation for physical, monetary, and emotional hardship, 29,601,000;
• Actual moving costs, 7,655,000;
• Cost of expediting the relocation sites, 17,099,000;
• Cost of preparing the relocation sites, 27,822,000;

- Cost of building schools, temples, clinics, public offices including a police station, etc., 27,128,000;
- Compensation for loss of livelihood until income can be generated in the relocated sites, 59,202,000;
- Administrative costs, 12,016,000;
- Roads to connect existing villages to relocation sites, 25,853,000;
- Miscellaneous costs, 5,457,000.

Total: US$626,400,000.

The above estimates are based on the following categories of the local inhabitants who would have to move:

1. People who might continue to live around the reservoir;
2. People who might move to the planned relocation sites;
3. People who might move to other places and buy their own land;
4. People who might move elsewhere, relying on their kin and acquaintances.

Of these, category 3 would involve the least expense and are the most likely to resettle successfully. After paying compensation to those in category 1, the government would have no further financial obligation to them once the dam was built. In addition, those who would move to the planned relocation sites were assumed to be mostly Laotians: few Thais were expected to relocate to the designated sites.

It was assumed that, with a maximum water level of 250 m, only 11 per cent of the total number of inhabitants would move to the planned relocation sites to be provided in the surrounding hills, and that 12 per cent would continue to live around the reservoir. Most (some 60 per cent) would move to other places and buy their own land.

Need to establish appropriate administrative organizations. The submergence caused by the construction of the Pa Mong Dam may cause major social problems on the national level, irrespective of the size of the dam. The greatest problems would be related to the administration established to deal with the submergence. Some 30 per cent of the total construction costs would have to be spent on compensation if the High Pa Mong Dam Project were realized. It is necessary, therefore, to strengthen the organization of both governments, particularly that of the Laotian government.

If the governments expect those evacuated to make a personal and unaided choice of livelihood at the new locations, they have to secure and assign suitable jobs to such people prior to their evacuation from the project site. The future prospects of the relocation sites and the surrounding region depend upon the type of work chosen by the evacuees after they move. In this respect, north-east Thailand is only one of many local districts, and is not seen as the most important region for the country's economy. On the other hand, the Vientiane Plain has been regarded as the most important area to the people of Laos, so that a plan to allow each of the 100,000 relocated people to select his or her own occupation can never be considered lightly in terms of the overall future of Laos.

For both governments, but particularly for the Laotian government, it will not necessarily be easy to make detailed plans and to carry them out accordingly. The expenditure for the relocation of the people should not "shrink" between donor and recipients. The appraisal of the required costs should be made decisively, and the evacuees should be paid in a timely and fair manner. Of course, this is easier said than done: as previously discussed, the submergence problems caused by the construction of tributary projects in the lower Mekong Basin (such as the Nam Pong Dam) were not always solved successfully. The issue of submergence caused by the Pa Mong Dam should be treated with the utmost care, keeping in mind the vital lessons learned from past unhappy experience.

All these matters are not merely domestic questions, but matters that should be borne in mind by the governments of assisting countries and the international lending agencies. All should cooperate with the local governments to form a consortium to establish a programme for dealing with problems of submergence.

Such an organization must take the following into account in order to minimize the negative effects and facilitate relocation:

1. Respect the individual, and give appropriate relocation advice if solicited.
2. Select the most appropriate relocation sites and adopt the most reasonable and feasible means of relocation acceptable to the governments and the people after frequent discussions. A completely satisfactory exchange of information, based on the most cordial understanding, should be expected on both sides. Details of the relocation project and implementation schedule should be clarified to gain a consensus.
3. Following consensus, start the construction of all required public facilities and private residences as soon as possible, and bring them to completion promptly.
4. Relocate people in a timely manner after all the public and private buildings and facilities are completed.
5. Begin all administrative services immediately.
6. Give appropriate advice and guidance to enable people to farm effectively at the relocation sites. It is important to inspire people to make efforts to increase agricultural production.
7. Help people who wish to become fishermen in the reservoir area. Begin adult education and professional training as soon as the submergence and relocation programme is finalized.

8. Provide ample funds and disburse them properly in accordance with the progress of the relocation programme.
9. Strive constantly to improve administrative organizations to ensure progress of the programme.

The organization would be expected to do the following in terms of compensation:

1. Pay compensation money as reasonably as possible, on the basis of a unified arrangement. A single set of criteria should govern all payments, regardless of the region involved.
2. Try to pay compensation before water impoundment begins. If this is impossible, the governments should explain the reason and make every effort to ease the evacuees' predicament. Separate payments should be avoided as much as possible.
3. Make open payment of all compensation.
4. Ensure that all regulation, and all that needs to be done in relation to compensation, does not harm the evacuees throughout their lives.

Once it becomes clear that some harmful effects will be caused by governmental decisions, the government should have the grace to apologize and award compensation. However, in regions like the Mekong Basin, where compensation often has to be paid to more than one riparian country (as is the case with the Pa Mong Dam), the author doubts that such ideals governing dam construction can be so easily realized.

Historical sites

Historical sites and dams

Careful consideration must be given to various factors in the development of river basins. It is considered the duty of modern society to pay special attention to preventing the destruction and inundation of valuable archaeological sites, the tools and implements of primitive human communities, and fossils from even earlier periods. In addition, efforts must be made to preserve intangible cultural assets, such as tribal customs and oral traditions which remain in various areas. In this connection, it is important to understand that the historical sites and cultural assets of developing countries in the tropical zones have not been adequately studied. In developing countries, unlike the advanced nations, the location and contents of such sites are not fully known, nor have they been publicized. In many cases, these areas have not been properly surveyed because of difficulties in transportation and communication. For instance, the site of an ancient civilization which flourished over two thousand years ago was suddenly discovered in north-east Thailand in the mid-1960s. As a result of this discovery, scholars for the first time became aware that their archaeological surveys of the lower Mekong Basin had been insufficient and realized that more exhaustive investigations would have to be conducted in the future.

It is unrealistic to expect a thorough archaeological survey to be undertaken promptly after the announcement of a dam-development project, especially in tropical regions. However, with regard to areas which will be inundated by a new dam, serious efforts should be made to begin collecting information immediately after the start of surveying work. If historical sites are discovered, this information should be widely publicized and funding should be sought for undertaking archaeological studies, as well as for the transfer or preservation of the discoveries. The necessary financial resources may not be available to developing countries, in which case it is important to seek the assistance of international organizations, such as UNESCO or the United Nations Development Programme (UNDP), or the governments and private organizations of the advanced nations.

The construction of the Aswan High Dam and preservation of the Abu Simbel Temple dramatically highlighted the inundation of historical sites by dams built in the tropical developing countries. Various similar cases have also attracted international attention: for instance, during the construction of the El Cajon Dam (completed in 1985) in Honduras in Central America, extensive excavation work was done on numerous Mayan sites in the affected area (Mekong Secretariat 1982); in the case of the Keban Dam in Turkey, as discussed in the following section, the government sought the cooperation of the University of the Middle East in the capital city of Ankara for the preservation of an ancient mosque and other historical assets. On the other hand, there are numerous cases in which no efforts were made to protect historical sites from inundation: for instance, in the case of the Cerron Dam in El Salvador in Central America, it is reported that numerous important historical sites lie at the bottom of the reservoir created by this dam project.

Loss and inundation of historical sites

The loss of historical sites can be viewed as an aspect of the overall problem of inundation. However, because of its special importance, the issue of the loss of historical sites is discussed separately and various previous instances are briefly described.

Keban Dam (Turkey). The Keban Dam in eastern Turkey was built between 1966 and 1973 by the Water Resources Development Agency (DSI) of the Turkish Government. This large dam (height 211 m, rockfill volume 15.332×10^6 m^3, installed capacity 1.24×10^6 kW) boasts a reservoir of 685 km^2 holding 30.6×10^9 m^3 of water. A total of 5,170 houses were submerged beneath the reservoir, affecting 30,000 people. In addition, 58 km of railroads and 90 km of highways were inundated. As a result, the cost of compensating the displaced population and facilities exceeded the US$100 × 10^6 cost of constructing the dam's power-generating plant.

This project also involved various problems related to the inundation of historical sites. The site of the Keban Dam is located at the uppermost section of the Euphrates River, and the town of Elazig, with a population of 110,000, is located in the vicinity of the dam. Referred to as Anatolia, or the land where the sun rises, this entire area is said to be the birthplace of an ancient Mesopotamian culture. In the latter half of the fourth millennium BC, this area was overrun by Indo-European tribes from the north, and a kingdom was established around 2000 BC – the kingdom of the Hittites. Armed with metal tools and horse-drawn chariots, Hittite warriors threatened the armies of ancient Egyptian pharaohs and constantly put pressure on the Babylonian region. The Hittites gradually extended their dominion over a large empire covering Anatolia and northern Syria. To this day, the ancient Hittite capital in Boazkyoi shows off its "lion gate," which is sculpted with the ancient Babylonian lion, bringing to mind the panoramic scene of these ancient people fighting with the many lions which once populated this region. The dam site of the Keban Dam looks to Boazkyoi in the west and is located in the northern region of the domain of the Mitanni culture which predated the Hittite influx into Anatolia. The Mitanni people are known as the conveyors of Babylonian culture to the Hittites. The Syrian town of Jalapurz, located downstream from the Keban Dam, is known for the epitaphs bearing Hittite hieroglyphics which were unearthed there. This area is also noted for later invasions by the Ottoman Turks and the Seljuks, who came out of Central Asia and overran Iran.

The long and ancient history of this region is richly represented by the fragments of bronze and iron implements which to this day are found in the earthwork in the vicinity of the Keban Dam. As soon as the Keban Dam Project was unveiled, archaeologists from Turkey, as well as from Europe and America, showed very keen interest in the progress of the work. In 1966, the University of the Middle East in Ankara, Turkey, announced that the Keban area was a veritable treasure trove of both buried and exposed historical sites. Following this announcement, archaeological teams from the University of Istanbul, the University of Chicago, and the University of Michigan were dispatched to Keban, while the British and German Archaeological Associations conducted a joint survey. A total of 38 excavations were undertaken in the eastern section of Elazig, and these studies were continued until the area was flooded by the reservoir. On the basis of recommendations by these teams, the Turkish government transferred an ancient mosque, which otherwise would have been lost to the lake. However, it is possible that numerous archaeological sites which were not uncovered by these efforts have now been lost for ever at the bottom of the Keban Dam's vast reservoir.

Historical sites in the lower Mekong Basin. Apart from the internationally known Angkor Wat Temple in Cambodia, there are many other historical sites in the lower Mekong Basin. However, most of the sites that are widely known date from after the tenth century. Two highly significant archaeological discoveries were made in north-east Thailand during the latter half of the 1960s.

The first discovery was made in the village of Non Nok Tha, located on the banks of the reservoir of the Nam Pong Dam. This site provided proof of rice cultivation dating back to 3500–4000 BC. In addition, evidence was found of the use of bronze implements dating back to 2000 BC or earlier. Shortly after this, a second discovery was made in the village of Ban Chiang, situated roughly 50 km east of the city of Udon Thani (often abbreviated to Udon), which was used as an air-strike base by the American forces during the Viet Nam War. This discovery featured numerous pieces of earthenware (bearing elegant surrealistic designs) and bronze-age tools and implements which were unearthed together with human remains. The members of the secretariat of the Mekong Committee, of whom the author was one, were very surprised by the news of this discovery. At the time, the international community was deeply involved in the efforts to save the historical sites threatened by Egypt's Aswan High Dam, but this discovery brought with it a very clear and strong awareness that the Mekong Basin was one of the ancient homes of human civilization. We were made painfully aware of the critical need and importance of conducting archaeological investigations at dam construction sites.

The Ban Chiang site has been carefully preserved and is now open to the public. In addition, the government of Thailand has sent some of the valuable findings from this site to various museums throughout the world, including the Japanese National Museum in Tokyo. The author has had the opportunity to visit Ban Chiang and was deeply impressed, and has also seen artefacts from this site on display in museums in Tokyo and Singapore. These artefacts are in an excellent state of preservation and are very impressive.

Following these important archaeological findings, a significant number of archaeological surveys were conducted in the region of the planned Pa Mong Project. The first study was headed by T. K. Marsh of the United States Bureau of Reclamation in 1969 (Mekong Secretariat 1976) and was entitled "Archeological Survey of the Na Klang District of Udon province (southern part of Nam Mong to be covered by the reservoir of the High Pa Mong dam)." The two archaeological discoveries noted above were located in the proximity of Na Klang which is situated at the south-ernmost extent of the reservoir of the High Pa Mong Dam (the villages of Non Nok Tha and Ban Chiang are situated 50 km south and 100 km east of Na Klang, respectively). This study was undertaken because it was feared that valuable historical sites would be lost to the planned new reservoir. In the course of his research, Marsh identified 22 sites of ar-chaeological interest and undertook pit excavations to the depth of 1 m at two places. Following the Marsh study, the National Energy Authority of the government of Thailand (currently EGAT) commissioned a study in the Thai villages situated within the area of the Pa Mong Dam's projected reservoir. The actual study, consisting of interviews in the villages affected, was implemented by the Fine Arts Agency of the Thai government. In 1973, using funds made available by the Ford Foundation of the United States, the Mekong Committee commissioned the Archaeology Faculty of Otago University in New Zealand to undertake a study in this region.

During the course of this study, some excavation work was carried out in the proximity of Na Klang. In addition, the entire area was investigated and data relating to the following factors was catalogued to determine the location of archaeological sites: topography, geology, vegetation, air tem-perature, rainfall, evidence of floods and drought, changes in river route, land use by local farmers, and historical evidence of population move-ments and settlement. As a result of this study, the following inferences have been made regarding the historical transitions in this region.

Human settlements in the lower Mekong Basin date back at least 2,000 years in two areas in this region. The first area consists of the basins of the Mun and Chi rivers in north-east Thailand (the two villages mentioned above both line the upper tributaries of the Chi River), and the Mekong Delta. Part of the second area does not belong to the Mekong Basin, but consists of settlements in the delta of the Chao Phraya River which flows through central Thailand. In this delta, primitive peoples populated the areas between the dense jungles lining the rivers and the highlands, and the savannas (including the barren salt lands) which spread out between the jungles. In the Mun–Chi Basin, people chose to settle in the sparsely wooded, weathered tablelands, which were flooded during the rainy sea-son. At this time, villages constituted independent communities with little interaction between them. However, the villagers held a common belief in a form of Indian Buddhism (as evidenced by the excavation of Buddhist icons). They dug canals and developed reservoirs of water for drinking and agricultural purposes, enjoying a farming life of relative abundance for that period. These settlements reached the peak of their development in the seventh century and gradually declined thereafter. By the tenth century, they had been displaced by the Angkor civilization, which had developed other agricultural techniques.

The Angkor civilization of the medieval period was the civilization of the northern shore of the Great Lake. During this period, the sparsely wooded and low-lying tablelands surrounding the Great Lake and in the Mun–Chi Basin were drained and reclaimed using a ring levee method. Buckets were used for irrigation, and reservoirs were developed to hold subterranean water for drinking purposes. Settlements were situated in the vicinity of these reservoirs. This represented a significant departure from the extensive farming patterns of primitive times, as these people had developed small-lot farming methods centred around individual reservoirs. They grew rice by transplanting seedlings to irrigated paddies, as they still do today. With the gradual growth in population, there was an increase in the number of small villages and some large towns appeared. However, these settlements, which flourished during the Angkor period, eventually disappeared. Although the reasons for this decline are not known, these settlements were affected by changing climatic conditions, changes in the flooding and drought patterns, and the incursion of the Thai tribes. Much of the cultivated lands eventually returned to forests.

The agricultural societies which once flourished in the lower sections of the Mekong Basin thus died out. With the start of international trade in rice in the nineteenth century, rice cultivation was resurrected in both the Chao Phraya and Mekong deltas. However, north-east Thailand has remained untouched by these developments to the present day.

Having established these historical developments in the agricultural societies which populated the river basins in this area, the archaeological team from Otago University proceeded to investigate the Loei district in north-east Thailand. Of the 38 sites which were investigated in this district, 30 per cent yielded prehistoric remains which could be related to Indian culture and religion. The other 70 per cent of sites were from more recent times, as evidenced by the discovery of stoneware, bricks, tiles, pipes, and pottery. Based on these findings, and on the agricultural practices of the past 2,000 years, the Otago University team concluded that there were no archaeological sites in the heavily wooded areas of the lower Mekong Basin. However, the team pointed to the possibility of sites existing in the ribbon areas of the Mekong and in the sparsely wooded tablelands, and concluded that investigations should be undertaken in these areas.

At all events, the important archaeological discoveries made in the latter half of the 1960s in north-east Thailand gave ample warning that any future development of the Mekong Basin should be undertaken with due caution. Needless to say, investigation of the upper reaches of the Mekong is of no less interest than investigation of the lower reaches. At present, China is actively developing the upper reaches of the Mekong, and we earnestly hope that the Chinese government will give its full attention to archaeological considerations as it proceeds with its dam-construction plans in this region.

Development of tourism resources and recreation

Tourism and developing countries

For the developing countries in the tropical regions, tourists from developed countries represent an important source of foreign exchange. In such developing countries there are several instances where dam sites have been identified as primary tourist attractions: well-known examples in Africa are the Aswan High Dam in Egypt, the Kariba Dam and Reservoir located between Zambia and Zimbabwe, and the Volta Dam and Reservoir in Ghana. The countries of the Mekong Basin may also be able to use future dams as important tourist attractions. The potential tourism aspects of dams and the use of reservoirs for recreational purposes merits further discussion.

The construction of a dam effectively improves the means of transportation between the urban areas and remote regions, and contributes to the promotion of various forms of exchange between these areas. Tourists will welcome the opportunity of seeing wild animals and plants in a natural setting and enjoying the beautiful scenery of reservoirs. Reservoirs can also provide various recreational opportunities for visitors.

On the other hand, it is no doubt true that the construction of a dam may lead to the disruption of the natural ecology by human intruders. In extreme cases, the building of a dam may result in the extinction of rare species. Both developed and developing countries must therefore exercise due caution in undertaking dam-construction projects in tropical regions. In Tanzania, where the author spent 2 years, visitors to the world-renowned Serengeti Natural Park near Mount Kilimanjaro can view the annual seasonal migration of large herds of animals in search of food and water. An ill-considered railway project intersecting this migratory course may well threaten the ecosystem which supports these wild herds; therefore, very careful consideration must be given to the construction of transportation routes for supporting dam projects in such areas.

Another instructive example can be found in Lake Victoria in Uganda. At one time in the past, plans were being drawn up for the construction of a dam at Murchison Falls on the northern shore of Lake Victoria. Standing at the source of the Nile River, this dam would have threatened with extinction a large number of species in this area, including crocodiles, hippopotami, aquatic plants, and birds which inhabit the waters immediately below the falls. In addition, the dam might have spoilt a very beautiful natural view. Finally, the construction of this dam was cancelled, and the author believes this is one plan that will never be implemented.

The point is that dam-construction projects must not focus solely on the economic benefits to be derived from the creation of a new tourist attraction; rather, full consideration must be given to cultural, social, humanitarian, and ecological factors. It is important to realize that due caution will sometimes dictate the cancellation of a project. However, it is regrettable that there are very few cases in which dam-construction plans in the tropical and torrid regions have taken into full consideration the effects of a projected dam on the above factors. The fact is that there is no record of how construction plans may have been modified to accommodate these important concerns.

Projections of tourism at the planned Pa Mong Dam

The vast majority of the people of Thailand – a developing country – have not had the economic power to own a car or take a sightseeing bus to enjoy a scene of natural beauty until relatively recently. With the passage of half a century since the end of the Second World War, conditions in Thailand are rapidly changing: the inhabitants have more stable and richer lifestyles and are following the Western example of using their holidays for recreation. For this reason, the beaches and resort areas located close to Bangkok – such as Pataya, Oahin, and Kaoyai – are becoming more crowded every year. The people of South-East Asia are light-hearted and family-oriented and have a preference for picnics and camping; the Thais are more prosaic people, appearing to prefer recreational activities to the appreciation of natural beauty.

The Nam Pong Dam in north-east Thailand was completed in the mid-1960s. In the first 3 years after its completion in 1965, a total of 210,000 tourists visited the dam. However, a closer look reveals that 85 per cent of the visitors were local residents and that tourists from Bangkok and other distant places accounted for only 15 per cent of the visitors. These figures probably reflect the economic conditions of the times, as well as poor road conditions and the scarcity of automobiles. Taking these figures into consideration – and bearing in mind the later changes in population trends, economic conditions, and the preferences of the Thai people – the National Energy Authority of the Thai government (currently the Energy Development Promotion Agency) in 1973 (2 years before the end of the Viet Nam War) undertook to estimate the number of tourists who would visit the High Pa Mong Dam if construction were to be started immediately. This study indicated that the total number of visitors to the Pa Mong Dam and Reservoir from Laos and Thailand would exceed 100,000 in 1990 (when the project was due to be completed if started immediately). The annual number of visitors was projected to increase to 150,000 by the year 2000, to 200,000 by 2010, and to peak at 370,000 in 2030. The study concluded that the number of visitors would stabilize at this level

Table 5.17   Classification of water-related diseases

Category	Examples of disease	Preventive strategy
I. Water-borne		Improve water quality
(a) Classical	Typhoid, cholera	Prevent casual use of other unimproved sources
(b) Non-classical	Infectious hepatitis	
II. Water-washed		Improve water quality
(a) Superficial	Trachoma, scabies	Improve water accessibility
(b) Intestinal	Shigella dysentery	Improve hygiene
III. Water-based		Decrease need for water contact
(a) Water-multiplied percutaneous	Schistosomiasis	Control snail populations Improve water quality
(b) Ingested	Guinea worm	
IV. Water-related vectors		Improve surface-water management
(a) Water-biting	Malaria, Gambian sleeping sickness	Destroy breeding sites of insects Decrease need to visit breeding sites
(b) Water-breeding	Onchocerciasis	

Source: Schiller (1984).

in the years thereafter. On the basis of these projections, the National Energy Authority calculated that net income from tourists in the year 2030 would amount to US$90,000 (Mekong Secretariat 1976). Although this appears to be a conservative estimate, the construction of the High Pa Mong Dam was never undertaken, and it would not be meaningful to comment further on these figures. Nevertheless, as a society becomes richer and more comfortable, no doubt its members will show greater interest in recreational activities and the appreciation of the beauty of nature. This means that the formulation of appropriate plans for tourism and recreational facilities should be included from the initial design stages of any dam-construction prject.

Water-related diseases

Tropical diseases, including water-related diseases, are characterized by unique regional and biological features and are not amenable to a single non-specific treatment (tables 5.17 and 5.18).

Onchocerciasis (river blindness), borne by *Simulium* spp., was at one time endemic to the area of the Volta Dam and Reservoir. With the completion of the dam, the incidence of this disease was reduced because

Table 5.18 Diseases associated with, or transmitted in, water

(A)	Through food intake having been in contact with contaminated waters (Vegetal irrigation, washing of edibles, or use of water in their industrial management) Example: Salmonellosis, Colibacillosis.
(B)	Through direct intake of contaminated waters. Example: Dysentery, Salmonellosis, Hepatitis, Enterovirosis, Amoebiasis.
(C)	Through contact with waters where pathogens are developing. Example: Schistosomiasis
(D)	Through transmission of infections or parasitosis through vectors bred in water. Example: Malaria, Onchocerciasis, Filariasis.

Source: Carlos Adlerstein. "Water, Environment and Health." *IWRA Water International* Vol. 6-3 IWRA, 1981

the *Simulium* fly (carrier of onchocerciasis) will deposit its eggs only in fast-flowing streams with a high oxygen content. On the other hand, as initially expected, there was a major increase in the number of *Simulium* larvae in the river below the dam. The reservoir itself witnessed a significant increase in snails which are vectors for schistosomes; as a result, it is reported that this area experienced major outbreaks of bilharzia.

The growth of *Simulium* larvae in the river below the dam can be attributed to the increased flow of water released from the power station. In such areas, frequent releasing of excess water from the dam will create excellent conditions for *Simulium* spp. to multiply. Therefore, in areas in which onchocerciasis is endemic, it is necessary to pay careful attention to the design and use of spillways and water outlets.

The construction of a dam can also lead to the creation of marshlands with a dense growth of algae around the reservoir. These marshlands provide ideal breeding grounds for malaria-carrying mosquitoes. In the United States, the Tennessee Valley Authority (TVA) acted to avoid this problem by building levees and drainage facilities to prevent marshes from forming in the vicinity of the reservoir. Furthermore, the water level of the reservoir was raised and lowered on a weekly basis to dry up the areas where mosquitoes laid their eggs. This was designed to take advantage of the malarial mosquito's habit of depositing its eggs in quiet waters (minimal vertical motion) with a flow velocity of less than 30 cm/s. (The TVA has stopped raising and lowering the water level since the disappearance of malaria in this area.) Irrigation canals should be sloped to ensure a flow velocity of more than 30 cm/s as a countermeasure to malaria. However, in the case of the Volta Dam, the presence of fast-flowing water in the river below the dam led to the generation of a larger *Simulium* population and the occurrence of onchocerciasis.

The situation is somewhat different in India because onchocerciasis and schistosomiasis do not exist there. Designers of dams and irrigation projects in that country therefore need only worry about countermeasures for malaria. This is a salutary reminder that the factors that must be taken into consideration in dam construction in tropical areas differ significantly from place to place, and in some cases the proper measures to be taken may be totally at variance with those in others.

The construction of dams may contribute to the increased incidence of bronchial asthma and other conditions. This is because the freshwater fish which are released in the reservoir can become infected with parasites and chironomids such as *Cladotangt arsus*. The author himself suffered from food poisoning after eating a freshwater fish in a first-rate restaurant in Rwanda in central Africa. Although these diseases are not contagious, diarrhoea and other stomach problems caused by poisonous fish are extremely common in this area.

The construction of new dams often gives rise to a large population of displaced persons. If these people come from areas which are not infested by tropical diseases, they will be prone to disease when they settle in areas with an inferior health environment. Needless to say, the construction of a dam involves the influx of a large working population from various areas. Under these conditions, the outbreak of a communicable disease at the dam site is sure to claim the lives of many victims. No doubt, any such outbreak will jeopardize the completion of the project, as evidenced by the notorious failure of Ferdinand de Lesseps in building the Panama Canal. It is thus important to note that, after completion of the dam, there is always the danger of the outbreak of epidemic diseases among immigrant populations and fishing communities in the area of the reservoir. In addition, tourists who stay for an extended period in the vicinity of the dam may be exposed to these diseases.

We have already touched on the issue of contagious diseases in irrigated areas located downstream from dams. Another problem that must be noted is that, in tropical areas, the soil tends to become very moist, providing an ideal environment for the spread of intestinal parasites. Parasites such as duodenal worms and dung roundworms enter the human body either orally or through the skin. The larvae are discharged in the faeces of the infected person and grow to maturity in the soil. Irrigation canals and ponds provide very good breeding grounds for snails which serve as intermediate hosts for schistosomes. The case of the irrigated areas of the Aswan High Dam is widely known; however, this problem is not unique to the Nile River. The irrigation area of the Kainji Dam in Nigeria has also been associated with serious outbreaks of schistosomiasis, underlining the fact that the spread of disease in irrigated areas constitutes one of the negative aspects of dam construction.

Malaria

Epidemic areas. It is reported that one-third of the world's population is living in malaria-infested regions. Serious malaria epidemics are also recorded in the history of Japan. In the post-war period, many returning soldiers brought the infection back with them from abroad, and the rice fields in an extensive portion of the country became infested with malaria-bearing mosquitoes. Fortunately, by the 1960s, malaria had naturally disappeared from Japan. The malarial vector in Japan was *Anopheles sinensis*, a mosquito found in swamps and paddy fields, which was driven to extinction by draining the swamps and by improving the management of rice fields. Another malarial vector was the mosquito *Anopheles minimus*, found on the islands of Iriomote and Ishigaki in Okinawa; malaria used to claim the lives of several thousand people every year in this area, but this vector has also disappeared. More recently, there have been reports in the Japanese newspapers of a new and highly resistant strain of malaria, the origins and details of which remain unknown.

Malaria is also carried by mosquitoes which live in the paddy fields of China. The World Health Organization (WHO) is assisting in a project to bring this problem under control using water-control methods for serial flooding and drought.

Malaria in the lower Mekong Basin. Malaria constitutes the most serious water-related disease in the lower Mekong Basin. It is reported that 90 per cent of the 60 million people in the four riparian countries of the lower Mekong Basin live under the threat of malaria. The risk of malaria thus covers the entire lower basin of the Mekong today.

In Thailand, which is reported to have the most aggressive antimalaria campaign among these four countries, malaria annually claimed the lives of 201.5/100,000 until 1949; this figure had fallen to 30.2 by 1960. With the start of a full-scale campaign of insecticide spraying and medical treatment in 1965, the mortality rate dropped further to 10–12/100,000 by the early 1970s. A survey conducted in north-east Thailand during 1971 and 1972 showed a malarial incidence of 1.2–5.2 per cent. (Much of this was clustered in the foothills of the forest areas.) Shortly after this, Thailand began to experience an increase in the number of cases of malaria, which had fallen below 124,000 reported cases in 1970. From this lowest level, the incidence of malaria more than doubled to 287,000 cases in 1974; this is believed to be the result of the spread of DDT-resistant mosquitoes.

In the case of Laos, almost the entire country, with the exception of the capital city of Vientiane, is designated as a risk zone. Although a mosquito-extermination campaign was mounted during the late 1950s, all efforts were halted after 1960 because of the war. Both Viet Nam and

Cambodia also experienced an increase in the number of malaria victims during the war years.

Nam Ngum Dam. The Nam Ngum Dam Project in Laos provides an instructive example of failure to control malaria. Construction work began in 1969 in this malaria-infested area. At the start of the project, no effective measures had been taken to prevent an outbreak of malaria (Mekong Secretariat 1976; 1972). As a result, shortly after the start of the work, 30 per cent of the work force became infected, including a consultant engineer who died in hospital. At this point, the Mekong Committee and the WHO were called in to bring the problem under control. All houses at the dam-site village were sprayed, the jungle adjoining the construction camp was cleared, and prophylactic drugs were routinely distributed. The construction schedule was curtailed, and various efforts were made to speed up the pace of the work, so that the dam was completed by 1971.

Nam Pong Dam. The Nam Pong Dam in Thailand was completed in 1966. The region of the dam is noted for its fertile and expansive rice fields, which were relatively free of malaria before the building of the dam. It is reported that appropriate countermeasures were taken during and after the construction of the dam. Preventive measures taken included careful surveillance, the promotion of public health programmes, and education of local residents.

Nam Phron Dam. The Nam Phron Dam, also located in north-east Thailand, was constructed in an area of malaria-infested hills and forests. These forests, and the stagnant ponds which appeared during the dry season, provided the ideal conditions for the spread of malaria-carrying mosquitoes. Although it is unclear how great the effect of the reservoir has been, a significant increase in malaria patients has been reported in the vicinity of the dam site. As a result, surveillance programmes have been put into effect to monitor the entire basin and the surroundings of the reservoir, as well as the villages of the displaced people.

Habits of malaria-carrying mosquitoes in the lower Mekong Basin. The *Anopheles* mosquito which is found in the lower Mekong Basin will not deposit its eggs in clear water or irrigated areas which are exposed to direct sunlight (Mekong Secretariat 1972). This mosquito, therefore, generally prefers ponds in dark, wooded areas. Although this has led to the contention that no direct relation may exist between water-resources development projects and malaria epidemics, we cannot afford to lower our guard. It is widely recognized that further research must be conducted into the response of the *Anopheles* mosquito to environmental changes, specifically to seasonal and regional changes.

There are two distinct species of *Anopheles* mosquito which inhabit the foothills and jungles of the lower Mekong Basin – *A. minimus* and

A. balabacensis. The former is found during the dry season between December and May in warm streams and in areas overgrown by water plants; the latter breeds during the rainy season and deposits its eggs in marshlands, ponds, and jungle swamps and can be generally found in the outer circumference of heavily wooded areas. Although the use of DDT has sharply reduced the number of *A. minimus* in the plains, the overall population of this vector remains large. With regard to *A. balabacensis*, it is reported that jungle loggers and all those with low immunity entering the jungle are particularly susceptible to infection by this mosquito.

Malaria prevention in the Mekong Basin. The experiences at the Nam Pong, Nam Phron, and Nam Ngum dams have impressed upon Thailand the importance of establishing a comprehensive system for malaria prevention at dam sites (table 5.19). Before the beginning of the construction work, a thorough survey of the area is undertaken to gauge the potential of a malaria epidemic, and appropriate control measures are formulated. These measures are implemented during the course of the construction work and are followed up after the project is completed. The essential measures include the application of insecticide sprays (DDT) and prophylactic drugs in the area of the dam site and breeding areas in the jungle, and the clearing of an area of radius 100 m around the camp site. In addition to these measures, other precautionary steps must be taken, such as not using non-immunized workers in deep jungle areas.

Particular measures must be taken to prevent the infection of people displaced by the construction of a dam. Rice farmers who have lived in the cultivated areas along the river have limited immunity against malaria and will be placed at risk if moved to the tablelands. On the other hand, people who have been living in malaria-infested areas may spread the disease to others when they are moved to low-immunity areas.

There are, thus, many factors to be taken into consideration when building a large dam, such as the High Pa Mong Dam. Possible ecological changes resulting from the construction of the dam must be carefully examined. Separate assessments must be made of the effects of these changes on the areas above and below the dam and the area of the reservoir, as well as the planned irrigation area. Before the start of construction, a "Malaria Control Council" should be created and the assistance of the WHO and other expert organizations should be sought.

Schistosomiasis

Symptoms. Although the number of malaria patients is declining on a global basis, the risk of malaria is being replaced by the growing threat of schistosomiasis, a disease carried by trematode worms (flukes). At least three species of schistosomes are known to cause schistosomiasis (bilharziasis) in humans: these include *Schistosoma japonicum* found through-

Table 5.19 Malaria-control measures for water-resource development projects

1. Surveys and investigations before construction
   Geographical reconnaissance
   Parasitological examination
   Entomological investigation
   Physical and hydrographic surveys
   Land cover, vegetation, forest
   Climate, rainfall, temperature
   Current malaria-control operations
2. Control measures
   Intensification of current control operations
   Source reduction, application of larvicides
   More frequent spraying
   Vegetation and jungle clearing
   Prophylactic treatment
   Establishment of health post
   Extension of spraying operations (5–10 km)
   Reservoir site clearing and shoreline reduction
   (Source reduction and larvicide application are justified particularly in view of
   the practice of night work in the open and the presence of outdoor biting
   mosquitoes)
3. Planning and siting construction camp
   Control of shanty construction near the camp
   Entomological and parasitological surveillance
   Screening of labourers
   Prophylactic treatment
   Frequent spraying of hutments as they are completed
   Routine monthly or bi-monthly spraying
4. After-construction services
   Observations and protection of work area residents
   Continuation of spraying as required in work area and for a radius of about
   5–10 km
5. Cost items
   Earthworks for source reduction and drainage
   Clearance
   Larvicide application
   Surveillance
   Health post
   Spraying
   Diagnosis and treatment

Source: Mekong Secretariat, Nam Pong Environmental Management Research
Project, Human Health and Nutrition, Diseases, Parasites and Disease Vectors,
Mekong Committee, June 1979

out East Asia, *S. mansoni*, found in Africa and Central and South
America, and *S. haematobium*, which is found in Africa. The former two
species are parasites which attach to the intestinal blood vessels, causing
bloody stools and cirrhosis of the liver; the third species enters the blood

vessels of the bladder and causes haematuria. The eggs of the flukes are passed through the urine and hatch in the water to grow as parasites on snails. Later, they detach themselves from the snail host as larvae (cercariae) swimming freely in the water, which ultimately enter the body of a human through the skin.

The island of Leyte in the Philippines has been notorious for its schistosomiasis epidemics, as have the Volta Dam in Ghana and the Kariba Dam on the Zambezi River. In the case of the Volta Dam, the epidemic was caused by the sharp increase in the population of snails which serve as the intermediate host in this disease. It is believed that the water discharged from the dam caused a significant topographical change in the area below the dam; this, in turn, pushed back the influx of sea water and opened up a greater area for snails to inhabit. To correct this problem, the canals in the delta have been dredged to increase the influx of sea water. By far the most widely known instance of schistosomiasis is the epidemic which occurred in the areas irrigated by the Aswan High Dam. The sad truth of the matter is that developing countries may be able to find financing for giant dams and irrigation projects, but often they do not have the financial resources necessary for mounting an effective preventive campaign against malaria and schistosomiasis.

Schistosomiasis in the lower Mekong Basin.   It is believed that 10 million people are infected today with schistosomiasis in the three countries of Laos, Cambodia, and Viet Nam. In cooperation with the governments of the four countries of the lower Mekong Basin, the Mekong Committee undertook an extensive study of schistosomiasis over the 5-year period from 1970 to 1974. This project was supported by the Smithsonian Institute and the United States and the USAID programme of the Department of the Interior. The study confirmed the existence of the snail *Lithoglyphopsis sperta*, a vector for schistosomiasis, in parts of the lower basin. Schistosomiasis itself was found to exist in the following three locations: Khone Island (Laos), located at the lower end of the middle section of the Mekong; Kratie (Cambodia), located in the lower basin; and in the village of Phibun Mangsahan, located in north-east Thailand at the confluence of the Mun and Mekong rivers. All three sites are located at strategic points in the basin and are relatively accessible. The parasite discovered in this study was named *Schistosoma japonicum* Mekong strain. This parasite is found in abundance between the end of March and early June, a period which overlaps with the end of the dry season and the beginning of the rainy season.

The schistosomiasis found at Khone Island infected only humans and dogs, whereas no trace of infection was found in cows and other domestic animals. At the time of the study, 15 per cent of the population of Khone

Island was infected. However, the infection rate was 30–40 per cent in children between the ages of 4 and 15 years. The total number of infected persons was estimated to be in the range of 10,000.

Snails have not been found in the section of the Mekong between the Myanmar border and Khemarat. For this reason, it is believed that the planned construction of the dams on the main stream of the Mekong at Pak Beng, Luang Prabang, Sayaboury, and Pa Mong, which are located above Khemarat, will not pose a problem. Snails breed during the dry season when river levels are low, but they are more successful breeders when the water level is relatively high during this season. Therefore, countermeasures for schistosomiasis may be necessary in the case of the High Pa Mong Dam where the low flow levels may be increased during the dry season.

The schistosomiasis epidemic which followed the completion of the Aswan High Dam has caused some concern that the development of the lower Mekong Basin may result in a similar outbreak. However, the two environments are completely different, and the types of snails found in these two settings also differ. The snail *Lithoglyphopsis sperta* of the lower Mekong Basin grows to a length of 3 mm. This is a mollusc which climbs onto the rocks, tree branches, water plants, and various decaying materials in the river during the dry season when the water level declines. It is reported that this snail cannot survive in reservoirs and irrigation canals. Its total population is relatively small and it is able to survive only under extremely restricted environmental conditions; for this reason, it is generally believed that this snail will not be able to inhabit reservoirs. However, it is conceivable that changes in the ecology may trigger changes in the snails. In any case, appropriate preventive measures should be taken, such as the draining of marshes, covering irrigation canals with concrete, regularly applying insecticides, cutting the grass on the levees, building public toilets in the nearby villages, preventing the pollution of the river, not bathing in the river, and promoting better water facilities to avoid contact with untreated river water for washing and cooking. (Such a water system would have to depend on uninfected water sources from other rivers.)

Other diseases of the lower Mekong Basin

In the lower Mekong Basin, outbreaks of dengue fever are occasionally seen in Bangkok; however, few people are affected and this disease does not pose a serious problem. On the other hand, enteritis caused by the use of tainted water is widespread. The economic conditions in the majority of the villages and hamlets of the lower Mekong Basin do not allow the digging of wells to reach subterranean water sources. As such, the main sources of drinking water are rainwater and small reservoirs which are fed from surface sources; this results in a high rate of infection among the villagers. People thus infected become too weak to work and are pushed toward further poverty, leading in turn to malnutrition and rendering the poor more susceptible to other diseases. This is the vicious circle which sadly affects many of the developing countries.

Among other water-related diseases, mention should be made of diseases caused by hookworms, which mature in the soil, and food poisoning caused by eating raw or poorly cooked freshwater fish. Poor and incomplete treatment of sewage and garbage will inevitably threaten the health of the inhabitants. In the final analysis, to combat these problems effectively, the general level of education must be raised and people must be given a better understanding of sanitation.

Notes

1. The average annual precipitation in the Tennessee River Basin is 1,300 m. An average of 43 per cent (560 mm) becomes surface run-off every year while the remaining 57 per cent evaporates. Although there are seasonal differences in the amounts of precipitation and evaporation, the annual figures are stable. From records of the period from 1935 to 1980, it has been estimated that 43 per cent of the annual average precipitation in the river basin before the start of the TVA Development became surface run-off. This figure (43 per cent) remained exactly the same after 1945 when the TVA reservoirs were completed. In other words, the effect of the TVA's construction of reservoirs was negligible (Elliot 1973).
2. Because of the Kariba Dam on the Zambezi River, the amount of precipitation in the arable land on the southern bank has increased and raised agricultural productivity (Mekong Secretariat 1982).
3. *The Water Quality Criteria Handbook*, issued in 1972 by the US Environmental Protection Agency, describes a case of a reservoir constructed near fluoride deposits in the state of Andhra Pradesh in southern India. The fluoride leaked into the groundwater and contaminated the reservoir, causing local residents who drank the water to contract unexpected illnesses.
4. The author visited the Keban Dam on the Murat River, a tributary in the upper Euphrates River Basin in Turkey, and confirmed that this dam is built on limestone. Water leakage has been clearly detected in the environs of the lower reaches of the reservoir.
5. D. S. Cardern was the first to notice the relationship between artificial lakes and the earth's strata. In 1945, he presented a paper stating that an earthquake in the strata from the impoundment of water from Lake Mead on the Hoover Dam was caused by the construction of the dam.
6. The Hoover Dam is an arch-gravity-type dam with a height of 221 m and a crest length of 390 m. Reservoir capacity is 37×10^9 m^3.
7. The Volta–Akosombo Dam is 94 m high and 1,240 m long at the crest. The reservoir is 400 m long and 25 km wide at its widest point. It has a gross capacity of 165×10^9 m^3 and an area of 8,730 km^2.
8. The Kariba Dam is 131 m high and 633 m long at the crest. The reservoir has a gross capacity of 185×10^9 m^3 and an area of 5,180 km^2.

9. A local earthquake with a magnitude of 5.3 was reported in November 1981.

10. This concrete, arch-type dam (height: 266 m) was constructed at a point 30 km upstream from the mouth of the Pieve River in northern Italy in November 1960. The Toc Hill at the left bank of the reservoir collapsed (300,000 m³) into the reservoir in October 1961. The water in the reservoir passed through the spillway (overflow height: over 200 m) and directly flowed into the village immediately downstream, killing all but 79 of the 1,269 inhabitants.

11. The Toc Hill, 1,320 m high, situated at the left bank of the reservoir, was made of dolomite, a type of sedimentary rock that is very unstable. For this reason, the local residents used to call the hill "The Walking Mountain." After the construction plans were announced, the local residents submitted a protest to stop the dam construction; nevertheless, the project was forced through.

12. Immediately prior to the catastrophe, the water level in the reservoir was lowered from full capacity at a speed of 20 cm/h (tests using models had shown that lowering the water level more than 1 m in 24 h caused instability).

13. Immediately before the completion of the dam, a crack with a width of 30 cm and a length of 2,500 m appeared near the top of the Toc Hill and 500,000 m³ of dirt and rock fell into the reservoir. After grouting in the crack and conduction model tests, it was judged that if the planned full-capacity water level were lowered from an elevation of 722.5 m to one of 699.5 m, even if land collapse happens, the depth of the flow which would be spilled over the dam crest due to waves caused by the land collapse would be less than 1.5 m, which is thought to be harmless.

14. Despite the fact that the land collapse had started before the major catastrophe, managers of the dam did not issue an evacuation order to people living by the reservoir until the day before the catastrophe. The residents of villages located directly downstream were not given an evacuation order until a mere 2 h before the great land collapse; thus, it was too late for most people to evacuate.

15. Of vegetable plankton grown in the Nasel Lake of the Aswan High Dam, the following are known: *Nitzschia*, *Melosira*, *Microcystis*, *Oscillatoria*, *Phormidium*, *Anabaena*, and *Oocystis* spp.

16. The two species of water hyacinth (rooted and free-floating) are further divided into various sub-species (Elliot 1973): (a) rooted macrophytes: cat-tails (*Typha* spp.), sedges (*Scirpus* spp.), lilies (*Nympha* spp.), *Elodea*; (b) free-floating macrophytes: water hyacinth (*Eichhornia crassipes*), water fern (*Salvinia auriculata*), water cabbage (*Pistia stratiotos*), duckweed (*Lemna* spp.).

17. Water for household and industrial use by the three large cities in the vicinity of the reservoir is discharged into the reservoir for the *15 de Noviembre* Dam in El Salvador. To compensate for this, water hyacinths have been actively propagated in the reservoir in the past 30 years since the reservoir was flooded. This situation may also arise with the High Pa Mong Dam.

18. West Malaysia has a total area of 132,000 km², of which forest occupies 66,000 km², or half. Of the remaining half, plantations take up 35,000 km², and 31,000 km² are cropland, marsh, lake, waterways, and city. The annual average concentration of earth and sand in rivers is 20–100 parts per million for rivers that have forested basins, whereas the annual average concentration for rivers in areas with human activity is 100–4,500 parts per million. Annual precipitation is 2,400 mm, which is much higher than the average in Japan. However, the volume of earth and sand efflux from major rivers is 145 t/km² of river per year or only about 85 m³/km³/year. This figure is calculated on the basis of the assumption that there are 1.7 t of earth, sand, and small stones in 1 m³ of river (Fatt 1985).

19. The sampling of the efflux of earth, sand, and silt from the main stream of the Mekong River was started in 1952 by Harza Engineering, a US construction consultancy. The

Mekong Committee took over the sampling activities in 1962 and conducted sampling throughout the lower basin. Later, the Provisional Mekong Committee took over the work and recommenced sampling in Thailand, Viet Nam, and Laos. These samplings were all conducted as part of the collection of hydrometeorological data. It is hoped that the new Mekong River Commission will focus increasing attention on such sampling activities.

20. The eastern boundary line of the Mekong River Basin is unclear during the flooding season. To the east of the Preyveng–Gokong branch of the Mekong River are the Vaico Oriental and the Vaico Occidental, tributaries of the Saigon River next to the Mekong Basin. When flooding reaches its peak, water from the Mekong Basin flows into the Vaico River Basin, creating a single basin.

21. The Joncs Plain is low-lying ground composed of clayey alluvial sediment that was severely affected by sea water in the sedimentation process, giving it a high sulphate content. For this reason, reeds that can withstand salt water flourish and the organic matter content is high. It is said that this is the cause of the acid, sulphate soil.

22. "Floating" rice grows to a height of 1.5–3 m. Every March and April, the rice stubble left from the previous year is burned. The soil is then tilled, and seeds are planted. The growth of the rice is usually faster than the increase in water level during the flood season. In years when the rice does not keep pace with the rising flood-water levels, damage is caused over a wide area.

23. Irrigation that delivers water at a rate of approximately 1–1.1 litre/s/ha is required to cultivate rice in the Mekong Delta from the dry season to the beginning of the rainy season.

24. Drainage is problematic in some areas of the Mekong Delta. In areas where drainage is easy, gravity drainage is used, taking advantage of the difference between high and low tides. In areas enclosed by embankments and areas where drainage is problematic, drainage is achieved by pumping. Water must be drained off at a rate of approximately 2–4 litres/s/ha. Rice varieties that give a high yield require more drainage than native species.

25. According to the fishing industry specialist Dr Pantulu, a former member of the Mekong Secretariat, the conditions in the lower reaches of the Mekong River are similar to those of the Ganges and Brahmaputra rivers in India.

26. In recent years, evaluation of fish based solely on economic viability has come to an end. Indeed, the Thai government's Fisheries Bureau embarked on a Pla Buk breeding project for the sake of conservation in 1983 and has been promoting propagation of the species since then. We are now living in an age when all countries pay serious attention to the issue of environmental conservation.

27. Before the Viet Nam War, the total annual catch from the Great Lake was reported as 35,000–70,000 t.

28. As mentioned in chapter 6, some planners on the Secretariat of the Mekong Committee have recently pointed out the advantages of increasing the amount of water in the delta during the dry season by constructing a large regulating weir at the entrance to the Great Lake. France also supports this plan. However, there are insufficient studies regarding concerns about the effect on fish, build-up of sediment, and other environmental changes; these should be investigated thoroughly.

29. Changes in the water level of the dam lake have both positive and negative effects on fish (Begg 1973). For example, the number of cichlids caught in the lake of the Kariba Dam in South Africa is intimately linked to fluctuations in the reservoir water level: when the water level rises, the fish move to a wider, newly expanded area of water in search of food, leading to an increase in the number of fish caught using gill nets; on the other hand, when the water level falls, the fish descend to deeper levels, reducing the catch. Another cause of the reduction in the catch when the water level falls is that the

extended shoreline of the dam lake becomes shortened and the lakeside area decreases; since this location is very important for the catch, the volume of the latter goes down. When the water level rises in the Kariba Dam lake, young fish of the species *Tilapia* move to the lake shore to feed; when the water level falls, they go down to deeper levels.

30. North-east Thailand was densely wooded and inhabited by many wild animals before 1961, when the first economic development plan began. After 30 years of rural development, however, the forested area of north-east Thailand now covers only somewhat less than 2.2×10^6 ha or 12.9 per cent of the total area of the north-east – 16.9×10^6 ha. Forest depletion in this area is the worst: the forested areas of northern, central, and southern Thailand have shrunk to respectively 45.5, 23.4, and 19.0 per cent of total land area.

31. At one point after water began to fill the Nam Ngum Dam Reservoir, the Smithsonian Institute reported that it was cooperating with the Laotian government to carry out a study on the impact of the dam's construction on wildlife; however, the study was interrupted by the war and remains incomplete (Mekong Secretariat 1976).

BIBLIOGRAPHY

Akinluyi, T. O. (1984). "Human Waste Disposal and Faecal Pollution of the River Niger Delta Waters." *Water International* Vol. 9(1).

Aksoy, S. (1970). *River Bed Degradation of Large Dam*. ICOLD Q38, R43, Paris.

Aluares, M. (1976). *Solid Material Contributed by a Watershed and Its Variation when Modified*. ICOLD Q47, R6, Paris, April.

Armstrong E. L. (1973). *Dam Construction and the Environment*. ICOLD Q40, R16, Paris.

Azpilman, A. (1976). *The Effect of Landslide of Furnas Reservoir*. ICOLD Q47, R26, Paris.

Bandeira de Mello, J. A. (1985). *Dams and the Environment in Tropical, Sub-Tropical and Arid Regions*. ICOLD Bulletin 50, Paris.

Begg, G. W. (1973). *The Biological Consequences of Discharge above and below Kariba Dam*. ICOLD Q40, R29, Paris, June.

Bellier, C. (1979). *Nile Water Study*. Vol. 1, Main Report. Ministry of Irrigation, Government of Sudan.

Biswas, A. K. (1982). *Environment and Sustainable Water Development: Water for Consumption*. IWRA, Urbana.

Buttling, S. (1973). *Predicting the Rate and Pattern of Storage Loss in Reservoirs*. ICOLD Q40, R37, Paris, June.

Catakli, O. (1973). *Problems of Water Loss: On Some International Aspects of Dam Construction*. ICOLD Q40, R2, Paris, June 1973.

Chaduick, W. L. (1976). *The Effect on Dams and Reservoirs of Some Environmental Factors*. ICOLD Q74, Paris, April.

Chand, K. (1973). *Some Water Resources Problems and Human Environment, IWRA Water for Human Environment*. IWRA, Urbana, September.

Choudhury, G. R. (1984). "Management of Sedimentation in Bangladesh." *Water International*, Vol. 9(4).

Conn, W. V. (1970). *Inspection and Observation of Completed Dams on Karst Foundation*. ICOLD Q38, R26, Paris.

Doelhomid, S. (1982). "Sediment Control of the Brantas River." *Water International*, Vol. 7(3).

Economic Commission for Asia and the Far East (ECAFE) (1957). *Development of Water Resources in the Lower Basin*. Flood Control Series No. 12. United Nations, Bangkok.

Economic and Social Commission for Asia and the Pacific (ESCAP) Secretariat (1983). "Effect of Dams on Basin Population." *UN Water Journal*.

Elliot, A. (1973). *Consequences on the Environment of the TVA Reservoir System*. International Commission on Large Dams (ICOLD) Q40, R15, Paris, June.

Fatt, C. S. (1985). "Sediment Problems and Their Management in Malaysia." *Water International* Vol. 10(1).

Food and Agriculture Organization (FAO)/OSRO (1983). *Report of Mission to Cambodia*. FAO, Rome.

Futa, A. B. (1983). "Water Resources Development Organization of a Resettlement Programme (a Case Study of the Kpong Resettlement Programme in Ghana)." *Water International*, Vol. 8(3).

Gangardt, G. G. (1976). *Effects of Environmental Factors of Water Storage Reservoirs, Causes of Water Bloom and its Control*. ICOLD Q47, R18, Paris, April.

Gruner, E. (1973). *Classification of Risk*. ICOLD Q40, R6, Paris, June.

Guadiana, J. (1976). *Water Hyacinth in Mexico, Problems and Solutions*. ICOLD Q47, R12, Paris, April.

Hao, L., and X. Mai-Ding (1976). *Regulation of Sediments in Some Medium and Small-sized Reservoirs on Heavily Silted Streams in China*. ICOLD Q47, R37, Paris, April.

Harclow, W (1956). *The Volta River Project I, Report of the Preparatory Commission*. Governments of UK and Gold Coast.

Hori, H. (1991). *Environmental Influences of Dam Construction on Nature and Society*. ICOLD Q64, Paris.

ICOLD (1981). *Earthquakes, Dam Projects and Environmental Success*. ICOLD Bulletin 37. ICOLD, Paris.

International Council of Scientific Unions (1972). *Man-made Lakes as Modified Ecosystems*. International Council of Scientific Unions, Paris.

Japan Dam Association (1963). "Volta River Project." *Topmost Dams of the World*. Japan Dam Association, Tokyo, October.

——— (1977). "Volta River Project." *World Dams Today '77*. Japan Dam Association, Tokyo.

Johnson, S. H. III, and S. Kolavalli (1984). "Physical and Economic Impacts of Sedimentation on Fishing Activities: Nam Pong Basin, Northeast Thailand." *Water International*, Vol. 9(4), December.

Jovanovic, D. (1973). *Some Effects on the Environment of the Building of Dams in East Africa*. ICOLD Q40, R7, Paris, June.

Kall, R. (1970). *Impounding of Manicougan 5 Reservoirs as Possible Trigger Cause of Local Earthquakes*. ICOLD Q38, R41, Paris.

Kellerhals, R. (1973). *Observed and Potential Downstream Effect of Large Storage Projects in North Canada*. ICOLD Q40, R46, Paris, June.

Kinawy, I. Z. (1973). *Some Effects of the High Dam on the Environment*. ICOLD Q40, R59, Paris, June.

Kivinen, P. (1976). *The Consequences on the Environmental Factors of Water Storage Reservoirs, Causes of Water Bloom and its Control.* ICOLD Q47, R18, Paris, April.

Kordas, B. (1976). *Hydraulic Method of Sedimentation Forecasting in Reservoirs.* ICOLD Q47, R25, Paris, April.

Kumi, E. N. (1973). *Environmental Effects of the Volta River Project.* ICOLD, Q40, R56, Paris, June.

Lane, R. G. T. (1970). *Major Problems in the Operation and Maintenance of Dams and Reservoirs.* ICOLD Q51, R36, Paris.

Larson, C. L. (1984). "Controlling Erosion and Sedimentation in the Panama Canal Watershed." *Water International* Vol. 9(4).

Lightfoot, R. P. (1981). *Problems of Resettlement in the Development of River Basins in Thailand, River Basin Planning.* John Wiley & Sons, Chichester.

Mekong Secretariat (1970). *Fishery Resources, Report on Indicative Basin Plan.* Mekong Committee, Bangkok.

―――― (1970). *Indicative Basin Plan Report.* Mekong Committee, Bangkok.

―――― (1972). *Fisheries in the Lower Mekong Basin: Summary Report.* Mekong Committee, Bangkok, May.

―――― (1972). *Fish and the Mekong Project.* Mekong Committee, Bangkok, September.

―――― (1972). *Public Health and the Mekong Project.* Mekong Committee, Bangkok, December.

―――― (1975). *Silt Balance in Kampuchea.* Mekong Committee, Bangkok, July.

―――― (1976). *The Water Situation in the Lower Mekong Basin.* Interim Mekong Committee, Bangkok, April.

―――― (1976). *Wildlife and the Mekong Project.* Mekong Committee, Bangkok, April.

―――― (1976). *Environmental Effects of Pa Mong.* Mekong Committee, Bangkok, August.

―――― (1978–1980). *Nam Pong Environmental Management Research Project.* Mekong Committee, Bangkok.

―――― (1979). *Delta Agricultural Development Program, Viet Nam.* Interim Mekong Committee, Bangkok.

―――― (1979). *Project, Human Health and Nutrition, Diseases, Parasites and Disease Vectors.* Mekong Committee, Bangkok, June.

―――― (1979). *Water Quality Studies in Delta.* Interim Mekong Committee, Bangkok, June.

―――― (1980). *Development of Marshes and Tidal Land in Delta.* Interim Mekong Committee, Bangkok.

―――― (1980). *Geomorphology of Deltas.* Interim Mekong Committee, Bangkok.

―――― (1980). *Survey of Weeds, Nam Pong Environmental Management Research Project No. 9.* Interim Mekong Committee, Bangkok, July.

―――― (1981). *Mekong Aquacultural Achievements in the Lao PDR.* MKG/91. Interim Mekong Committee, Bangkok, January.

―――― (1980). *Water Weeds and Studies on Fish, Nam Pong Environmental Management Research Project.* Interim Mekong Committee, Bangkok, April.

————— (1981–1992). *Studies of Salinity Intrusion in the Viet Nam Mekong Delta.* Interim Mekong Committee, Australia.

————— (1982). *Environmental Impact Assessment.* Interim Mekong Committee, Bangkok.

————— (1982). *Geomorphology of Mekong Delta.* Interim Mekong Committee, Bangkok.

————— (1983). *Annual Report 1982.* Interim Mekong Committee, Bangkok.

————— (1983). *Environmental Investigation in Mekong Delta.* Interim Mekong Committee, Bangkok, October.

————— (1983). *Environmental Program in Viet Nam.* Interim Mekong Committee, Bangkok.

————— (1984). *Environmental Survey of Delta.* Interim Mekong Committee, Bangkok.

————— (1984). *Fishery Development Activities, Annual Report, 1984.* Mekong Committee, Bangkok.

————— (1984). *Preliminary Study on Salinity.* Interim Mekong Committee, Bangkok.

————— (1985). *Annual Report 1984.* Interim Mekong Committee, Bangkok.

————— (1985). *Geological Investigations for Feasibility Study of Nam Theun 2 Project.* Interim Mekong Committee, Bangkok, April.

————— (1985). *Hydroelectric Development of the Nam Theun Basin, Prefeasibility Report.* Interim Mekong Committee, Bangkok, April.

————— (1985). *Environmental Investigation of the Development of Water and Land Resources in the Mekong Delta, Viet Nam.* Interim Mekong Committee, Bangkok, June.

————— (1992). *Preparatory Environmental Study for the Lower Pa Mong Project: Main Report.* Interim Mekong Committee, Bangkok, July.

Minshull, J. L. (1973). *The Establishment of Fishery among the Resettled Batonka People at Lake Kariba, Rhodesia.* ICOLD Q40, R28, Paris, June.

Murray, J. A. (1976). *Nizamsagar Project: Problems of Sedimentation, Effects on Irrigated Area and Remedial Measures.* ICOLD Q47, R14, Paris, April.

Murthy, B. N. (1970). *Sedimentation in D. V. C. Reservoirs, India.* ICOLD Q38, R58, Paris.

Murti, N. G. K. (1970). *Supervision of Bhakra Dam and Reservoir.* ICOLD, Q38, R57, Paris.

NEDECO (Dutch Government Aiding Agency) (1993). *Draft Master Plan for the Mekong Delta in Viet Nam Ho Chi Minh City.* NEDECO, Viet Nam, June.

Nelson, M. (1984). "Economic/Institutional Issues in Sedimentation Management in Developing Countries." *Water International* Vol. 9(4).

Ohtake, M. (1984). *Seismicity Change Associated with Impounding of Major Artificial Reservoirs in Japan.* National Research Center for Disaster Prevention, Tokyo, October.

Okamoto, S. (1984). "Earthquake Resistance of Concrete Gravity Dams." *Introduction to Earthquake Engineering*, 2nd edition. Tokyo University Press, Tokyo.

Oliver, H. (1977). "The Kariba Dam." *World Dams Today '77.* The Japan Dam Foundation, Tokyo.

Pantulu, V. P., and J. E. Bardach (1969). *Fisheries Aspects of Pa Mong Project.* Mekong Committee, Bangkok.

Parfi, R. (1976). *Quantitative Analyses of Reservoir Sedimentation.* ICOLD Q47, R17, Paris, April.

Penella, M. P. (1973). *Improvement of River Pollution: Positive Effects of Reservoirs on the Environment.* ICOLD Q40, R55, Paris, June.

Phelines, R. E. (1973). *Some Biological Consequences of the Damming of the Pongola River.* ICOLD Q40, R14, Paris, June.

Pillips, V., and L. Georgeson (1973). *Environmental Considerations of Dam Construction and Operation in Seismically Active Urban Areas.* ICOLD Q40, R18, Paris, June.

Rooseboom, A. (1976). *Reservoir Sediment Deposition Rates.* ICOLD Q47, R21, Paris, April.

Roynolds, P. J. (1981). "Ecology and Environment, Beauty and the Beast." *Water International,* Vol. 6(2).

Saha, S. K., and C. J. Barrow (1981). *River Basin Planning.* John Wiley and Sons, New York.

Schiller E. J. (1984). "Water and Health in Africa." *Water International,* Vol. 9(2).

SCOPE/UNEP (1976). *Environment and Development.* SCOPE/UNEP, Paris, December 1976.

Senturk, F. (1970). *Interpretation of Piezometric Indication in a Dam Resting on Permeable Foundation.* ICOLD Q38, R44, Paris.

Shahin, M. M. A. (1986). "Discussion of the Paper Entitled 'Ethiopian Interests in the Division of the Nile River Waters'." *Water International,* Vol. 11(1), March.

Simons, D. B., and H. W. Shen (1973). *Review of Pa Mong Phase II Report.* Vol. 1(2), Supplement to *Main Report.* Mekong Committee, Bangkok.

Sluiter, L. (1992). *The Mekong Currency, Project for Ecological Recovery.* TERRA, Bangkok.

Smith, H. A. (1973). *Estimated Impact on Fish Resource, A Detrimental Effect of Dams on Environmental Nitrogen Supersaturation.* ICOLD Q40, R17, Paris, June.

Snowy Mountains Engineering Corporation (1979). *Pa Mong Multipurpose Project, Summary of Engineering and Economic Feasibility Studies.* Interim Mekong Committee, Bangkok, June.

——— (1979). *Water Quality, Pa Mong Multipurpose Project.* Interim Mekong Committee, Bangkok, June.

Sutadipradia, E., and H. Hardjorwitzitro (1984). "Watershed Rehabilitation Program Related to the Management of River and Reservoir Sedimentation in Indonesia." *Water International,* Vol. 9(4).

Taki, Y. (1978). *An Analytical Study of the Fish Fauna of the Mekong Basin as a Biological Production System in Nature.* Special Publication No. 1. Research Institute of Revolutionary Biology, Tokyo.

Tejwani, K. G. (1984). "Reservoir Sedimentation in India." *Water International,* Vol. 9(4).

UN ESCAP (1973). "Proliferation of Aquatic Weeds on Dams, UK." *Water Journal,* Bangkok.

University of Hawaii (1985). *The Composition of Mekong River Silt and Its Possible Role as a Source of Plant Nutrients in Delta Soils.* Mekong Committee, Bangkok.

US Bureau of Reclamation (1956). *Lower Mekong River Basin.* International Cooperation Administration (ICA), Washington D.C., March.

Vlastos, K. (1976). *Problems from Siltation and Floating Debris at Certain Reservoirs of the Public Power Corporation of Greece.* ICOLD Q47, R29, Paris, April.

Vladut T. "Environmental Aspects of Reservoir-induced Seismicity, Water Power and Dam Construction," May 1993.

Watershed (1995). *People's Forum on Ecology, Burma, Cambodia, Laos, Thailand & Viet Nam,* Vol. 1, No. 1, TERRA, Bangkok, June.

White, G. F. (1977). *Environmental Effects of Complex River Development.* Westview Press, Boulder, Colorado.

Whitmore, J. S. (1976). *The Influence of Changing Land Use on Inflow to Reservoirs.* ICOLD Q47, R22, Paris, April.

Yaguek, A. G. (1979). *Spanish Experiences of Earthquakes Induced by Reservoirs.* ICOLD Q51, R36, Paris.

Yiicel, O. (1976). *Model Investigations of Reservoir Sedimentation.* ICOLD, Q47, R11, Paris, April.

Zein, S. L., and R. T. Lane (1973). *Engineering Implications of the Environmental Study of Some Dams in Africa with Particular Reference to Seismicity.* ICOLD Q40, R58, Paris, June.

6

Creation of the Mekong River Commission and future outlook

This chapter begins with a review of the events leading to the creation of the new Mekong River Commission, followed by a discussion of "dam-development on the main stream of the Mekong," one of the most important issues on the agenda of the new Commission. In this connection, comments are made on the policies which the Secretariat of the former Mekong Committee has been attempting to pursue in recent years. Next, recent issues related to the "domestic development of the Mekong tributaries" in the four countries in the lower Mekong Basin are considered briefly. Then, certain aspects of the Greater Mekong Subregional Development Plan advocated by the Asian Development Bank (ADB) are explained. Finally, the author presents his own proposals or recommendations regarding how the new Commission should cope with the various issues and challenges it faces today.

Chapter 6 concludes with a discussion of the need for intra-regional cooperation among the riparian countries, which is an indispensable element in the future development of the Mekong Basin. In closing, the author outlines his thoughts on how he would like to see the Japanese government, private Japanese companies, and non-governmental organizations (NGOs) promote and support this "need for effective mutual cooperation among the governments of the six riparian countries."

Events leading to the creation of the Mekong River Commission

There was widespread international interest and support for large-scale dams and other large-scale construction projects during the 1950s and 1960s. This interest derived from the belief that large-scale projects had much to offer in terms of economies of scale. This view was also prevalent in the area of water-resources development, and it was generally believed that the economic benefits of a man-made lake would increase with its size. Larger reservoirs held out the promise of being a more effective and efficient means for flood control, hydroelectricity generation, irrigation, and the supply of drinking water and water for industrial uses. For a world hoping for a rapid recovery from the exhaustion and destruction of the Second World War, this vision was both attractive and convincing.

However, the world gradually became more aware of the negative aspects of building large dams. Some of the more acute problems involved the inundation of large tracts of land, and social problems such as the need to relocate large numbers of residents. At the same time, the international community was becoming more conscious of the negative environmental impact of these structures. Thus, the people who opposed the construction of large dams because of the inundation of their lands were, ironically enough, joined by sections of the public who had reaped economic benefits from large-dam construction.

By the beginning of the 1970s, the development of the lower Mekong Basin began to be affected by these negative sentiments. In particular, there was a heated debate in Thailand over the pros and cons of new undertakings, based on past experiences with dam construction. As the 1970s wore on, the public tended to hear only the negative aspects of dam construction. Before long, not only the residents directly affected by inundation but also substantial sections of local populations who were traditionally acquiescent to all government decisions, began to voice active opposition to large-dam construction.

New dam projects were further jeopardized by the intensification of the war in Viet Nam. When Cambodia became engulfed in the fighting, it, in effect, quit the Mekong Committee. The absence of Cambodia made it impossible to pursue any new dam-construction projects in the lower Mekong Basin. The Mekong Committee thus virtually lost its ability to function and was reduced to a mere facade.

To remedy this situation, the Committee decided in 1978 to make a new start, with the creation of an Interim Mekong Committee formed without the participation of Cambodia. By this time, however, the governments of Thailand, Laos, and Viet Nam had neither the enthusiasm nor the capacity actively to promote water-resources development as a part of "creating a foundation for lasting peace and stability in the lower basin area." Furthermore, the Western countries which had enthusiastically supported the work of the Mekong Committee suspended their financial assistance when the three countries of Indo-China adopted socialist systems. Assistance to Thailand continued, but this increasingly took the form of direct bilateral aid, targeting specific economic and technological fields which reflected the individual positions and preferences of the donor countries. On the other hand, the ADB was steadily increasing its influence in this region. Originally formed in 1966 with the support of the Economic Commission for Asia and the Far East (ECAFE), the ADB had accumulated the financial resources and executive capabilities to make its presence felt among Asian nations. Following this development, ECAFE and the Mekong Committee gradually gave up their earlier positions of leadership and retreated into the background. In order to revive its influence, the Interim Mekong Committee decided in 1988 to formulate a revised plan for the comprehensive development of the lower Mekong Basin; however, donor nations remained cool toward this revised plan and adopted a "wait-and-see" attitude on the grounds that the post-war political situation in Viet Nam and Laos had not changed appreciably. In addition, any realistic discussion of a comprehensive development plan for the lower basin area was virtually impossible without the participation of Cambodia.

The "international development projects," which should have been the highlights of the revised plan, were inevitably quite conservative compared with those proposed in the wide-ranging original Indicative Basin Plan report announced by the Mekong Committee in 1970, mainly because of inundation and environmental concerns. What remained of the international development projects in the revised basin plan published in 1988 were some limited schemes which were mere shadows of the original concept.

In the revised version, two large-scale projects – the construction of the Low Pa Mong Dam and the Nam Theun 2 Dam on a Mekong tributary – were to be started before the year 2000. (The Pa Mong Project aimed at minimizing the displaced population; the Nam Theun Project involved a relatively low-risk structure aimed at minimal population displacement while generating at a high capacity.) Moreover, even in the area of domestic development, only a few small-scale water-resources development projects were to be undertaken in the three countries excluding Cambodia. Thus, the 1988 development plan itself lacked any outstanding or innovative features.

Nevertheless, no action has actually been taken for the implementation of the revised 1988 plan, and almost all the proposed projects have remained shelved for all these years. This conservative plan, which focused on the development of the central segments of the lower basin, was derailed by a combination of factors and events. The countries of the lower Mekong Basin were each caught up in their own problems, while it became increasingly difficult to gain funding from donor countries. Furthermore, global concern for environmental protection and strident criticism of the harm done by inundation became obstacles to implementation. (As of early September 1995, the only portion of the international project which is moving forward, albeit slowly, is the development of the Nam Theun River in Laos, where a strict study of the environmental impact, discussed below, has been completed.)

In June 1991, King Sihanouk of Cambodia announced that his country intended to rejoin the Mekong Committee. This represented a historic change of far-reaching significance. Responding to this statement, a peace conference was convened in Paris in October of the same year and the reinstatement of Cambodia was formally discussed in the plenary meeting of the Interim Mekong Committee held in November. However, Cambodia's return was not ratified on this occasion because of disagreements which arose between Thailand and Viet Nam in the process of drafting a formal agreement.

The United Nations Development Plan (UNDP) headquarters in New York reacted by issuing a general call to the four riparian countries in the lower Mekong Basin, including Cambodia. Responding to this invitation, the four countries met in Hong Kong in the fall of 1992 to discuss the formation of a new Mekong River Commission. This was followed by a meeting in Kuala Lumpur, where an agreement was reached to form a working group for deliberating on the details of re-forming the Commission. Several factors prompted the progress of this work: first of all, at this time, the countries of this region were becoming incorporated into the Association of South-East Asian Nations (ASEAN) framework, and the Asia-Pacific region was beginning to emerge as an economic bloc. Secondly, China had launched the construction of the Manwan Dam in the upper reaches of the Mekong (Lancang Jiang) in Yunnan Province. These developments impressed upon these four countries the need to create a new organization for restarting the development of the lower Mekong Basin.

The working group extensively discussed the "effective and equitable development and use of the waters of the Mekong River," with an agreement finally being reached in Hanoi at the end of November 1994. The formal signing of this historic document which enunciated the high-minded ideals of "cooperation for the sustainable development of the

Mekong Basin" was carried out in early April 1995 in the northern Thai city of Chiang Rai (see the Appendix on pp. 373–385). Observers from both China and Myanmar were present at this signing.

Mekong River Commission

The newly formed body, hitherto referred to as the Mekong Committee, was named the Mekong River Commission. In the earlier format, the member nations were represented by sub-ministerial officials, with a Secretariat operating under them. In the new organization, nations are represented by cabinet ministers, or their equivalent, who comprise the Council. On the next level of the organization, the Joint Committee functions as an agency of the Council and is comprised of national representatives of the bureau-chief level. The role of the Joint Committee is to implement the policy directives and decisions of the Council. Finally, the Commission has a Secretariat which provides technical and management support services to the Council and the Joint Committee.

The Mekong River Commission has thus been transformed from a two-tiered, middle-level organization to a high-class, triple-tiered organization. In other words, it has been invested with significantly greater power and authority than its predecessor.

In the earlier format, the appointment of the chief of the Secretariat was by recommendation of the UNDP headquarters, which was unconditionally adopted. This practice has been changed in the new organization: the UNDP now provides a list of nominees from third-party countries from which the Joint Committee makes its selection, subject to the final approval of the Council. In the recent selection process, more than 400 nominees from all over the world were considered. The position went to a Japanese high-level specialist attached to the Japanese Ministry of Agriculture, Forestry, and Fisheries, who assumed his post in September 1995.

The Mekong River Commission was formed as an agency for international cooperation for the sustainable development, use, management, and preservation of the water resources of the Mekong Basin. However, the formulation of a broad-based, long-term, international programme for the use of these resources in a manner which is equitable, rational, and acceptable to all member countries is very difficult to achieve, particularly because the areas capable of supplying water and energy needs in the Mekong Basin are separated from the areas consuming water and energy by significant distances and national boundaries. This geographic imbalance between supply and demand is sure to make the task of the newly formed Commission very difficult indeed.

Regarding the "need to undertake 'comprehensive' development, use, preservation, and management of water resources in the river basin," the Mekong River Commission made the following statement at its inception:

If the efforts of all of our countries are to be integrated into a unified effort for achieving sustainable development, then the Mekong Basin must be viewed as a single entity. For this reason, the Commission has determined to adopt a comprehensive and all-inclusive approach to the treatment of the whole Mekong Basin ... The protection and preservation of the environment and the ecosystem of the river basin is a prerequisite to sustainable development. Therefore, we must seek to maintain and to preserve the subtle ecological balance in the entire Mekong River system. Henceforth, environmental preservation will stand as the first priority while we promote the development and management of the water resources not only of the lower basin but of the entire Mekong Basin.

Nevertheless, China and Myanmar have not sought full membership of the Mekong River Commission and have remained as observers. However, judging from the historical and socio-economic backgrounds of the countries of this region, there is no guarantee that the induction of the two upper-basin countries would add to the longevity of any agreement for the comprehensive development, use, management, and preservation of the water resources of the entire Mekong Basin. The possibility of a future split among these six nations and the abandonment of the lofty ideal of sustainable development cannot be discounted. In any case, the principle that the "basin is one" rings true, and the author fervently hopes that the six nations can permanently maintain good relations in the future.

The Mekong River Commission has decided to expand the scope of the development plan for the lower basin and has initiated the process for "formulating a comprehensive development plan for the entire Mekong Basin." The Commission's aim is not simply to formulate a concrete master plan but "to achieve the balanced and sustainable development and use of the water resources and other related resources of the entire river basin while paying careful attention to the environmental and ecological preservation." This comprehensive development plan is to encompass "all aspects of the natural, social, economic, and cultural environments." In addition, due care must be taken to ensure the "lifestyles and comfort of the indigenous populations." Furthermore, the Commission has confirmed the following caveat:

The comprehensive development plan shall be formulated in view of these requirements, and its contents shall be subject to repeated reconsideration and modification whenever rendered necessary by new developments and changes in the general environment.

Future dam projects in the lower Mekong Basin

The Mekong River Commission may experience various difficulties in the future, but the question remains as to what specific development projects it will choose to promote. This section examines the outlook for the Commission's future activities, dividing the main water-resources (and related-resources) development projects being actively discussed or promoted (as of mid-1995) into projects for the main stream and those for the tributaries.

New development proposal for serial dams on the main stream and suggestions concerning the problem of displaced populations

The first proposal for serial dams on the main stream of the lower Mekong was presented by ECAFE in the 1950s. This proposal went through numerous modifications to emerge as the Indicative Basin Plan report announced in 1970. This was further modified in the 1988 proposal. At this point, with the exception of the lowering of the height of the Pa Mong Dam, the plan remained essentially the same as the original Japanese proposal presented in 1961 for a series of large dams.

However, a sudden shift in direction began to occur in 1989 when the Secretariat of the Interim Mekong Committee presented a plan to build a series of low dams. Although this was not a formal proposal, many people began to draw the conclusion that the construction of large dams was a thing of the past. Following this, the Secretariat commissioned a study by a joint group of French and Canadian consultants who altered the plan entirely into a run-of-the-river type concept. It is reported that the working group of the three member nations of the Interim Committee approved this plan, indicating that a seminal turning point had been reached.

However, there is no evidence that a definite decision has been made. It is unclear whether the Mekong River Commission will, in fact, adopt this as its final plan, and the decision-making process is just now beginning to enter its crucial stage. The details of the proposals made after 1989 are discussed below.

Main stream development plan announced by the Secretariat in 1989

In September 1989, the Secretariat acquired funding from the UNDP for a short-term consultancy contract for a 3-month study leading to a new development proposal to be presented by the Secretariat. In preparation for this study, the Secretariat had estimated the total demand for electric power in the lower basin area over the next 30 years: total demand was projected at 19×10^6 kW (of which 14×10^6 kW was for Thailand) in the year 2000, and 57×10^6 kW in 2020 (of which 38×10^6 kW was for Thailand). On the basis of these gross electric-power projections, the mainstream dam projects included in the 1988 proposal were individually re-examined.

The conclusion of this re-examination was that the Low Pa Mong Dam (full-reservoir water elevation: 205–210 m) should be constructed by the year 2000. Regarding the other proposed mainstream dams, the study called for the revision and the lowering of the full water elevation of the reservoirs of the Ban Koum Dam (130 m) and the Stung Treng Dam (88 m) to 125 and 80 m, respectively. Thus, the overall conclusion was that it would be desirable to lower the height of the planned dams.

Another notable feature of the Secretariat's 1989 proposal was the recommendation for building a low dam with locks which provide gates at the exit and entrance of the Great Lake. These locks would be used to adjust the volume of flow, the lake being maintained at maximum capacity until mid-January, and released thereafter. It was indicated that this scheme would add to the natural flow volume in the Mekong Delta during the dry season and would help to ensure the volume of water needed for stable agricultural activity during the dry season.

This brings to mind the report submitted in 1970 by the French firm of consultants, SOGREAH, which concluded that the building of a large weir providing entrance and exit locks on the Great Lake would make only a limited contribution to flood control in the delta. Following the adoption of this report, the Mekong Committee had completely abandoned the idea of building locks on the Great Lake. The Secretariat was now actively promoting a plan for building massive locks to benefit agriculture in the dry season.

Run-of-the-river type development plan provided by the Secretariat in 1994

Following the Secretariat's 1989 proposal, the consultant services of the Compagnie Nationale du Rhône from France and the Acres Company from Canada were provided with the support of the French government and the UNDP to undertake a new study in order to conceive all the mainstream dam projects on the lower Mekong as low "run-of-the-river type hydroelectric power generating dams." In November 1994, this plan (fig. 6.1) was submitted to a technical working group panel representing three member countries and was approved. (Cambodia did not participate.) In accordance with this proposed plan (Compagnie Nationale du Rhône et al. 1994), the participants agreed on the following projects: the Dong Sahong Hydroelectric Project near the Khone Waterfalls in Laos (full water elevation: 70–72 m; 240,000 kW), the Ban Koum Hydroelectric Project at the Thai–Laos border (full water elevation: 120 m;

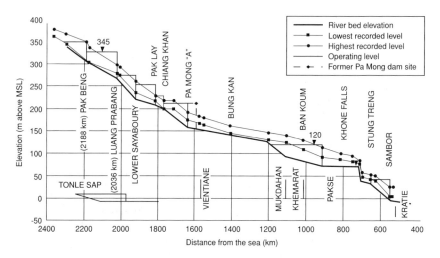

Figure 6.1. Mainstream cascade dams planned in 1994
Source: Mekong Secretariat 1994

2.33×10^6 kW), the Stung Treng Low Dam in Cambodia (full water elevation: 55 m; 980,000 kW), the Sambor Hydroelectric Project in Cambodia (full water elevation: 40 m; 3.3×10^6 kW), and the entrance and exit weir at the Great Lake (full water elevation: 10 m; 140,000 kW) (table 6.1). It was also decided at this time that feasibility studies would be undertaken first for the Dong Sahong, Ban Koum, and Sambor projects.

If this course of action is followed, it is quite likely that the Mekong River Commission will adopt the proposal of the consultants as it stands now, to construct low, run-of-the-river dams throughout the length of the main stream of the lower Mekong.

The new proposal for low, run-of-the-river dams fundamentally differs from the original development concept in that the efficacy of reservoir-type projects has been totally negated. If this plan is implemented, the advantages, not only of the much-awaited irrigation of north-east Thailand and other places along the main stream but also of flood control in the delta and other areas, will be forfeited.

Will the Mekong River Commission be able to abandon all its past plans and give its unconditional approval to the implementation of this new proposal?

Suggestions regarding the problem of population resettlement

The construction of large dams always brings with it the difficult problem of relocating the people affected by inundation. However, there are some cases in which resettlement has succeeded when countries (or regions) have combined this issue with other regional development projects. The

Table 6.1 Comparative study of potential cascade-dam projects

Projects and sites	Distance from sea (km)	Capacity (MW)	Energy (GWh/year)	Population displaced	Land area flooded (km²)	Internal Rate of Return (%)
RUN-OF-RIVER PROJECTS						
First-category projects						
Dong Sahong	719	240	1,640	0	0	14.6
Ban Koum	928	2,330	10,200	2,573	90	13.9
Sayaboury	1,930	1,260	5,990	1,720	20	13.9
Pak Beng	2,188	1,230	5,670	1,670	50	11.7
Subtotal		5,060	23,500	5,963	160	
Second-category projects						
Sambor	560	3,300	14,900	5,120	590	14.6
Luang Prabang LPA10	2,036	970	5,650	5,200	85	12.9
Subtotal		4,270	20,550	10,320	675	
Third-category projects						
Pak Lay PLB10	1,818	1,010	4,840	8,710	70	12.4
Fourth-category projects						
Pa Mong "A"	1,651	2,030	8,870	23,260	40	13.9
Least attractive						
Stung Treng	670	980	4,870	9,160	480	8.8
Grand total	9 projects	13,350	62,630	57,413	1,425	
1989 MAINSTREAM DEVELOPMENT SCENARIO	6 or 7 projects	19,000	93,000	310,000	7,600	

Source: Compagnie Nationale du Rhône, France, et al. (1994).

337

cases presented below may serve as points of reference for the future development of the Mekong Basin.

Prompt preparatory measures must be taken to cope with the problem of resettling populations as soon as a dam-construction project is proposed. The residents of the area to be affected by inundation must be surveyed at the outset and an appropriate resettlement site identified. If this resettlement site is already populated, careful attention must be given to the question of whether the two populations will be able to live harmoniously. The authorities must create opportunities for consultation between the two populations and should endeavour to satisfy both sides. Necessary job training must be provided and all the needs of the resettled population must be investigated and provided for. Some members of the resettled population will insist on staying put; the authorities must be prepared for such contingencies, and could, for example, assist such people to resettle near the reservoir to take up fishing or some other profession. In all cases, a "regional development plan" (for the resettlement site) should be formulated which fully utilizes the energies and capabilities of the resettled population. The construction of any dam involves a very long-term project which must go through the successive stages of proposal, investigation, planning, and construction. Necessary measures for coping with resettled populations must be undertaken in tandem with these successive development stages. The regional development of resettlement sites should also proceed in parallel. Finally, all dam-project designers, builders, executive officers, and others involved in dam projects must respect the basic human rights of resettled populations and the residents of such regions.

Towards the end of 1994, the author inspected the Lancang development projects in Yunnan Province, China, and was told that the Chinese government had taken measures to satisfy the above requirements, not only on the Lancang but also in all of its major river and dam developments. According to what the author was told by government officials about their resettlement planning, China could proudly call itself an "advanced country."

Another successful example is the Sennar Dam in Sudan constructed on the Blue Nile in 1925. This is an irrigation dam located 350 km upstream from the capital city of Khartoum, boasting a 160 km^2 reservoir with a capacity of 1×10^9 m^3. The building of a major new city in conjunction with the dam project was started earlier and completed in 1914. It is said that the city flourishes today as one of the important urban areas of the Sudan. The Roseires Dam is located a further 270 km upstream from the Sennar Dam. A total area of 290 km^2 was inundated by the Roseires Dam. Together with the construction of this dam, the city of

Damazin was built as the resettlement site for the population displaced by this dam, and has become a well-developed city.

An example of failure is the Khashm El Girbe Dam, which was built to provide for the water needs and irrigation of the resettlement site established for the people displaced by the Aswan High Dam. A total of 17,000 nomadic Nubians were moved into this area. However, erosion of the land caused by over-grazing threatened the dam with heavy silt build-up. This case shows how difficult the resettlement of large populations can be, not only for the resettled people themselves but also for the government authorities who must provide for their needs.

Regarding the success story cited above of the new city built to accommodate the construction of a dam on the Blue Nile, the author himself has not inspected the site and knows of it only through his research. Several years ago the author had the opportunity to meet with the Minister of Water Resources of the Sudan, who confirmed the truth of this success story. Of course, the Mekong Basin cannot be compared to the Sudan or Egypt, but these cases provide instructive lessons in how to cope with the very difficult problem of inundation.

Some notable development projects on the lower Mekong tributaries

Since the formation of the Mekong River Commission, the four countries of the lower Mekong Basin have been expected to undertake a number of dam- and river-development projects on the Mekong tributaries within their borders. This section outlines some of these projects, commenting on the necessity and scale of these projects based on the rather limited information available.

Meeting the demand for energy and water in Thailand

As mentioned above (p. 334), the 1989 proposal estimated that the demand for electric power in Thailand would stand at 14×10^6 kW in 2000 and would rise to 38×10^6 kW by 2020. EGAT has estimated higher levels of demand, indicating in its 1993 projections that demand in 2000 would reach 17.7×10^6 kW, rising to 31.7×10^6 kW by 2010.

Thailand currently has lignite- and oil-fired thermal generating plants in operation. However, the government would prefer not to build new thermal plants because this would exacerbate the already serious problem of atmospheric pollution. Nuclear generating plants have been studied but some hesitation remains in pursuing this line of development. Thailand has already exhausted most of its opportunities for domestic water-resources development and there are very few promising dam sites left to be developed. These problems are compounded by widespread environmental concerns and opposition to inundation. Thailand is therefore pinning its hopes on development projects in the neighbouring countries of Myanmar and Laos.

In the past, Thailand maintained a cautious stance toward the importation of electric power from neighbouring countries with different political systems. However, power demand is growing in Bangkok and its environs, as well as in the industrial zones on the eastern coast, and supply conditions have become very tight. Demand for water is also rising and the capacity of the Chao Phraya Basin to meet the demands for irrigation and for urban and industrial water use is being strained. On the other hand, north-east Thailand requires substantial water resources for irrigation during the dry season.

The members of EGAT are turning their attention to the rivers of Myanmar. By jointly building two dams on the Salween River, which flows from north to south through eastern Myanmar, up to 6×10^6 kW of power can be generated. In this connection, a 2-year feasibility study was launched in January 1994. Thailand has also been studying various plans for jointly developing four rivers to generate energy in the area of the Thai–Myanmar border and has called for leaders of the Myanmar government to consult on these plans. Furthermore, plans are being studied for building a long-distance canal system linking the Salween River to Thailand's second-largest hydropower dam, the Bhumibol (constructed on the Ping River, an upper-basin tributary of the Chao Phraya River), to augment the installed capacity of this dam. This project would also contribute to improving the irrigating capacity of the lower Chao Phraya River during the dry season, while also helping to meet the water needs of Bangkok and adjacent industrial areas.

Thailand is also looking to cooperate with Laos in the development of water resources in that country. Thailand is hoping that the installed capacity of the Nam Ngum Dam in Laos will be increased from the current level of 150,000 kW to 190,000 kW.[1] In addition to the development of the Nam Tha River in northern Laos (discussed in detail below), Thailand is looking forward to the small- and medium-scale development projects on the Nam Beng, Nam Suong, and Nam Man rivers and the large-scale development project of the Nam Theun River.

In view of Thailand's rapidly growing water demand, it is reported that Laos has put forward a plan to sell water to Thailand by diverting the flow of its domestic Mekong tributaries to northern and north-east Thailand.

The plan to augment the power-generating capacities of the two hydroelectric power dams in Thailand – the Silikit Dam (gross reservoir capacity: 6.66×10^9 m^3) and the Bhumibol Dam (gross reservoir capacity: 12.2×10^9 m^3) – by drawing additional water from Myanmar and Laos represents a desperate measure on the part of EGAT. In response to the growing opposition to the development of new dam projects, the King of Thailand issued a special statement in early 1994 calling on these opposition groups to exercise greater restraint so that the government could build the dams required for ensuring the energy needs and the irrigation and drinking-water needs of the country. Notwithstanding this statement, EGAT, which is responsible for the construction of hydropower dams in Thailand, was forced to make a public pledge that no new hydropower dams would be built in the next 15 years. This provides an indication of the strident opposition to new dam projects which has spread throughout the Thai public.

In early 1994, the Thai government unveiled the Kok–Ing–Nan Rivers Water Diversion Plan, which entails drawing approximately 1×10^9 m^3 of water per year from the Nam Kok River, a Mekong tributary in northern Thailand, and diverting this through the Nam Ing River (also a Mekong tributary) and on to the uppermost region of the Chao Phraya River. The government has stated that the feasibility study would be completed within 2 years. The implementation of this plan would affect the flow volume of the Mekong by reducing the inflow from the two tributaries; by contrast, the flow volume of the Nan River, a tributary of the Chao Phraya River, would increase and would result in the augmentation of the installed capacity of the Silikit Dam. In addition, the volume of water supplied to Bangkok for its urban and industrial needs would be increased. Finally, the paddy fields of the entire lower basin of the Chao Phraya River would receive more water.

A similar plan exists in north-east Thailand, where water shortages during the dry season are a problem. To alleviate this shortage, this plan calls for pumping water from the Mekong in the vicinity of Nong Khai and diverting this to the Chi and Mun rivers via an aqueduct 200 km long. This is referred to as the Kong–Chi–Mun Plan. In 1992, the Thai government revealed the outline of this wide-ranging plan, which will cost an estimated US40×10^9 and take 42 years to complete. The idea is that water will be pumped from the Mekong during the rainy season only, and will be stored in the Mun and Chi rivers, to be released during the dry season to irrigate the fields in the basin of these two rivers. It is reported that, in the future, Thailand wants to divert as much as 300 m^3/s from the Mekong (and this during the dry season).

These two plans for diverting the flow of the Mekong – the Kok–Ing–Nan and the Kong–Chi–Mun dams – have aroused the anxiety of the countries of the lower Mekong Basin. In particular, Viet Nam has voiced strong opposition on the grounds that these projects will have a major negative impact on the influx of salt water into the delta (during the dry season). Even Laos has expressed concern on the grounds that the diversion of 300 m^3/s would affect mainstream navigation in the lower basin during the dry season. Regarding the Kok–Ing–Nan project, the Thai government can take the position that it can rightfully divert the flow of a domestic tributary in northern Thailand to feed the Chao Phraya River. However, the Kong–Chi–Mun Plan, which threatens directly to reduce the flow volume of the Mekong, can be fairly described as appalling. It is said that, in early 1992, the Vietnamese government issued a strong protest and demanded that Thailand abandon its plan. As we have seen, the Executive Agent of the Secretariat of the Interim Mekong Committee, Mr Chuck Lankester, who tried to mediate the case, was said to have been forced to resign as a result of these events. Although this problem is directly related to the issue of water use which provided the impetus for the creation of the newly formed Mekong River Commission, no resolution is in sight.

The very difficult task of negotiating and adjusting these differences will devolve upon the Mekong River Commission, and, in the worst-case scenario, there is the risk that this issue may lead the Commission into an intractable deadlock at an early stage in its existence.

Laotian plans for development of the Mekong tributaries

Table 6.2, released by the government of Laos in August 1993, presents a list of existing and planned development projects in the Mekong tributaries in Laos. The dates in the column headed "Project completion" indicate tentative start-up dates for power generation. As there are only limited prospects of growth in domestic power demand in Laos, most of the power to be generated is intended for export to Thailand or other neighbouring countries.

Development plan for the Nam Theun River.   In February 1995, the ADB released its "List of Energy Development Projects" at the Forum for the Comprehensive Development of Indo-China, held in Tokyo.

Of the projects listed, the Nam Theun 2 and Theun–Hin Bun projects are currently in progress. However, these two projects are mutually exclusive: that is to say, with the completion of the Nam Theun 2 Project, water from the Nam Theun River will be diverted to the neighbouring Se Bang Fai River, rendering the Theun–Hin Bun Project in the lower basin untenable.

The Nam Theun 2 Project (600,000 kW). Using funds provided by the World Bank, the Australian consultants, the Snowy Mountains Engineering Corporation (SMEC), completed a feasibility study in 1991. Following this, SMEC joined the Trans Field Company of Australia to form the Transmec Group, a development and consulting firm. In turn, the Transmec Group has teamed up with EDF, the French public power

Table 6.2 Hydroelectric power projects under development and under consideration in June 1997 (by EDL) in Laos

No.	Project	Developer (MOU signing date)	Status	Project completion	Capacity (MW)
1	Nam Theun Hinboun	EDL + MDX + NG (MOU 23/6/93)	PPA Signed	1998	210
2	Houay Ho	Daewoo + Loxley + EDL + (MOU 23/9/93)	PPA under negotiation	1999	150
3	Nam Theun 2	Transfield + EDL + EDF + Italian Thai (MOU 07/03/94)	PPA under negotiation	2001	681
4	Hongsa Lignite (Phase I)	Lao-Thai Power (MOU 30/12/93)	PPA under negotiation	2001	600
5	Xepian & Xenamnoy	Dong Ah (MOU 03/8/94)	F/S Submitted	2001	439
6	Xekaman 1	HECEC (MOU 06/06/94)	F/S Submitted	2001	468
7	Nam Ngum 3	MDX + GOL (MOU 16/3/94)	F/S Submitted	2002	400
8	Nam Ngum 2	Shlapak (MOU 16/1/91)	F/S Submitted	2002	465
9	Nam Theun 1	SUSCO (MOU 25/3/94)	F/S Submitted	2002	400
10	Nam Theun 3	Heard Energy Cop. (MOU 01/08/94)	F/S Submitted	–	273
11	Nam Khan 2	Hydro Quebec Int. (MOU 24/6/94)	F/S Submitted	–	126
12	Nam Ngiap 1	Shlapak (MOU 16/1/91)	Study pending	–	440
13	Nam Cha 1	HECEC (MOU 06/4/94) (withdrawn)	Study pending	–	115
14	Nam Cha 2	HECEC (MOU 06/4/94) (withdrawn)	Study pending	–	70
15	Nam Mang 3	Ch. Kanchang (MOU 21/1/94) (Cancelled)	F/S available	–	50
16	Sekatam 1	HydroPower Pty Ltd. (MOU 15/10/94)	F/S Submitted	–	130
17	Sekatam 2	HydroPower Pty Ltd. (MOU 15/10/94)	F/S Submitted	–	130
18	Nam Ou	Pacific Rim Energy Partner (MOU 7/12/93)	F/S Submitted	–	600
19	NamLik 1/2	SIT Enterprise Cop. (MOU 16/02/95)	F/S Submitted	–	100
20	NamNgiep 2	VKS (MOU 01/03/95)	Study pending	–	565
21	NamNgiep 3	VKS (MOU 01/03/95)	Study pending	–	565
22	NamSeuang	VKS (MOU 01/03/95)	Study pending	–	190
23	Se Kong 4	Modular (MOU 26/11/94) (Cancelled)	Study pending	2002	470
24	Nam Thal	S.P.B (MOU 07/10/95)	Study pending	–	230
	Total				7,867

Source: Given by EDL. Government of Laos in 1997
Abbreviations: EDL, Laotian State Power Company; GOL, Government of Laos; MOU, Memorandum of Understanding; PPA, Power Purchase Agreement.

company, and the Phatra Thanakit/Jasmine/Italthai Company of Thailand to jointly finance a US$$1.2 \times 10^9$ BOT (build, operate, and transfer) agreement which was formalized in June 1994. It is reported that work on this project is proceeding with a projected completion date in 1998. This constitutes the first BOT project to be undertaken in Laos.

In March 1995, an agreement was signed between EGAT of Thailand and the government of Laos for the sale of 600,000 kW of electric power to be generated by this project. Under this agreement, power will be supplied to Thailand for a period of 25 years. The power from this hydroelectric dam will be transmitted across the Mekong from Savannakhet in Laos to the Thai provinces of Mukdahan and Roi Et, where the power will be linked to the Thai domestic power grid.

The Theun–Hin Bun Project (210,000 kW). The Theun–Hin Bun Project involves the construction of a flow-through hydroelectric dam with an installed capacity of 210,000 kW, which is to be located midway between the Nam Theun 2 Dam, discussed above, and the Nam Theun 1 Dam, which stands at the lowest reaches of the Nam Theun River. (The site of the Theun–Hin Bun project is located approximately 100 km upstream from the confluence of the Nam Theun and Mekong rivers, and 40 km downstream from the Nam Theun 2 Dam.)

This Theun–Hin Bun project will cost an estimated US$280 million to complete. According to reports in Thai newspapers, the project will be undertaken by a joint venture company involving EDL, the Laotian state power company, Nor Hydroelectric of Norway, and MOX, a privately-owned power company in Thailand. The equity position of the three companies in the newly formed joint venture is to be 55, 25, and 20 per cent, respectively. The three partners will operate the project for 25 years after its completion, at which point the entire project will revert to the government of Laos. Construction is scheduled to be completed in 1998. Part of the power output will be supplied to domestic users in Laos.

The ADB has actively promoted the Theun–Hin Bun project and has provided the Laotian government with various types of legal and technical support. It is said that the ADB hopes to make this a showcase project for private and public sector cooperation in hydroelectric power development.

The problems of both projects – development and the environment. The two above-mentioned projects have been criticized for their expected negative impact on the environment. If the Nam Theun 2 Dam is completed as originally designed, it will inundate an area of 400 km^2, forcing the resettlement of 4,000 people (14 hamlets), including minority groups. If either of the two projects is completed, there is the danger that the reduced flow volume of the Nam Theun River may trigger significant ecological changes in the area. In the former project, the water passing through the generators will be diverted to the upper regions of the Se Bang Fai River, a neighbouring tributary; in the latter project, the water from the dam will be carried to the tributary river of Nam Hin Bun. In both cases, this diversion will result in a significant reduction in the flow volume in those parts of the Nam Theun River located below the dams, affecting a total of 27 downstream hamlets.

At present, the plan is not to discharge the outflow from dams to the downstream Nam Theun River during the dry season between January and April. Downstream from these two dams, there are, in total, 16 small natural lakes and 30 waterfalls which comprise an area that is very rich in fishery resources (including fish returning to spawn in this river system). Because of this, the Laotian people of this region are able to enjoy a diet of various types of fish every day. The dam projects are therefore a serious cause of concern in this area.

There are also other environmental problems to be considered. The Nakay Plateau is home to a wide range of wild animals, including tigers and elephants. The dam projects may well threaten their habitats, as well as forest environments.

The Mekong River Commission and the government of Laos must urgently review the five outstanding plans for dam-construction and river-diversion projects in the upper and lower regions of the Nam Theun River from an overall perspective. The projects to be implemented should be selected and carried out with due caution under the precept of "environmentally friendly development."

Se Kong River hydroelectric power dam-development project and the Irawaddy dolphin.   From the 1980s, the Laotian government changed its development policies which had, until then, focused on the development of the province of Vientiane and its environs. Under the revised plan, greater attention was given to the development of the outlying regions, with the highest priority assigned to the development of southern Laos made up of the six provinces of Se Kong, Champasak, Savannakhet, Saravan, Kammouan, and Attapu.

The hydroelectric power development projects for the Nam Theun River noted in the preceding section may be viewed as part of the general development plan for southern Laos. However, the focus of these projects is less on regional development and more on contributing to the considerable national income secured from the sale of electric power to Thailand.

The development projects for the southern provinces were initiated with two objectives in mind. The first objective was to build small- and medium-sized hydroelectric power projects for the electrification of the towns and villages of the southern provinces to support urbanization and the operation of small factories. The second objective was to use the roads built for the dam projects as a springboard for improving the land transportation and communications networks.

Parts of these development projects have already been completed. First, the installed capacity of the Selabam hydroelectric power station located at the lowest section of the Se Done River (which joins the Mekong at Pakse) was raised from 2,000 to 5,000 kW. Next, a new hydroelectric power station with an installed capacity of 45,000 was constructed at the well-known Se Set Falls (1991).

The two provinces of Se Kong and Attapu in southern Laos share borders with the central region of Viet Nam and north-east Cambodia and are close to north-east Thailand. Although this is a strategically important area from the viewpoint of national defence, because of its distance from the capital Vientiane, transportation and communication remain poorly developed, making this one of the most isolated regions in Laos. The Se Kong River and its tributaries which flow through these two provinces constitute one of the major tributary systems of the Mekong, and have the potential to yield a great deal of hydroelectric power. The government of Laos is hoping that the construction of a series of hydroelectric power dams on the Se Kong River system will provide electricity for export to Thailand. In addition, the government hopes to use part of this power for the development of forestry, mining, and processing industries in this region, stimulating dynamic economic progress in the two provinces. Naturally, the development of medium- and large-scale dams will have a significant positive impact on the transport and communications networks of this region. As is well known, this is a region rich in mineral resources which await exploitation. Furthermore, the famous Boloveng Plateau, extending along the eastern portion of the Se Kong Basin with the town of Pak Song at its centre, is noted for its cultivation of tea, coffee, and other products.

According to the prefeasibility study carried out by the Japan International Cooperation Agency (JICA) (1994), a total of 17 prospective projects, including some of relatively low priority, have been identified in the Laotian portion of the Se Kong Basin (the upper Se Kong Basin covering an area of 160,460 km^2, including the Se Piang Basin where a river-diversion project is planned). The total installed capacity of all the planned facilities comes to 2.175×10^6 kW with an estimated annual generated energy of 11,782 GWh.

The installed capacity of each of the candidate projects exceeds the current level of power demand in the two provinces of Se Kong and Attapu. Therefore, at least for the foreseeable future, most of the power generated by these projects will be fed to the power grids in Thailand, with smaller amounts being exported to central Viet Nam and north-east Cambodia.

According to the above-mentioned JICA report, the development of the Se Kong will reduce the volume of the floods effectively during the rainy season, while increasing the flow volume in the lower reaches of the river during the dry season. Furthermore, the report estimates that the total resettled population in the above-mentioned three projects will be in the range of 5,000 people. It also states that these development projects may trigger the spread of water-related diseases and may have an adverse effect on the habitats of aquatic life, particularly fish. However, the report makes no mention of the possible inundation of mineral resources.

Among the various development projects in the Se Kong Basin, the most active progress is being made in the study of the Se Kamane 1 Dam, a high dam being planned at the mouth of the Se Kamane River, a tributary of the Se Kong. However, the feasibility report (compiled by a Tasmanian hydroelectricity company) has drawn widespread criticism due to the many environmental problems which the project would entail. Most of these criticisms are directed at two issues – the problem of inundation and resettled populations, and the problem of the adverse effects on forests and fisheries.

The critics point out that, although the area of the Se Kamane Basin has been designated by the Laotian government as an environmentally protected zone, the planned dam would cause the inundation of 190 km^2 of land and forests and would require the resettlement of 2,000 people (400 families). In addition, the dam itself would cut off the migratory routes of fish in the area. The government of Laos remains unaffected by these criticisms and it is reported that it has signed a billion-dollar contract with Tasmanian Hydroelectric for the construction of the Se Kamane 1 dam and other dams.[2]

The Tasmanian report does not contain a clear assessment of the impact on aquatic life. However, the JICA report on the comprehensive development plan for the Se Kong basin includes certain projections on this matter, mentioning that the construction of the dams may have a negative impact on the dolphins which are seen from time to time in the Se Kong. The JICA report also expresses concern about the adverse effect on rare wildlife in this region, and states that the completion of the project will inundate and destroy forest areas which serve as habitats and feeding grounds for these rare species. The problem of obstructing the migratory routes of fish is also considered, and the report recommends that wildlife reserves should be established around the reservoir. Regarding these concerns, as mentioned in chapter 4, the Mekong Committee had already made a similar suggestion before 1976: it called for the establishment of wildlife reserves, including a large area of the Laotian part of the Se Kong Basin.

It appears that those in charge of studying these Se Kong Basin projects for the ADB and JICA have focused their attention solely on developing the hydroelectric power potential of this river system. The author believes that other uses of these water resources should be investigated concurrently. For instance, multi-purpose development possibilities for irrigation, release of freshwater fish, and related forestry and mining operations should be actively pursued. The town of Se Kong, located on the main stream of the river, is a newly built town established to serve as the centre for development in this region. The town was created by clearing an area in the forest and it remains unclear whether the productivity of the soil in this area is high enough to sustain agriculture. However, there are several towns further downstream, such as Attapu, which have the potential for diversified agricultural development in the future. This indicates that the dams on the Se Kong should not be used exclusively for hydroelectricity generation and that various schemes should be devised for the multi-purpose use of these water resources. There is no doubt that the dam projects in the Laotian part of the Se Kong will affect the flow volume of the river as it runs through north-east Cambodia.

The development of this river system should, therefore, be considered from the perspective of the multi-purpose and transregional utilization of its water resources and a comprehensive vision should be formulated for the future development of the entire Se Kong Basin which overlaps the territories of these two countries. In this context, the tendency to focus solely on the development of the hydroelectric potential of this river system is questionable.[3]

Dam-development projects on the Nam Leuk River (Laos). The Nam Leuk River development project features the construction of a rockfill hydroelectricity dam (600,000 kW) on the Nam Leuk River which flows north of the city of Vientiane. Power generated by this dam would serve domestic needs in Laos, with the excess being sold to Thailand. According to the environmental assessment study conducted by JICA, this project will not have any significant negative impact on the natural environment. For this project, the Japanese government will provide loans from the Overseas Economic Cooperation Funds (OECF) totalling $¥3.93 \times 10^9$ (30-year loan at 1 per cent rate of interest). From this amount, $¥100 \times 10^6$ has been earmarked for the protection of fish, forests, and wildlife.

Irrigation projects in the lower basin of the Se Bang Fai River (Laos). The Se Bang Fai River, another major tributary of the Mekong, flows through the southern Laotian provinces of Savannakhet and Kammouan. The lower basin of the river contains a plateau with an area of 90,000 ha. Although this river has a relatively high flow volume compared with its neighbouring rivers, its water has not been used at all for irrigating the paddy fields on the plains; instead, this water is the cause of frequent flooding in the river's lowest reaches.

Plans are now being made for hydroelectric power projects in the neighbouring Nam Theun and Se Kong rivers and it is highly likely that this area will have an abundant supply of electricity in the near future. For this reason, the Laotian government is hoping to improve the current conditions in this area by undertaking appropriate flood-control measures and by implementing low-pumping irrigation in the future. The government's position seems to be that, first and foremost, these plans must be devised in such a way that they are environmentally friendly and acceptable to the local people.

Development projects in the Cambodian tributaries

Cambodia finally appears to be making some headway towards economic reconstruction in Phnom Penh and its surrounding areas. However, the situation in the outlying areas remains unchanged, with poor security conditions and a large number of land-mines not yet removed. Safety is a very serious concern, not only in the river basins of the north-east and in the vicinity of the Great Lake, but even at the dam site of the planned Prek Thnot Project adjacent to Phnom Penh. Conditions are such that foreigners are unable to set foot in these areas.

However, there are two exceptional projects which can be approached with relative ease: these are the projects for the reconstruction of the Kirirom and Kamchay hydroelectric power stations. Kirirom is located in the mountainous area 120 km west of Phnom Penh. Completed in February 1968, this project, with an installed capacity of 10,000 kW, transmitted all the energy generated to Phnom Penh. However, all the facilities were destroyed during the Cambodian conflict. Recently, local newspapers have reported that Austria and Sweden will provide financial aid for its rehabilitation. The Kamchay, in the south, has the potential to generate about 100,000 kW, and also provides irrigation. The Japanese governmental organization JICA sent a survey team in April 1995, but it is reported that JICA has been taking a cautious attitude regarding the promotion of further surveys for development. Both of these sites are bounded by the Elephant Mountains and are located just outside the Mekong Basin. This is an area of very abundant rainfall (5,000 mm/year) and is conveniently located for transmission of power to Phnom Penh, Kampot, and the port facilities at Sihanoukville (Kampong Som).

As one of the promising planned projects for the development of the Cambodian tributaries of the Mekong, the Prek Thnot Dam Project near Phnom Penh has already been referred to on several occasions in this book. Construction work on this project began in 1968. Funding was made available through the official development assistance (ODA) of the Japanese government and the work was undertaken by a Japanese company. However, the site was abandoned in 1974 as a result of the war and deteriorating security conditions. The turbines for the hydroelectric power station remained in storage in Phnom Penh. The project was originally designed for the irrigation of 90,000 ha and a power output of 18,000 kW. An environmental assessment study was attempted in 1989 but was called off because of guerrilla activities. More recently, the project was reviewed by an Australian NGO. Although the government is keen to restart the project, no progress has been made because of the environmental issues related to the planned reservoir and the sharply diminishing economic benefits to be gained from irrigating this area. As the plan stands now, a hydroelectric power station of 18,000 kW capacity is to be built, as was originally intended. The electricity from this station will be sent to Phnom Penh using the above-mentioned Kirirom–Phnom Penh power lines and will contribute to alleviating the power shortage in the capital city. In early 1988, a concrete intake weir was constructed downstream from the project site by the local government to enhance irrigation. The implementation of the upstream Prek Thnot Project is therefore keenly awaited.

Development projects in the tributaries of the Mekong Basin in Viet Nam

Hydroelectric power projects outside the Mekong Basin.   Electricity supply conditions in Viet Nam appear to have been stabilized with the completion of the north–south power transmission lines (500 kV, total length 1,200 km) in 1994. As a further measure, the government is very interested in building a large-scale hydroelectric power station immediately upstream from the existing Hoa Binh Dam (2.1×10^6 kW capacity completed with Soviet assistance) which is located on the Black River, a major tributary of the Red River which flows through northern Viet Nam. Viet Nam has requested financial assistance from Japan for the construction of this newly planned Sonla Dam, which is projected to have an installed capacity of 3.6×10^6 kW. In the south, Viet Nam has also approached the Japanese government for financial support for the development of the abundant water resources of the Dong Nai River, which flows through the vicinity of Ho Chi Minh City. Regarding the latter project, a comprehensive study was conducted by Nippon Koei using funds provided by JICA. Meanwhile, the feasibility study for the Ham Tuan hydroelectric power project has recently been finalized by an electric-power development company at one of the tributaries of the Dong Nai River and its construction has been started.

Hydroelectric power project for the Yali Falls in the Mekong Basin.
Parallel to these hydroelectric power projects located outside the Mekong

Basin, the government of Viet Nam has been pursuing various development projects within the basin. For instance, feasibility studies have been
completed for the hydroelectric power station at the Yali Falls located in
the upper reaches of the Se San River, which flows through the central
plains. Construction of this project, for which various plans have existed
for many years, has been completed with the assistance of the government of the Ukraine.

The Ya Soup Project in the Mekong Basin.   Located within the same
Mekong Basin, the Darlac Plateau situated in the upper reaches of the
Srepok River, which flows through the central plains, has a population of
approximately a million people. The Ya Soup region, located in a corner
of the Darlac Plateau, is noted for its fertile soil and abundant rainfall.
For the development of this region, the government of Viet Nam peti-
tioned the Interim Mekong Committee for South Korean financial assistance. As a result of this petition, preliminary studies for this project were
completed in February 1995. The project calls for the construction of the
Ya Soup Reservoir and the irrigation of 8,210 ha (to be increased to
15,000 ha at the final stage of the project) of agricultural land, and the
generation of 1,450 kW of power. The reservoir facilities will also reduce
damage from flooding. The completion of this project is expected to make
a major contribution to the promotion of agriculture and regional development in Darlac Province.

Other projects in the Mekong Basin.   The foregoing are among the long-
standing projects of the Mekong Committee. It is feared that the actual
implementation of these projects may cause various changes in the water
quality of the rivers and subterranean water sources of this area, as well
as influencing the vegetation of the area. To monitor these changes and
to establish the conditions for long-term sustainable development in this
region, Denmark has recently embarked on a regional survey as a project
of the Mekong River Commission.

Author's recommendations on the tributary-development projects

The various projects for the development of the tributaries discussed in
this section are far more concrete than the development plans for dams
on the main stream of the Mekong.

Development projects for the tributaries should, of course, be undertaken in order of immediate and near-future needs. However, full consideration should be given to the dam projects on the main stream of the
Mekong which will be undertaken in the more distant future. Furthermore, as stated in the ADB recommendations discussed in the following
section, due attention should be paid to the optimal linkage of these
tributary-development projects to the various projects for the improve-

ment of transportation and communications, and the upgrading of hydropower generation and power transmission within and outside the Mekong
Basin. Before proceeding with construction, appropriate priorities should
be carefully established in order to achieve a well-balanced plan for
overall regional development.

*ADB recommendations for improving transportation,
communication and power transmission inside and outside the
Mekong Basin*

In December 1993, the Japanese government hosted a preparatory meeting in Tokyo for the Forum for the Comprehensive Development of Indo-
China. Following several additional meetings, the Forum for the Comprehensive Development of Indo-China was held in Tokyo in February 1995
with the participation of the representatives of 20 nations and the EC.

   The objective of the Forum for the Comprehensive Development of
Indo-China was to promote the development of export-oriented industries in the three countries of Indo-China and to nurture their stagnating
socialist economies so that they would be able to withstand the pressures
of the market economy and competitive practices. The following measures were to be taken for the achievement of this objective:
1. To provide political stimulus for the promotion of greater international economic assistance to the Mekong region;
2. To implement international development projects beyond national
   boundaries, and to facilitate the exchange of information among donor
   nations, aid-receiving nations, and international organizations in order
   to promote greater efficiency in these projects.

   Prior to the holding of the Forum, the Japanese government divided
all infrastructure-related activities into "hardware-oriented activities"
and "content-oriented activities," which were then assigned to the ADB
and the UNDP, respectively. In response to this assignment, the ADB
retained the services of consultants to review from its own perspective all
the outstanding development plans for the Mekong Basin which had
previously been studied by the Mekong Committee and others. Following
this review, the ADB formulated a set of plans focusing on the regional
development and improvement of transportation and energy generation.
The ADB approach was to enlarge the geographic area of the plan to
include Myanmar and the Chinese province of Yunnan on the grounds
that a more broadly based regional plan would be more effective. The
outcome of these efforts was the ADB's subregional infrastructure programme, which it presented in the form of a "Compendium of Project
Profiles" (ADB 1995).

   The contents of the individual projects mentioned in the ADB "List" are not examined here. Instead, referring to the contents of the List and to the 1994–1995 Annual Report of the Interim Mekong Committee (the last Annual Report of the Interim Committee), several of the projects described for the improvement of transportation and the upgrading of power-transmission networks are outlined below.

Plans for improvement of transportation in the upper Mekong Basin

The French government undertook a study for the development of river transportation in the Upper Mekong between the end of 1993 and the end of January 1994. This study examined transportation projects for linking Luang Prabang in Laos with Jinghong, located upstream on the Mekong in the southernmost section of Yunnan Province in China.

According to the French study, an annual volume of 98,000 t of cargo is being transported between the two areas of Luang Prabang and Jinghong. This is projected to reach 2×10^6 t by the year 2000. Although this 20-fold increase in transport volume over a 6-year period represents a very bold projection, the study estimates that the amount transported via the Mekong between the two points will not exceed 150,000 t in 2000. However, if both the current scenario for the development of the waterways in the upper Mekong and also related improvements in land transportation are, in fact, implemented, a dramatic increase in cargo movements can be expected after 2000. This would necessitate the dredging of waterways, the installation of navigation aids (nautical markers), and the development and improvement of river ports.

More specifically, the following improvements would be required. Ferry services should be inaugurated to link the northern Laotian towns along the Mekong, such as at Pak Beng. Navigation aids would be needed between Luang Prabang and the Golden Triangle area. The three ports at Houay Xay, Pak Beng, and Luang Prabang need to be improved. Furthermore, the French study points out that environmental assessment studies for the dam projects in the upper regions of the Golden Triangle, and various legal studies related to international navigation, should be undertaken as soon as possible.

Improving transportation in the Mekong Delta and the My Thuan Bridge Project

In the Mekong Delta are numerous canals which serve the irrigation and river-transport needs of this area. The delta is also rich in fisheries and marine resources. A study of navigation plans for the Mekong Delta and its branch, the Bassac River, is now about to be restarted after a lapse of three decades. In February 1993, the World Bank dispatched an investigative team to the delta, which concluded that priority should be given to the improvement of the port facilities at Can Tho near the Bassac estuary in order to promote navigation on the Bassac. In comparison, the World Bank study assigned a lower priority to developing navigation at the Mekong estuary. Following this study, Denmark signed an agreement with the Interim Mekong Committee to undertake a feasibility study for dredging the Bassac estuary and the improvement of navigation.

The dredging of the Bassac estuary and other improvements for promoting navigation do not, in themselves, pose any serious problems. However, a problem has arisen in connection with the development of land-transportation routes which are intimately related to improvement of navigation. The issue involves the construction of the My Thuan Bridge (a long-cherished plan of the Mekong Committee for constructing a bridge in the Mekong estuary). For Viet Nam, the My Thuan Bridge would constitute a very important land-transport link between Ho Chi Minh City and the several ports located to the west of the city: these include the ports at My Tho (located on the left bank of the Song Cua Dai River, a branch river in the Mekong Delta), Vinh Long (located on the right bank of the Cua Co Chien River, another branch of the Mekong), and Can Tho (located on the right bank of the Bassac estuary). The bridge is scheduled to be built immediately upstream from Vinh Long (on the Mekong).

In July 1993, the Australian government undertook a one-year feasibility study of this project. At the conclusion of the study, the Australian government announced that it would invest approximately A$100 × 10^6 for the construction of this bridge. (The project was to be jointly financed by the governments of Australia and Viet Nam at a ratio of 7:3.)

It was at this point that the Cambodian government began to voice its doubts concerning the construction of the My Thuan Bridge. The Cambodian position is that, under the Australian plan, the bridge will be too low. The Cambodians have argued that, unless the bridge is built to a minimum height of 37.5 m above water level, ocean-going ships weighing more than 10,000 t will be unable to pass under the bridge. This would effectively rob the Mekong of its value as an international shipping route.[4]

What changes would have to be made to accommodate the Cambodian position that the Mekong should be reserved as a shipping route for ocean-going vessels sailing up to Phnom Penh? First, the height of the My Thuan Bridge would have to be raised, and this would obviously increase the construction cost of the bridge. Secondly, because the Mekong is too shallow to allow the passage of ships over 3,000 t, an extensive dredging operation would be required. These are highly uneconomical propositions. The Cambodian position is even more uneconomical, considering the fact that the river route will add an entire day to transport time. Thus, from a logical perspective, this is an entirely irrational position.

Why does Cambodia insist on forwarding this totally uneconomical proposal which jeopardizes the agreement for economic and technical assistance between Viet Nam and Australia? Cambodia cites "historical and cultural factors" in demanding that the Mekong below Phnom Penh be preserved as an international shipping route. It is difficult to evaluate this position because these "historical and cultural factors" have not been satisfactorily explained. However, it can be easily seen that, unlike Viet Nam, Cambodian access to the open sea is limited. There is only one waterway to the port of Phnom Penh (restored and expanded by Japanese aid), and that is through the Mekong Delta. Other than this, Cambodia must rely totally on the port at Sihanoukville as its sole interface with international ocean-going trade. If this is the reason why the Cambodian government insists on upgrading the passageways of the Mekong Delta for use in international shipping, its position can be understood.

Development of an east–west transportation corridor

Although southern Laos languishes today as an underdeveloped region, its outstanding potential for hydroelectricity generation, forestry, agriculture, mining, and tourism are well known. This area also holds the promise of future development as an important junction in the east–west transportation corridor linking Bangkok to the central Vietnamese ports of Da Nang and Qui Nhon. Furthermore, southern Laos, which borders on both Viet Nam and Cambodia, occupies a highly strategic position in the defence of Laos.

As shown in figure 6.2, several east–west trade routes have been proposed for linking the three countries of Thailand, Laos, and Viet Nam. The construction of these routes will undoubtedly have a very significant effect on agriculture and water resources development in southern Laos.

The concept of east–west transportation corridors was formulated by the ADB using funds made available by the French government. This concept has its origins in the "Study of the Role of the Mekong River in the Development of Regional Transportation" undertaken by the Secretariat of the Interim Mekong Committee in 1991, and the "Feasibility Study of Mekong River Bridges in Southern Laos and North-east Thailand" completed in 1992 through a grant from the ADB.

The most important element in this concept consists of the international bridges to be built linking Thakhek (Laos) to Nakhon Phanom (Thailand) and Savannakhet (Laos) to Mukdahan (Thailand) on the two proposed routes. It is reported that the Laotian government favours the former route because it is closer to Vientiane, whereas the Thai government favours the latter route because it links north-east Thailand to the Vietnamese port of Da Nang. There is another link road further south of

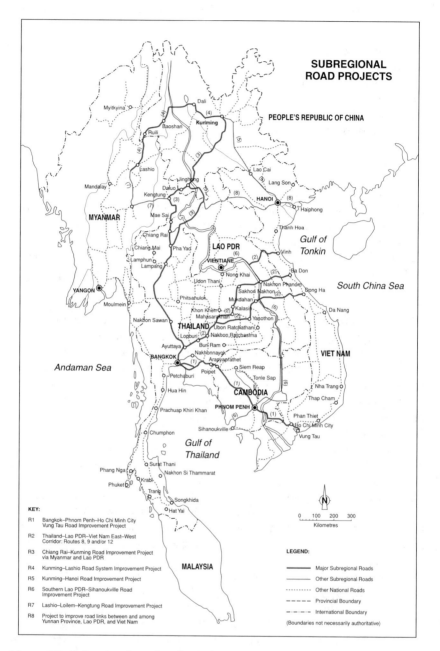

Figure 6.2. Subregional road projects planned by ADB
Source: ADB 1995

356

Savannakhet to Mukdahan: this is the route between Pakse (Laos) and the opposite bank, which fortunately belongs to Laos. Although it crosses the Mekong, the route is therefore considered domestic, so the Japanese government faces no problem concerning the provision of bilateral aid. JICA is considering building a long bridge at this section, and is also planning to build another Mekong Bridge (1,360 m long) at a site far downstream between Kampong Cham and its opposite side. Since both sides of the river at this point belong to Cambodia, there is no danger of international conflict.

At any rate, when either of the above-mentioned bridges in Laos is completed, the east–west corridor would intersect Laos's Highway 13 (the section running through Luang Prabang, Vientiane, Thakhek, Savannakhet, and Pakse). At the time of writing, Highway 13 is being re-paved and its bridges restored. ADB funds are being used for the northern section of the highway between Luang Prabang and Vientiane. The restoration of the section south of Vientiane through Thakhek is being financed approximately equally by the Swedish International Development Agency (SIDA) and Japan's JICA. Finally, the southern section of the highway linking Thakhek to Savannakhet is being financed by funds from the World Bank. The restoration of Highway 13 can make a significant contribution to the development of this region if it is thus linked to the east–west corridor plan.

However, the author is concerned that the east–west transportation corridor may jeopardize the independence of Laos as a small, land-locked nation. In other words, the southern section of this weak and small country sandwiched between the powerful nations of Thailand and Viet Nam may be used as a mere tool to serve the interests of economic development of the two larger countries. It is not the author's intent to fret about preserving due respect for national sovereignty or to exaggerate the dangers of regional and subregional rivalry and estrangement. Although the east–west corridor will, no doubt, bring many material benefits to the government and people of Laos, one cannot help but be concerned that it may also cause much disruption to the daily lives of these people and may lead to the loss of many intangible spiritual and cultural assets. Although the officers of the ADB and the members of the Japanese government ministries who participated in organizing the Forum are sure to say that these metaphysical concerns are no more than personal and irrational fears, the author nevertheless feels that these intangibles deserve consideration.

Upgrading the electric power transmission networks

As discussed previously, demand for electricity is growing rapidly in Thailand and Viet Nam. By linking the power grids of the two countries, significant energy losses could be avoided, while various advantages could be gained. This is the background for this proposed project.

The construction of hydroelectric dams in the lower Mekong Basin and the linking of power-transmission networks has constituted one of the most cherished dreams of the Secretariat ever since the earliest development projects for this region were drawn up. The representatives of the four riparian countries of the lower Mekong Basin (with the addition of the governments of China and Myanmar) who participated in the conference held in Chiang Mai in September 1994 gave their support to this plan. Japan is playing a leading role in undertaking a study (between June 1995 and September 1996) on the construction of power-transmission networks in Cambodia, Laos, Thailand, and Viet Nam. It should be noted that the concepts currently being studied extend beyond the limits of the lower Mekong Basin.

The trunk lines in the power-transmission network under study at present would carry a current of 500 kV, while the branch lines would carry 230 kV. The network would extend from China and Myanmar in the north to Malaysia in the south. This concept will certainly lend vigorous impetus to the hydroelectricity development projects currently being studied for the various Laotian tributaries of the Mekong. The realization of a broadly based regional power-transmission network and the implementation of the east–west transportation and communications corridors, and the combining of these three strategic projects, is sure to enhance the developmental potential of the six riparian countries of the Mekong Basin. By the same token, the hopes and dreams of all who have been involved in these projects are now expanding to take in new possibilities. However, many years will have to pass before the current pace of economic development inside and outside the Mekong Basin gives rise to demand which would be capable of supporting these massive power-transmission networks. What is important to remember at this point is that no decisions have been made on the most crucial aspect of this entire dream: that is to say, nothing has been decided on the scale or the format of the dams to be constructed on the main stream of the Mekong.

Main tasks of the New Mekong River Commission

The New Mekong River Commission was established by the riparian governments in the lower Mekong Basin on the basis of lofty ideals. There are a number of difficult tasks which the Commission must perform, of which the author considers the following to be the three most important:
1. To observe whether the riparian countries foster and maintain reasonable and fair utilization of water resources, as promised;

2. To maintain the existing quantity and quality of river water in the main Mekong;
3. To work out a new comprehensive development plan for the river basin.

Construction of mainstream dams

As described in the previous chapter, the Secretariat of the Interim Mekong Committee provided two separate yet similar development plans between 1989 and 1994 for mainstream dam projects. As an alternative to the previously proposed large-scale dam construction, the plans are to build low dams in succession on the main stream in the lower basin.

Will the new Commission really discard the long-cherished dream of large-scale dams on the main stream, favoured since the 1950s, and embrace instead the development of small, low dams? Let us examine the various arguments.

With regard to this question, it may be worth while to introduce remarks made in Japan by Mr Daniel Beard, Director of the Bureau of Reclamation of the US Department of the Interior.[5] The Japanese people who are involved in water development pay great attention to Mr Beard's remarks because the Bureau is widely recognized as one of the leading promoters of dam construction, not only in the United States but also, until recently, overseas. After visiting Japan, the Director made the same speech in Thailand. His remarks apparently represent the Bureau's most recent official view on dam construction. The speech amounted to an abandonment of the Bureau's past conviction of the superior merits of large-dam construction, which gave great encouragement to those who have opposed large-scale projects all along. Such opponents have gained influence since the 1970s. Beard said the new outlook was due to the changed realities under which the Bureau has been working recently. His remarks, which the author heard at his public address, are outlined below.

The past ideal, to "tame rivers," is no longer applicable to the people of the United States. The Bureau has decided to withdraw from the development of dams in the United States. Nowadays, nobody in the United States doubts that large-dam construction has brought about several adverse effects. The thinking that prevailed in the United States until the 1970s has undergone a complete about-face, partly because nearly all the water resources have been developed but mostly because of broad recognition of the negative side-effects caused by past dam construction.

Nowadays, people have become sensitive to environmental concerns, including the endangering of wildlife by dam construction. Awareness of water pollution has also become very great. From now on, when a dam is constructed, environmental impact will be a major concern. At the same time, spending is under severe constraints, including the Bureau's budget. No dam project is launched without all information regarding it being made public from the earliest stages of planning and development through actual construction. Although this is not generally known, construction costs are often found to exceed their budgets by more than 50 per cent. The main reason is that costs originally estimated for related items, such as construction of access roads and compensation to relocated people, are often found to have been too low.

What is the alternative to dam construction? The answer is to solve the expected future problems of water and energy shortages in each region, viewed and evaluated comprehensively. As a first step, we should control and manage water and energy demands in the region. It is recommended that we abandon the past view of regarding river basins as tools to produce food, energy, and transportation; we should, instead, take a new attitude, controlling such impulses and attaching importance to less tangible, more spiritual, values.

Mr Beard's position is drawn from the current and future circumstances in the United States, where tight budgets oblige the Bureau to carry out so-called "restructuring." Needless to say, the development and utilization of water resources should be made according to the particular circumstances of each country or region. (The Bureau has recently declined China's request to assist in the construction of the Three Gorges Dam on the Yangtze River. It is hard to tell whether the international political situation influenced this decision.)

For developing countries, it is recommended that the planning and implementation of development projects should be made with the least impact on the environment, opting for cheaper and safer implementation as much as possible. On the other hand, developing countries should take care to screen out schemers and profiteers who involve themselves only for quick gain.

Most Japanese accepted Mr Beard's remarks, since both Japan and the USA are at similar stages of development, as well as under similar environmental constraints. However, they did so with slight bewilderment at the outset, since it marked such a radical departure. At the end of the day, there seemed to be many Japanese who felt that his views were not only valuable but also applicable to Japanese society.

However, countries still in the early stages of development, particularly the riparian governments in Indo-China, seem unlikely to heed Beard's advice. National pride, and the desire to catch up with perceived development in industrialized countries, influence their judgment. Unfortunately, rather than profiting from the adverse experience of advanced countries, they tend to overreact with such statements as "we need to develop our own country," "we have to retain our freedom to develop," or "don't poke your nose into our affairs." It would seem, therefore, that the Japanese government is likely to be all the more perplexed when invited to participate in future Indo-Chinese development projects. It is not clear whether the Chinese government has heard Mr Beard's views, but they would probably disregard them, even if they have.

In December 1994, the author visited several dam sites on the Lancang River in Yunnan Province (see chapter 4) and inquired how the Chinese government felt about the current worldwide criticism of the development of large dams. They answered thus: "We must make our country as powerful and rich as we can, as soon as possible. Therefore, we would never agree with such scepticism. We are carrying out the development of the Lancang in consultation with downstream countries. Moreover, we have been handling the relocation of local people with the greatest caution. We say with pride that our way of moving residents from the project sites should be considered exemplary by all riparian governments in the lower Mekong Basin."

The Chinese government, then, is very proud of its large-scale construction of dams such as the Three Gorges Dam and the cascade dams on the main stream of the Lancang. It seems that the general public has supported this construction, and perhaps the Chinese government would not show any interest in Mr Beard's views.

The author has not asked people in Thailand for their reaction to Mr Beard's statement, but their response would probably be fairly complicated. As we have seen, the Thai government recently enforced the construction of the Pak Mun Dam at the mouth of the Mun River in north-east Thailand. In its official report, "Perspectives for Mekong Development," this project was chosen by the Interim Mekong Committee in 1988 as one of the projects to be implemented before the year 2000. In the report, the Committee estimated that it would result in 8,100 ha of irrigation and 136,000 kW of power generation. Those who had grave concerns over its environmental impact and hardship caused to people due to inundation from the reservoir vehemently opposed construction, but the government ignored them.

As mentioned in the section on notable development projects on the lower Mekong tributaries (pp. 339–342), when an anti-dam movement threatened to sweep Thailand, the King issued this statement: "As the development of dams is necessary to fulfil water and energy needs, the movement to oppose construction will damage national interests."[6] However, even if the government wishes to promote dams to supply electricity, irrigation, and industrial and domestic water, there are only a few appropriate sites left for construction.

Regarding the plan to build the Low Pa Mong Dam on the main Mekong, the Thai government at one point indicated its intention to promote it but soon backed off when the Laotian government opposed the project. Therefore, although the Thai government does not wish to rely on foreign countries, owing to concerns for national security, it now expects to get help for its water-resources development from Laos, China, and Myanmar, as described in the section on meeting the demand for energy and water in Thailand (pp. 339–342). Following the government's lead, the Thai private sector has shown great interest in promoting projects in neighbouring countries.

When Mr Beard visited Thailand in November 1994, the Thai press featured his remarks as front-page news. Because of that, the government will henceforth develop dams more discreetly, while showing more interest in the development in neighbouring countries. However, of the journalists who opposed the construction of dams within Thailand, some have now focused their criticism on dam construction by Thailand in neighbouring countries. Although they do not seem keen to criticize the development in the Lancang Basin upstream, they are opposed to the development of some of the large dams in Laos: for example, they criticized the Nam Hin Bun Project in Laos over the preservation of forested lands there. They also expressed scepticism over the need to preserve the environment while expanding agricultural production in the Mekong Delta, which the Interim Mekong Committee has been keen to do for a long time, even though the delta is a long way from Thailand.

So far, however, it seems that very few people in Thailand are interested in events occurring outside their country. Nevertheless, there must be influential Thai citizens concerned about Thailand's long-range welfare who might seriously oppose the plan to build consecutive cascade low dams on the main Mekong. This said, it must be acknowledged that the majority of them may feel that the dam sites are in too remote an area to merit overriding visionary concern.

Unlike the Chinese government, the Thai government seems unable to decide clearly whether it should further promote the planning and implementation of dams on the main stream of the lower Mekong because it has to consider various serious consequences. On the other hand, the three Indo-Chinese governments are currently too busy to rehabilitate or improve their own countries. Therefore, unless some international lending agency urges them to act, none will initiate such development.

It was reported a few years ago that a team of experts of SIDA, who visited the Interim Mekong Committee just before it was reorganized, criticized the plan to develop the Low Pa Mong Dam Project, saying that even the low-dam plan would eventually cause the inundation of a large number of local inhabitants' houses and broad farmlands and was, therefore, unrealistic. Such a statement is easy for outsiders to make: they are not responsible for satisfying the future demands for water and energy of the countries in question by promoting development while protecting the natural and social environments, or for solving the resettlement problems; nor do they sympathize with the officials of the riparian governments involved, who must shoulder these responsibilities.

Considering both the overall circumstances of development in the lower Mekong Basin and the probable results of the development of large dams, the author proposes the shelving of all the development plans for mainstream dams in the lower basin for the time being. Even the latest plan to build low cascade dams and run-of-the-river power stations on the main stream, which was recommended by French consultants and tentatively approved by engineers of three riparian governments (excluding Cambodia), seems to lack sufficient study of the impact. Moreover, it is quite certain that once such a low-dam plan is authorized by the Mekong River Commission and dams are built on the main stream as planned, it would no longer be possible to mitigate floods in the Mekong Delta or to promote large-scale irrigation schemes in north-east Thailand. Until recently, both of these had long been hoped for by all the people concerned. A careful study of the most appropriate size of the dams to be built on the main stream is required.

When the Mekong Committee was established in the 1950s, emphasis was put on flood control in the delta and the improvement of navigation. Some time later, the merits of power generation and irrigation were stressed. Serious plans were made to achieve these aims by building large-scale dams; however, owing to consideration of the resettlement problems anticipated with the construction of large dams, some alternative plans to build a series of lower dams were proposed, mainly to generate power.

As outlined above, the purpose of developing dams on the main stream of the lower Mekong has changed over the past 40 years. In fact, nothing concrete has yet been achieved. The author feels that this inaction has helped to maintain the happiness and tranquillity of the inhabitants in the basin and that, in view of this, the Mekong River Commission should re-examine all the past mainstream development plans, aiming for the true happiness and well-being of the inhabitants in the entire river basin by means of sustainable development.

The Commission should take time to hold widespread public hearings to relieve conflicts between development and environmental conservation, as well as to resolve the contradiction between the suffering of people who must be moved from the project sites and the welfare of those who would benefit from the development. If such hearings could forge a consensus to build a series of dams on the main stream in the lower basin, construction could begin, with the aim of minimizing the adverse effects of construction. If no consensus is achieved as a result of the hearings, development may have to be postponed until some agreement is reached, which could take a long time. As a cautionary reminder, although the regions and surrounding circumstances are different, it ought to be kept in mind that, in the Amazon Basin, dams have been built only on the tributaries; no dam has ever been built on the main stream of the Amazon.

The author is neither a naturalist nor a romanticist; nevertheless, remembering the unspoiled beauty of the Mekong and the peaceful and contented nature of the people in the riparian countries, he would urge that the development of dams on the main stream be carried out not merely from the viewpoint of energy generation but from a comprehensive viewpoint that pays attention to other merits, including flood control. Mainstream dam development should be initiated with the utmost deliberation and caution.

Necessity to plan and implement the future development of tributaries in the lower basin for multi-purpose use

The section on notable development projects on the lower Mekong tributaries (pp. 339–352) above outlines several notable tributary projects, the development of which is currently being carried out, or is likely to be initiated in the very near future. The author would also point to the Subregional Infrastructure Development Programme in Indo-China and the Greater Mekong Area, which has been promoted mostly by the ADB with the support of the Japanese government.

At the Third Subregional Economic Cooperation Conference held in Hanoi in April 1994, the ADB presented the following six proposals for the development of the generation and transmission of power in Cambodia, China, Laos, Myanmar, Thailand, and Viet Nam:

1. The hydroelectric power development study on the Se Kong and Se San rivers (both of which are major tributaries in the lower Mekong Basin and flow through Viet Nam, Laos, and Cambodia), including the power-transmission interconnection between these countries and Thailand;
2. The prefeasibility study of the Nam Tha hydroelectric power project in Laos (which includes the transmission line to Thailand);
3. The feasibility study of the power-transmission interconnection between the Jinhong hydroelectric power station in Yunnan and Thailand;
4. The Nam Theun Basin hydroelectric power development study in Laos, including a transmission interconnection between Thailand and Viet Nam;
5. The hydroelectric power development study of the Tanlween Basin located between Myanmar and Thailand;
6. The implementation of the Theun–Hin Bun hydroelectric power

project in Laos, including the power-transmission interconnection with Thailand.

Following the above conference, the Forum for the Comprehensive Development of Indo-China was held in Tokyo in February 1995, in which the ADB presented a list of future projects. On the list, all of the above-mentioned projects were included except the Tanlween development project. The following four projects were added, in accordance with the recommendations of the Mekong Secretariat:

1. The study on micro-hydroelectricity for rural electrification (basin-wide);
2. The run-of-the-river hydroelectric power possibilities on the Mekong main stream (the author feels that this addition shows the depth of the Secretariat's attachment to the idea of developing cascade low dams on the main stream);
3. The review and assessment of water resources for hydroelectric power, and identification of priority projects in Cambodia;
4. The prefeasibility study on the Kamchay hydroelectric power project in Cambodia.

The list presented by the ADB includes the Son La hydroelectric power project on the Black River – a major tributary of the Red River in Viet Nam – and its transmission interconnections, the long-term sub-regional generation and transmission development study, the subregional strategy study for the utilization of natural gas, and the gas pipeline from Viet Nam and Cambodia to Thailand. In addition, a related institutional project aimed at strengthening the private sector's participation in hydro-electric power development (basin wide) was proposed by the Mekong Secretariat, while the ESCAP proposed the initiation of a study on re-gional cooperation on wind-energy development and utilization.

Reading the explanations provided by the ADB for the above pro-posals and the welcoming comments by the representatives of the ripar-ian governments, the author was impressed that the future development of energy in the Mekong River Basin would be promoted on the basis of what the ADB has proposed together with the Mekong Secretariat and ESCAP. Each of the proposed projects mentioned above should, there-fore, be examined carefully.

The author respects the excellent work done by all the experts em-ployed by the ADB who created the Mekong Development proposal, all the more so since it was completed in less time than is customary. Never-theless, one fundamental point must be questioned: all the proposed projects related to water resources have been announced solely as power projects. The Japanese government asked the ADB to list all the desir-able, important projects related to the improvement of transportation, communication, and energy supply and transmission in the subregion so that there would be no reason to question in the future whether it treated projects solely as potential hydroelectric power projects. Nevertheless, the list includes a number of projects which should have been appraised as multi-purpose dam projects (for irrigation, flood control, fishing, navigation, water supply, etc., in addition to power generation). For example, in the study of the development of the Se San and the Se Kong basins, the people concerned should see beyond the dams' value as a power-generation resource, viewing them comprehensively for the many benefits they will bring to inhabitants of the basins.

The author is disturbed by the attitude of the ADB, which seems concerned solely with the project's economic value – understandable though this is as the viewpoint of a regional development bank. However, if the development of a river basin is planned only with the purpose of power generation and implemented from the economic viewpoint alone, the river may suffer.

The ADB planned the development of the basin in accordance with the instructions of the government of Japan and, being a bank, it focused primarily on economic concerns. However, the Mekong River Commission should promote the macro-development of both the main stream and the tributaries, with the overall aim of achieving sustainable development for people living in the entire region, without being entrapped by the well-meaning but limited perspective of a financial lending agency.

How the Mekong River Commission can fulfil its mandate

The following description was made by the author in December 1994 at the International Symposium on Regional Infrastructure for the Mekong River Basin, held by the Japan Global Infrastructure Funds (GIF) in the United Nations University. The symposium was attended by officials from the four riparian governments. Fortunately, the GIF symposium was held immediately after the Working Group Meeting of the Interim Mekong Committee, at which the group unanimously agreed to initiate the new Mekong River Commission.

In light of this news, the author changed his prepared speech to one more suitable to the new circumstances. The revised content suggested the next steps based on the understanding that the dispute between Thailand and Viet Nam over the sharing of Mekong River water had finally ended, and that the four riparian governments had agreed in Hanoi to cooperate in the development of the Mekong Basin.

As the participants in the symposium were executive-level officials, not those involved in discussing fundamental policy, the author chose more concrete and practical issues for the speech, starting with a briefing on the past contribution of the UN ECAFE (now ESCAP – the United Nations Economic and Social Commission for Asia and the Pacific) and of several donor countries, particularly Japan, for comprehensive development planning and for the execution of some of the recommended development projects in the lower basin. The author then made several suggestions. More than a year has passed since the inauguration of the Commission, making it unlikely that much of it is still applicable; nevertheless, the suggestions were as set out below.

Establishing networks to monitor the Mekong flow.   In order to prevent future water disputes, the author suggests a review of the ongoing system of the hydrometeorological network in the lower basin, which has been set up after long, hard efforts by the Mekong Committee, to mitigate flood damage and drought in the lower basin.

The current system aims mainly to monitor the quantity of the river flow and the water level, and to transmit the results to the monitoring centre in the Secretariat's office. This system should be improved and converted to a flow-monitoring network by which constant surveillance on both the quantity and quality of water can be conducted on the interaction of the entire watershed and the rivers. The resulting data on the activities of upstream countries should be transmitted immediately to the Secretariat of the Mekong River Commission. Upon receiving any unusual data, the Secretariat would convey these to the concerned riparian government(s), requesting an examination of the cause. If the cause is found to be activities by people upstream, with no countermeasures being taken by the upstream government(s), the Secretariat should report this to the top levels of the Commission, seeking solutions.

Achieving comprehensive development planning for the entire basin.   In the new agreement, it was stipulated that a comprehensive development plan should be worked within the lower basin. However, in order to enable all the riparian neighbours to cooperate fully for sustainable development, it will be necessary to revise the plan later to meet the needs of development of the whole basin, including the Yunnan region (China and Myanmar), as soon as possible.

In such a comprehensive plan to develop the entire basin, it would, of course, be necessary to take into account the available subregional development plan of the Mekong prepared by the ADB, as well as the results of the study by the UNDP on human development in the Mekong Basin. However, care should be taken to avoid allowing too great an influence from the ADB plan, which seems to be biased in its focus on the development of communications, transportation, and energy as its priority areas; instead, it should be well balanced in all aspects. Development planners should take into account the wishes of donor countries and put queries to the Commission regarding total development, including agriforestry and fisheries development as well as the environmental impact, until a clear understanding is achieved. They should also make efforts to gather the opinions of reputable experts to help design a viable development strategy, emphasizing the importance of preserving the natural and social environments.

Unfortunately, recently formulated development plans for the lower Mekong Basin were made by a closed group of planners consisting of members of the Secretariat in collaboration with the staff of consultant engineering firms. They compiled an overall development plan, gathering information from a number of individual development projects, without adequate consideration being given to the environment; they did not pay much attention to the circumstances of the local people. It was unfortunate that the riparian governments readily approved the plan without a detailed critique of its contents.

In the future, the riparian governments should exercise their own independent judgement and, when any of them feels it necessary to revise or improve the draft, they should insist on it and make efforts to do so. To make a satisfactory comprehensive development plan, and to succeed in the actual future development of the basin, the author makes the following suggestions.

An international advisory group of experts.   In addition to the above-mentioned suggestions, it is absolutely necessary to set up an advisory group composed of selected high-level international experts. These experts would be attached to the new Commission to assist it in fulfilling its intended task of smoothly solving all important problems – such as advising on the selection and adoption of the elements of the comprehensive development plan and on the selection and promotion of major individual development projects. This advisory group should be on call as needed by the Commission.

It would be desirable for these advisory experts to have wide experience of comprehensive development planning of international river basins of this type and size. They should also be conversant with social and economic affairs, the various environmental and ecological studies, and the legal and institutional studies on river-basin development. Members of the group should be impartial and objective. Meetings should take place more than twice a year, at the discretion of the Commission. To ensure continuity, the term of the advisory group should be at least 3 years. To fulfil their tasks, they must be given ample time before meetings to examine data, references, and drafts. Their opinions and comments should be summarized by the Secretariat and conveyed to the Commission without embellishment or omission. The executive officer of the Secretariat should help organize the group and attend every meeting.

Respect towards the new Mekong River Commission. The above three sections are the author's suggestions to the Commission. However, what is perhaps more important is the attitude of the riparian governments, donor governments, international lending agencies, and NGOs towards the new Commission.

When the Commission was inaugurated in April 1995, some people expressed their misgivings over the relationship between the ADB's development plan, *Subregional Infrastructure Projects in Indochina and the Greater Mekong Area*, and the future activities of the Mekong River Commission. The author shared this anxiety. A director responsible for the ADB plan reportedly said: "The ADB and the Commission are just like the two wheels of a cart; they are inseparable in regard to the Mekong activities. Since they will help each other, there should be no cause for concern." At best, this comment seems ambiguous. The ADB will no doubt gain an advantage over the Commission, for it has already achieved numerous instances of *fait accompli* in the development of projects all over Asia, both in their planning and their execution. The ADB should, therefore, clearly and publicly express its views, now that the Commission is the highest authority concerning the study and planning of the development of water resources and all other categories directly related to their development, at least in the lower Mekong Basin. These include transportation, communication, energy development, and environmental impact. This may be difficult for the ADB, but unless it does so, confusion and conflict will almost certainly prevail.

On the other hand, the Mekong River Commission should make its position clear on any matter without hesitation and whenever necessary to any assisting government or international lending agency, such as the ADB and the World Bank, from which it is expecting to draw the funds necessary for the development of the region. Needless to say, the Commission should not rely too much on any particular aid-giving country or organization in promoting projects, even if that country or organization offers to finance preliminary studies and the execution of the project fully. The Commission should not take a passive attitude in choosing and determining its priorities in the development of its own projects. The only thing the Commission should rely on is the discreet judgement of its own Standing International Advisory Group, whose members are selected by the Commission itself from among the world's top experts and who are expected to have the courage to voice their independent opinions.

All the above comments described above were included in the author's speech at the United Nations University in December 1994.

Author's expectations of development assistance by the Japanese government and suggestions to initiate periodic meetings with the Mekong River Commission

In August 1995, the United States decided to resume diplomatic relations with Viet Nam; moreover, the United States now enjoys full trade with China. US participation could have a marked influence on the development of the Mekong Basin; however, it may take more time to embark upon assistance for the development of the river basin. Until then, a large part of the development of the Mekong will be promoted by official and private aid from other donor countries, including Japan, in addition to the aid from international lending agencies.

After the Forum for Comprehensive Development of Indo-China held in Tokyo in February 1995, the Japanese government asked the ADB to assist in the planning and development of the infrastructure projects in the Greater Mekong, which includes Indo-China. It is, therefore, quite natural that Japan has shown much respect and little opposition towards the ADB's dynamic proposal and somewhat impetuous action to develop the region, which immediately followed the proposal. Perhaps this posture will be maintained by the Japanese government for the time being; however, as far as the lower Mekong Basin is concerned, Japan should understand that the final responsibility for all studies, planning, and development lies with the Commission, not the ADB.

After the Viet Nam War ended in April 1975, the Japanese government was restrained from rendering any significant assistance to the Indo-Chinese countries. In recent years, however, Japan has been providing assistance mostly on a bilateral basis, not through the Interim Mekong Committee. From now on, the Japanese government should render its technical and financial assistance more positively through the Mekong River Commission, the ADB, the World Bank, and the UNDP, while continuing with its bilateral assistance to the area. The author hopes that the government will initiate periodic consultative meetings with the Commission at least once a year, in addition to the designated regular donor-consultative meetings with the Commission. Through such periodic meetings, the assistance programme could be carefully examined by both sides and all the necessary adjustments be made smoothly.

Author's expectations of the private sector and NGOs

In April 1995, the Japanese government convened the Indo-China Forum, an explanatory meeting held in Bangkok. At the Forum, an appeal was made by the government to induce the private sector to be a more active force in the area.

There is no doubt that a number of private enterprises from foreign countries, including many from Japan, could make a significant contribution with their intense vitality – in the form of "cooperation" – in parallel with the official assistance aimed at developing the Mekong Basin. However, a recent article in a leading Japanese newspaper discussed the Mekong in terms of new trading opportunities.[7] In this regard, the author would ask all Japanese enterprises to reconsider their past record in South-East Asia, reflecting on their aggressive and self-centred actions carried out in the name of business. The author fervently hopes that the interested parties will not be profiteers but will, rather, look to extend cooperation in order to foster the long-term well-being of the people of the basin.

Nowadays, we expect NGOs to offer wholehearted and effective assistance to the developing world. As NGOs are goodwill organizations by nature, the author hopes that they will do all they can to ensure that the delicate relationship between development and the conservation of the environment is preserved, and that they give appropriate and timely advice to all related organizations and individuals providing cooperative services for the development of the Mekong Basin. Meanwhile, the riparian governments, donor governments, and international lending agencies should endeavour to publicize development information without delay in order to assist NGOs that are sincerely concerned about the environment and the people's well-being.

Development for the future, spiritual unification, and responsibility of organizations and individuals

The author has described many of his own convictions and hopes. All efforts regarding the development of the Mekong River Basin should be based on the wish to formulate plans and carry them out so that the sustainable development of the basin can be continued into the twenty-first century and beyond.

The achievement of "sustainable development" depends on whether those involved in development value the right of the people in the river basin, and their descendants, to create the most desirable relationship between development and the environment. This requires a careful examination of the close relationship between causes and effects from many viewpoints. In order to do this, those involved in development must have the capacity for impartial scientific judgement. However, the author strongly believes that one of the guiding principles of the aid-giving organizations and riparian governments should be a sincere fellow-feeling for the people living in the river basin. The development of the Mekong Basin must not be viewed simply as a business opportunity. It is strongly hoped that all plans, executions, and management are made on the basis of genuine concern for the inhabitants of the basin.

The author would like to end the chapter by quoting from his speech at the United Nations University in December 1994:

Currently, a number of suggestions have been made for future development. However, I believe it absolutely essential that each riparian government refrains from taking an egoistic and wilful attitude by aiming solely at its own economic benefit, and instead chooses the best way to create a well-balanced and stable society. Each government should try to maintain a calm and peaceful approach, conveying to its people that the most important thing is to live richer individual lives. What should be sought is the "spiritual unification" of the six riparian governments, aiming at the common goal of attaining sustainable water-resources development.

The religion of most of the people of the six countries is Buddhism. Although there may be some differences between Hinayana Buddhism and Mahayana Buddhism, the basis of belief is the same. As the author understands it, the guiding spirit of Buddhism is the wish to foster a deeper harmony between the mind and the surrounding world, while treating each other with love and forgiveness.

If the six riparian countries are to proceed together for the betterment of their people's livelihood through a spirit of love and mercy, they must come closer to the idea behind the creation of the new Mekong River Commission – sustainable development.

Finally, the author would like to point out that the majority of the Japanese are Buddhists, while many people in the other countries aiding the development of the Mekong are Buddhists, Christians, Hindus, Muslims, and followers of other religions. All of them esteem mercy and mutual forgiveness, and everyone who would aid the Mekong should keep these ethical principles uppermost in their minds. Only by serving the people in the Mekong on the basis of these guiding principles will the countries and individuals who offer assistance be regarded as real friends and be respected by the local governments and inhabitants in the twenty-first century.

Notes

1. The volume of water in the reservoir of the Nam Ngum Dam has been declining in recent years, perhaps as a result of deforestation in the upstream areas. A proposal has been made to pump water into the reservoir from neighbouring rivers. In addition, the construction of a second dam on the Nam Ngum River, to be operated jointly with the existing dam, is being seriously considered.

2. According to an article from the 23 December 1994 edition of *The Nations*, a Thai newspaper.

3. The fact that both the ADB and JICA are viewing the Se Kong River development projects solely from the perspective of energy development can be verified from the "Compendium of Project Profiles" (released in Tokyo in February 1995).

4. This description appeared in the February issue of *The Nations*, published in Bangkok.

5. The Bureau of Reclamation of the US Department of the Interior was established in 1902 for the purpose of developing water resources in the western United States, providing irrigation for farmland, and stimulating economic development focused on the farming, mining, and manufacturing industries. Today, the Bureau of Reclamation is the leading provider of water throughout America and is the sixth-largest provider of hydroelectric power. It oversees 45 per cent of all surface water in the western United States. Flood control and marine transport are under the jurisdiction of the US Army Corps of Engineers; the Bureau of Reclamation is not involved in these matters.

6. According to *The Nations* in February 1995.

7. "The Mekong Basin: Companies Aim to Promote Business Negotiations, Taking Advantage of the Upturn in the Political Situation to Expand Trade" in the 6 August 1995 edition of *Asahi Shinbun*, a leading newspaper in Japan.

Appendix

AGREEMENT

ON THE COOPERATION FOR THE SUSTAINABLE DEVELOPMENT OF THE MEKONG RIVER BASIN

The Governments of The Kingdom of Cambodia, The Lao People's Democratic Republic, The Kingdom of Thailand, and The Socialist Republic of Viet Nam, being equally desirous of continuing to cooperate in a constructive and equally beneficial manner for sustainable development, utilization, conservation and management of the Mekong River Basin water and related resources, have resolved to conclude this Agreement setting forth the framework for cooperation acceptable to parties hereto to accomplish these ends, and for that purpose have appointed their respective plenipotentiaries:

The Kingdom of Cambodia:
H. E. Mr. Ing Kieth
Deputy Prime Minister and Minister of Public Works and Transport

The Lao People's Democratic Republic:
H. E. Mr. Somsavat Lengsavad
Minister of Foreign Affairs

The Kingdom of Thailand:
H. E. Dr. Krasae Chanawongse
Minister of Foreign Affairs

The Socialist Republic of Viet Nam:
H. E. Mr. Nguyen Manh Cam
Minister of Foreign Affairs

Who, having communicated to each other their respective full papers and having found them in good and due form, have agreed to the following:

CHAPTER I. PREAMBLE

RECALLING the establishment of the Committee for the Coordination of Investigations of the Lower Mekong Basin on 17 September 1957 by the Governments of these countries by Statute endorsed by the United Nations,

NOTING the unique spirit of cooperation and mutual assistance that inspired the work of the Committee for the Coordination of Investigations of the Lower Mekong Basin and the many accomplishments that have been achieved through its efforts,

ACKNOWLEDGING the great political, economic and social changes that have taken place in these countries of the region during this period of time which necessitate these efforts to re-assess, re-define and establish the future framework for cooperation,

RECOGNIZING that the Mekong River Basin and the related natural resources and environment are natural assets of immense value to all the riparian countries for the economic and social well-being and living standards of their peoples,

REAFFIRMING the determination to continue to cooperate and promote in a constructive and mutually beneficial manner in the sustainable development, utilization, conservation and management of the Mekong River Basin water and related resources for navigational and non-navigational purposes, for social and economic development and the well-being of all riparian States, consistent with the needs to protect, preserve, enhance and manage the environmental and aquatic conditions and maintenance of the ecological balance exceptional to this river basin,

AFFIRMING to promote or assist in the promotion of interdependent sub-regional growth and cooperation among the community of Mekong nations, taking into account the regional benefits that could be derived and/or detriments that could be avoided or mitigated from activities within the Mekong River Basin undertaken by this framework of cooperation,

REALIZING the necessity to provide an adequate, efficient and functional joint organizational structure to implement this Agreement and the projects, programs and activities taken thereunder in cooperation and coordination with each member and the international community, and to address and resolve issues and problems that may arise from the use and development of the Mekong River Basin water and related resources in an amicable, timely and good neighbourly manner,

PROCLAIMING further the following specific objectives, principles, institutional framework and ancillary provisions in conformity with the objectives and principles of the Charter of the United Nations and international law:

CHAPTER II. DEFINITIONS OF TERMS

For the purposes of this Agreement, it shall be understood that the following meanings to the underlined terms shall apply except where otherwise inconsistent with the context:

Agreement under Article 5: A decision of the Joint Committee resulting from *prior consultation* and evaluation on any *proposed use* for inter-basin diversions during the wet season from the mainstream as well as for intra-basin use or inter-basin diversions of these waters during the dry season. The objective of this *agreement* is to achieve an optimum use and prevention of waste of the waters through a dynamic and practical consensus in conformity with the Rules for Water Utilization and Inter-Basin Diversions set forth in Article 26.

Acceptable minimum monthly natural flow: The acceptable minimum monthly natural flow during each month of the dry season.

Acceptable natural reverse flow: The wet season flow level in the Mekong River at Kratie that allows the reverse flow of the Tonle Sap to an agreed upon optimum level of the Great Lake.

Basin Development Plan: The general planning tool and process that the Joint Committee would use as a blueprint to identify, categorize and prioritize the projects and programs to seek assistance for and to implement the plan at the basin level.

Environment: The conditions of water and land resources, air, flora, and fauna that exists in a particular region.

Notification: Timely providing information by a riparian to the Joint Committee on its *proposed use* of water according to the format, content and procedures set forth in the Rules for Water Utilization and Inter-Basin Diversions under Article 26.

Prior consultation: Timely *notification* plus additional data and information to the Joint Committee as provided in the Rules for Water Utilization and Inter-Basin Diversion under Article 26, that would allow the other member riparians to discuss and evaluate the impact of the *proposed use* upon their uses of water and any other effects, which is the basis for arriving at an agreement. *Prior consultation* is neither a right to veto the use nor unilateral right to use water by any riparian without taking into account other riparians' rights.

Proposed use: Any proposal for a definite use of the waters of the Mekong River system by any riparian, excluding domestic and minor uses of water not having a significant impact on mainstream flows.

CHAPTER III. OBJECTIVES AND PRINCIPLES OF COOPERATION

The parties agree:

Article 1. Areas of Cooperation

To cooperate in all fields of sustainable development, utilization, management and conservation of the water and related resources of the Mekong River Basin including, but not limited to irrigation, hydro-power, navigation, flood control, fisheries, timber floating, recreation and tourism, in a manner to optimize the multiple-use and mutual benefits of all riparians and to minimize the harmful effects that might result from natural occurrences and man-made activities.

Article 2. Projects, Programs and Planning

To promote, support, cooperate and coordinate in the development of the full potential of sustainable benefits to all riparian States and the prevention of wasteful use of Mekong River Basin waters, with emphasis and preference on joint and/

or basin-wide development projects and basin programs through the formulation of a basin development plan, that would be used to identify, categorize and prioritize the projects and programs to seek assistance for and to implement at the basin level.

Article 3. Protection of the Environment and Ecological Balance

To protect the environment, natural resources, aquatic life and conditions, and ecological balance of the Mekong River Basin from pollution or other harmful effects resulting from any development plans and uses of water and related resources in the Basin.

Article 4. Sovereign Equality and Territorial Integrity

To cooperate on the basis of sovereign equality and territorial integrity in the utilization and protection of the water resources of the Mekong River Basin.

Article 5. Reasonable and Equitable Utilization

To utilize the waters of the Mekong River system in a reasonable and equitable manner in their respective territories, pursuant to all relevant factors and circumstances, the Rules for Water Utilization and Inter-basin Diversion provided for under Article 26 and the provisions of A and B below:

A. On tributaries of the Mekong River, including Tonle Sap, intra-basin uses and inter-basin diversions shall be subject to notification to the Joint Committee.

B. On the mainstream of the Mekong River:

1. During the wet season:

a) Intra-basin use shall be subject to notification to the Joint Committee.

b) Inter-basin diversion shall be subject to prior consultation which aims at arriving at an agreement by the Joint Committee.

2. During the dry season:

a) Intra-basin use shall be subject to prior consultation which aims at arriving at an agreement by the Joint Committee.

b) Any inter-basin diversion project shall be agreed upon by the Joint Committee through a specific agreement for each project prior to any proposed diversion. However, should there be a surplus quantity of water available in excess of the proposed uses of all parties in any dry season, verified and unanimously confirmed as such by the Joint Committee, an inter-basin diversion of the surplus could be made subject to prior consultation.

Article 6. Maintenance of Flows on the Mainstream

To cooperate in the maintenance of the flows on the mainstream from diversions, storage releases, or other actions of a permanent nature; except in the cases of historically severe droughts and/or floods:

A. Of not less than the acceptable minimum monthly natural flow during each month of the dry season;

B. To enable the acceptable natural reverse flow of the Tonle Sap to take place during the wet season; and,

C. To prevent average daily peak flows greater than what naturally occur on the average during the flood season.

The Joint Committee shall adopt guidelines for the locations and levels of the flows, and monitor and take action necessary for their maintenance as provided in Article 26.

Article 7. Prevention and Cessation of Harmful Effects

To make every effort to avoid, minimize and mitigate harmful effects that might occur to the environment, especially the water quantity and quality, the aquatic (eco-system) conditions, and ecological balance of the river system, from the development and use of the Mekong River Basin water resources or discharge of wastes and return flows. Where one or more States is notified with proper and valid evidence that it is causing substantial damage to one or more riparians from the use of and/or discharge to water of the Mekong River, that State or States shall cease immediately the alleged cause of harm until such cause of harm is determined in accordance with Article 8.

Article 8. State Responsibility for Damages

Where harmful effects cause substantial damage to one or more riparians from the use of and/or discharge to waters of the Mekong River by any riparian State, the party(ies) concerned shall determine all relative factors, the cause, extent of damage and responsibility for damages caused by that State in conformity with the principles of international law relating to state responsibility, and to address and resolve all issues, differences and disputes in an amicable and timely manner by peaceful means as provided in Articles 34 and 35 of this Agreement, and in conformity with the Charter of the United Nations.

Article 9. Freedom of Navigation

On the basis of equality of right, freedom of navigation shall be accorded throughout the mainstream of the Mekong River without regard to the territorial boundaries, for transportation and communication to promote regional cooperation and to satisfactorily implement projects under this Agreement. The Mekong River shall be kept free from obstructions, measures, conduct and actions that might directly or indirectly impair navigability, interfere with this right or permanently make it more difficult. Navigational uses are not assured any priority over other uses, but will be incorporated into any mainstream project. Riparians may issue regulations for the portions of the Mekong River within their territories, particularly in sanitary, customs and immigration matters, police and general security.

Article 10. Emergency Situations

Whenever a Party becomes aware of any special water quantity or quality problems constituting an emergency that requires an immediate response, it shall notify and consult directly with the party(ies) concerned and the Joint Committee without delay in order to take appropriate remedial action.

CHAPTER IV. INSTITUTIONAL FRAMEWORK
A. MEKONG RIVER COMMISSION
Article 11. Status

The institutional framework for cooperation in the Mekong River Basin under this Agreement shall be called the Mekong River Commission and shall, for the purpose of the exercise of its functions, enjoy the status of an international body, including entering into agreements and obligations with the donor or international community.

Article 12. Structure of Mekong River Commission

The Commission shall consist of three permanent bodies:
– Council
– Joint Committee, and
– Secretariat

Article 13. Assumption of Assets, Obligations and Rights

The Commission shall assume all the assets, rights and obligations of the Committee for the Coordination of Investigations of the Lower Mekong Basin (Mekong Committee/Interim Mekong Committee) and Mekong Secretariat.

Article 14. Budget of the Mekong River Commission

The budget of the Commission shall be drawn up by the Joint Committee and approved by the Council and shall consist of contributions from member countries on an equal basis unless otherwise decided by the Council, from the international community (donor countries), and from other sources.

B. COUNCIL
Article 15. Composition of Council

The Council shall be composed of one member from each participating riparian State at the Ministerial and Cabinet level, (no less than Vice-Minister level) who would be empowered to make policy decisions on behalf of his/her government.

Article 16. Chairmanship of Council

The Chairmanship of the Council shall be for a term of one year and rotate according to the alphabetical listing of the participating countries.

Article 17. Sessions of Council

The Council shall convene at least one regular session every year and may convene special sessions whenever it considers it necessary or upon the request of a member State. It may invite observers to its sessions as it deems appropriate.

Article 18. Functions of Council

The functions of the Council are:

A. To make policies and decisions and provide other necessary guidance concerning the promotion, support, cooperation and coordination in joint activities and projects in a constructive and mutually beneficial manner for the sustainable development, utilization, conservation and management of the Mekong River Basin waters and related resources, and protection of the environment and aquatic conditions in the Basin as provided for under this Agreement;

B. To decide any other policy-making matters and make decisions necessary to successfully implement this Agreement, including but not limited to approval of the Rules of Procedures of the Joint Committee under Article 25, Rules of Water Utilization and Inter-Basin Diversions proposed by the Joint Committee under Article 26, and the basin development plan and major component projects/programs; to establish guidelines for financial and technical assistance of development projects and programs; and if considered necessary, to invite the donors to coordinate their support through a Donor Consultative Group; and,

C. To entertain, address and resolve issues, differences and disputes referred to it by any Council member, the Joint Committee, or any member State on matters arising under this Agreement.

Article 19. Rules of Procedures

The Council shall adopt its own Rules of Procedures, and may seek technical advisory services as it deems necessary.

Article 20. Decisions of Council

Decisions of the Council shall be by unanimous vote except as otherwise provided for in its Rules of Procedures.

C. JOINT COMMITTEE

Article 21. Composition of Joint Committee

The Joint Committee shall be composed of one member from each participating riparian State at no less than Head of Department level.

Article 22. Chairmanship of Joint Committee

The Chairmanship of the Joint Committee will rotate according to the reverse alphabetical listing of the member countries and the Chairperson shall serve a term of one year.

Article 23. Sessions of Joint Committee

The Joint Committee shall convene at least two regular sessions every year and may convene special sessions whenever it considers it necessary or upon the request of a member State. It may invite observers to its sessions as it deems appropriate.

Article 24. Functions of Joint Committee

The functions of the Joint Committee are:

A. To implement the policies and decisions of the Council and such other tasks as may be assigned by the Council.

B. To formulate a basin development plan, which would be periodically reviewed and revised as necessary; to submit to the Council for approval the basin development plan and joint development projects/programs to be implemented in connection with it; and to confer with donors, directly or through their consultative group, to obtain the financial and technical support necessary for project/program implementation.

C. To regularly obtain, update and exchange information and data necessary to implement this Agreement.

D. To conduct appropriate studies and assessments for the protection of the environment and maintenance of the ecological balance of the Mekong River Basin.

E. To assign tasks and supervise the activities of the Secretariat as is required to implement this Agreement and the policies, decisions, projects and programs adopted thereunder, including the maintenance of databases and information necessary for the Council and Joint Committee to perform their functions, and approval of the annual work program prepared by the Secretariat.

F. To address and make every effort to resolve issues and differences that may arise between regular sessions of the Council, referred to it by any Joint Committee member or member state on matters arising under this Agreement, and when necessary to refer the matter to the Council.

G. To review and approve studies and training for the personnel of the riparian member countries involved in Mekong River Basin activities as appropriate and necessary to strengthen the capability to implement this Agreement.

H. To make recommendations to the Council for approval on the organizational structure, modifications and restructuring of the Secretariat.

Article 25. Rules of Procedures

The Joint Committee shall propose its own Rules of Procedures to be approved by the Council. It may form ad hoc and/or permanent sub-committees or working groups as considered necessary, and may seek technical advisory services except as may be provided for in the Council's Rules of Procedures or decisions.

Article 26. Rules for Water Utilization and Inter-Basin Diversions

The Joint Committee shall prepare and propose for approval of the Council, inter alia, Rules for Water Utilization and Inter-Basin Diversions pursuant to Articles 5 and 6, including but not limited to: 1) establishing the time frame for the wet and dry seasons; 2) establishing the location of hydrological stations, and determining and maintaining the flow level requirements at each station; 3) setting out criteria for determining surplus quantities of water during the dry season on the mainstream; 4) improving upon the mechanism to monitor intra-basin use; and, 5) setting up a mechanism to monitor inter-basin diversions from the mainstream.

Article 27. Decisions of the Joint Committee

Decisions of the Joint Committee shall be by unanimous vote except as otherwise provided for in its Rules of Procedures.

D. SECRETARIAT

Article 28. Purpose of Secretariat

The Secretariat shall render technical and administrative services to the Council and Joint Committee, and be under the supervision of the Joint Committee.

Article 29. Location of Secretariat

The location and structure of the permanent office of the Secretariat shall be decided by the Council, and if necessary, a headquarters agreement shall be negotiated and entered into with the host government.

Article 30. Functions of the Secretariat

The functions and duties of the Secretariat will be to:

A. Carry out the decisions and tasks assigned by the Council and Joint Committee under the direction of and directly responsible to the Joint Committee;

B. Provide technical services and financial administration and advise as requested by the Council and Joint Committee;

C. Formulate the annual work program, and prepare all other plans, project and program documents, studies and assessments as may be required;

D. Assist the Joint Committee in the implementation and management of projects and programs as requested;

E. Maintain databases of information as directed;

F. Make preparations for sessions of the Council and Joint Committee; and,

G. Carry out all other assignments as may be requested.

Article 31. Chief Executive Officer

The Secretariat shall be under the direction of a Chief Executive Officer (CEO), who shall be appointed by the Council from a short-list of qualified candidates selected by the Joint Committee. The Terms of Reference of the CEO shall be prepared by the Joint Committee and approved by the Council.

Article 32. Assistant Chief Executive Officer

There will be one Assistant to the CEO, nominated by the CEO and approved by the Chairman of the Joint Committee. Such Assistant will be of the same nationality as the Chairman of the Joint Committee and shall serve for a co-terminus one-year term.

Article 33. Riparian Staff

Riparian technical staff of the Secretariat are to be recruited on a basis of technical competence, and the number of posts shall be assigned on an equal basis among the members. Riparian technical staff shall be assigned to the Secretariat for no more than two three-year terms, except as otherwise decided by the Joint Committee.

CHAPTER V. ADDRESSING DIFFERENCES AND DISPUTES

Article 34. Resolution by Mekong River Commission

Whenever any difference or dispute may arise between two or more parties to this Agreement regarding any matters covered by this Agreement and/or actions taken by the implementing organization through its various bodies, particularly as to the interpretations of the Agreement and the legal rights of the parties, the Commission shall first make every effort to resolve the issue as provided in Articles 18.C and 24.F.

Article 35. Resolution by Governments

In the event the Commission is unable to resolve the difference or dispute within a timely manner, the issue shall be referred to the Governments to take cognizance of the matter for resolution by negotiation through diplomatic channels within a timely manner, and may communicate their decision to the Council for further proceedings as may be necessary to carry out such decision. Should the Governments find it necessary or beneficial to facilitate the resolution of the matter, they may, by mutual agreement, request the assistance of mediation through an entity or party mutually agreed upon, and thereafter to proceed according to the principles of international law.

CHAPTER VI. FINAL PROVISIONS

Article 36. Entry Into Force and Prior Agreements

This Agreement shall:

A. Enter into force among all parties, with no retroactive effect upon activities and projects previously existing, on the date of signature by the appointed plenipotentiaries.

B. Replace the Statute of the Committee for Coordination of Investigations of the Lower Mekong Basin of 1957 as amended, the Joint Declaration of Principles for Utilization of the Waters of the Lower Mekong Basin of 1975, the Declaration Concerning the Interim Committee for Coordination of Investigations of the Lower Mekong Basin of 1978, and all Rules of Procedures adopted under such agreements. This Agreement shall not replace or take precedence over any other treaties, acts or agreements entered into by and among any of the parties hereto, except that where a conflict in terms, areas of jurisdiction of subject matter or operation of any entities created under existing agreements occurs with any provisions of this Agreement, the issues shall be submitted to the respective governments to address and resolve.

Article 37. Amendments, Modification, Supersession and Termination

This Agreement may be amended, modified, superceded or terminated by the mutual agreement of all parties hereto at the time of such action.

Article 38. Scope of Agreement

This Agreement shall consist of the Preamble and all provisions thereafter and amendments thereto, the Annexes, and all other agreements entered into by the

Parties under this Agreement. Parties may enter into bi- or multi-lateral special agreements or arrangements for implementation and management of any programs and projects to be undertaken within the framework of this Agreement, which agreements shall not be in conflict with this Agreement and shall not confer any rights or obligations upon the parties not signatories thereto, except as otherwise conferred under this Agreement.

Article 39. Additional Parties to Agreement

Any other riparian State, accepting the rights and obligations under this Agreement, may become a party with the consent of the parties.

Article 40. Suspension and Withdrawal

Any party to this Agreement may withdraw or suspend their participation under present Agreement by giving written notice to the Chairman of the Council of the Mekong River Commission, who shall acknowledge receipt thereof and immediately communicate it to the Council representatives of all remaining parties. Such notice of withdrawal or suspension shall take effect one year after the date of acknowledgment or receipt unless such notice is withdrawn beforehand or the parties mutually agree otherwise. Unless mutually agreed upon to the contrary by all remaining parties to this Agreement, such notice shall not be prejudicial to nor relieve the noticing party of any commitments entered into concerning programs, projects, studies or other recognized rights and interests of any riparians, or under international law.

Article 41. United Nations and International Community Involvement

The member countries to this Agreement acknowledge the important contribution in the assistance and guidance of the United Nations, donors and the international community and wish to continue the relationship under this Agreement.

Article 42. Registration of Agreement

This Agreement shall be registered and deposited, in English and French, with the Secretary General of the United Nations.

IN WITNESS WHEREOF, the undersigned, duly authorized by their respective governments have signed this Agreement.

DONE on 5 April 1995 at Chiang Rai, Thailand, in English and French, both texts being equally authentic. In the case of any inconsistency, the text in the English language, in which language the Agreement was drawn up, shall prevail.

For The Kingdom of Cambodia:

Ing Kieth
Deputy Prime Minister and Minister of Public Works and Transport

For The Lao People's Democratic Republic:

Somsavat Lengsavad
Minister of Foreign Affairs

For The Kingdom of Thailand:

Krasae Chanawongse
Minister of Foreign Affairs

For The Socialist Republic of Viet Nam:

Nguyen-Manh Cam
Minister of Foreign Affairs

PROTOCOL
TO THE AGREEMENT ON THE COOPERATION
FOR THE SUSTAINABLE DEVELOPMENT
OF THE MEKONG RIVER BASIN
FOR THE ESTABLISHMENT AND COMMENCEMENT
OF THE MEKONG RIVER COMMISSION

The Governments of the Kingdom of Cambodia, Lao People's Democratic Republic, Kingdom of Thailand, and Socialist Republic of Viet Nam, have signed on this day the AGREEMENT ON THE COOPERATION FOR THE SUSTAINABLE DEVELOPMENT OF THE MEKONG RIVER BASIN.

Said AGREEMENT provides for in Chapter IV the establishment of the Mekong River Commission as the institutional framework through which the AGREEMENT will be implemented.

BY THIS PROTOCOL, the signatory parties to the AGREEMENT do hereby declare the establishment and commencement of the MEKONG RIVER COMMISSION, consisting of three permanent bodies, the COUNCIL, JOINT COMMITTEE and SECRETARIAT, effective on this date with the full authority and responsibility set forth under the AGREEMENT.

IN WITNESS WHEREOF, the undersigned, duly authorized by their respective governments have signed this Protocol.

DONE on 5 April 1995 at Chiang Rai, Thailand.

For the Kingdom of Cambodia:

Ing Kieth
Deputy Prime Minister and
Minister of Public Works and Transport

For the Lao People's
Democratic Republic

Somsavat Lengsavad
Minister of Foreign Affairs

For the Kingdom of Thailand:

Dr. Krasae Chanawongse
Minister of Foreign Affairs

For the Socialist Republic of
Viet Nam:

Nguyen Manh Cam
Minister of Foreign Affairs

BIBLIOGRAPHY

Asian Development Bank (ADB) (1994). *Economic Cooperation in the Greater Mekong Subregion: Towards Implementation.* Proceedings of the 3rd Conference of Subregional Economic Corporation. ADB, Hanoi, April.
—— (1995). *Subregional Infrastructure Projects in Indochina and the Greater*

Mekong Area. A Compendium of Project Profiles for Forum for Comprehensive Development of Indochina. ADB, February.

Bangkok Post (newspaper), Thailand 1990–1995.

Compagnie Nationale du Rhône, France; Acres International, Canada; and Mekong Secretariat (1994). *Mekong Mainstream Run-of-River Hydropower*. Interim Mekong Committee, Bangkok, December.

JICA (1991). *Feasibility Study on Xe Katham Small Hydroelectric Power Development Project*. JICA, Laos, December.

——— (1994). *Se Kong Basin Hydroelectric Power Development Prefeasibility Study*. JICA, Tokyo.

Mekong River Commission Secretariat (1995). *Mekong Work Programme 1996*. Mekong River Commission Secretariat, Bangkok, October.

Mekong Secretariat (1970). *Inventory of Promising Tributary Projects in Laos*. Mekong Committee, Bangkok, September.

——— (1978). *Pa Mong Optimization and Downstream Effects Study*. Interim Mekong Committee, Bangkok, June.

——— (1985). *Hydroelectric Development of the Nam Theun Basin: Prefeasibility Study*. Interim Mekong Committee, Bangkok, April.

——— (1989). *Planning on Mekong Mainstream Development Scenarios*. Interim Mekong Committee, Bangkok, December.

——— (1993). *Mekong Work Programme 1994*. Mekong Committee, Bangkok, November.

——— (1994). *Mekong Work Programme 1994/95*. Mekong Committee, Bangkok, December.

——— (to September 1995). *Mekong News*. Mekong Committee, Bangkok.

——— (to 1995). *Mekong Review* Mekong Committee, Bangkok.

Nations (newspaper), Thailand, 1990–1995.

New Frontiers (magazine), Thailand, July 1995.

Snowy Mountains Engineering Corporation (SMEC) (1990). *Nam Theun Hydroelectric Project (Feasibility Report)*. SMEC, Australia, November.

——— (1991). *Reappraisal Report on Prek Thnot Multipurpose Project*. SMEC, Australia, December.

TATA Consulting Engineers (1991). *Final Report of Provincial Grid Integration Project (SPEC II)*. TATA Consulting Engineers, Laos/India, May.

TEAM Consulting Engineers (1992). *Preparatory Environmental Study for the Lower Pa Mong Project*. Interim Mekong Committee, Bangkok, July.

WAPCO (1984). *Summary of Projects Possibilities of Lower Mekong Water Resources Inventory*. WAPCO, The Netherlands, September.

Author

Hiroshi Hori was born in December, 1919, in Tokyo. He graduated with a Master of Science degree from the Department of Civil Engineering, University of Tokyo, in 1944 and became Doctor of Engineering in 1987.

After the Second World War, the author worked on river development for the Nissan Construction Company in Japan. From 1955 until 1956 he studied as a Fulbright scholar at the Graduate School of the University of Illinois, and worked for Harza Engineering Co., Chicago, on overseas dam development.

Returning to Japan, Dr Hori joined the Electric Power Development Co. From 1964 until 1969 he was seconded to the United Nations Mekong Committee in Bangkok where he worked on the Indicative Mekong River Basin Plan, and from 1969 until 1971 to Dar es Salaam, Tanzania, where he served as Chief Executive of the National Water Resources Council.

In 1971 Dr Hori was appointed Senior Resident Consultant at the Asian Development Bank in Manila. In 1973 he became Deputy Resident Representative of the United Nations Development Programme (UNDP) in Ankara, Turkey. He was then transferred to UNDP headquarters in New York where he served as Senior Technical Advisor until late 1975 when he returned to Japan.

From 1976 on, Dr Hori worked on overseas projects as Managing Director of Pacific Consultants International, Tokyo, and for other companies. He chaired various domestic and international academic societies, as well as lecturing at several well-known universities in Tokyo, including the University of Tokyo, Waseda University, and Hosei University.

At present, Hiroshi Hori is Honorary Chairman of the National Committee, International Water Resources Association (IWRA) and Chairman of the Mekong Study Committee, Japan International Cooperation Agency (JICA).

Hiroshi Hori has co-authored several books and published numerous papers in Japanese. To academic books and journals in English he has contributed several articles including: "Report of Study on Viet Nam Power Development," Institute of Energy Economics, Japan (1989); "Macro-Engineering – A View from Japan," *Technology Society*, Vol. 12, Number 1, Pergamon Press, USA (1990) and "Conflicts and Opportunities Concerning Development and the Environment in the Mekong Basin," *Water International*, IWRA, USA (1998).

Index